Springer Proceedings in Complexity

Springer Complexity

Springer Complexity is an interdisciplinary program publishing the best research and academic-level teaching on both fundamental and applied aspects of complex systems—cutting across all traditional disciplines of the natural and life sciences, engineering, economics, medicine, neuroscience, social, and computer science.

Complex Systems are systems that comprise many interacting parts with the ability to generate a new quality of macroscopic collective behavior the manifestations of which are the spontaneous formation of distinctive temporal, spatial, or functional structures. Models of such systems can be successfully mapped onto quite diverse "real-life" situations like the climate, the coherent emission of light from lasers, chemical reaction-diffusion systems, biological cellular networks, the dynamics of stock markets and of the Internet, earthquake statistics and prediction, freeway traffic, the human brain, or the formation of opinions in social systems, to name just some of the popular applications.

Although their scope and methodologies overlap somewhat, one can distinguish the following main concepts and tools: self-organization, nonlinear dynamics, synergetics, turbulence, dynamical systems, catastrophes, instabilities, stochastic processes, chaos, graphs and networks, cellular automata, adaptive systems, genetic algorithms, and computational intelligence.

The three major book publication platforms of the Springer Complexity program are the monograph series "Understanding Complex Systems" focusing on the various applications of complexity, the "Springer Series in Synergetics", which is devoted to the quantitative theoretical and methodological foundations, and the "SpringerBriefs in Complexity" which are concise and topical working reports, case-studies, surveys, essays, and lecture notes of relevance to the field. In addition to the books in these two core series, the program also incorporates individual titles ranging from textbooks to major reference works.

More information about this series at http://www.springer.com/series/11637

Shu-Heng Chen • Ying-Fang Kao
Ragupathy Venkatachalam • Ye-Rong Du
Editors

Complex Systems Modeling and Simulation in Economics and Finance

Editors
Shu-Heng Chen
AI-ECON Research Center, Department of
Economics
National Chengchi University
Taipei, Taiwan

Ying-Fang Kao
AI-ECON Research Center, Department of
Economics
National Chengchi University
Taipei, Taiwan

Ragupathy Venkatachalam
Institute of Management Studies
Goldsmiths, University of London
London, UK

Ye-Rong Du
Regional Development Research Center
Taiwan Institute of Economic Research
Taipei, Taiwan

ISSN 2213-8684 ISSN 2213-8692 (electronic)
Springer Proceedings in Complexity
ISBN 978-3-319-99622-6 ISBN 978-3-319-99624-0 (eBook)
https://doi.org/10.1007/978-3-319-99624-0

Library of Congress Control Number: 2018960945

© Springer Nature Switzerland AG 2018
This work is subject to copyright. All rights are reserved by the Publisher, whether the whole or part of the material is concerned, specifically the rights of translation, reprinting, reuse of illustrations, recitation, broadcasting, reproduction on microfilms or in any other physical way, and transmission or information storage and retrieval, electronic adaptation, computer software, or by similar or dissimilar methodology now known or hereafter developed.
The use of general descriptive names, registered names, trademarks, service marks, etc. in this publication does not imply, even in the absence of a specific statement, that such names are exempt from the relevant protective laws and regulations and therefore free for general use.
The publisher, the authors, and the editors are safe to assume that the advice and information in this book are believed to be true and accurate at the date of publication. Neither the publisher nor the authors or the editors give a warranty, express or implied, with respect to the material contained herein or for any errors or omissions that may have been made. The publisher remains neutral with regard to jurisdictional claims in published maps and institutional affiliations.

This Springer imprint is published by the registered company Springer Nature Switzerland AG
The registered company address is: Gewerbestrasse 11, 6330 Cham, Switzerland

Contents

On Complex Economic Dynamics: Agent-Based Computational Modeling and Beyond... 1
Shu-Heng Chen, Ye-Rong Du, Ying-Fang Kao, Ragupathy Venkatachalam, and Tina Yu

Part I Agent-Based Computational Economics

Dark Pool Usage and Equity Market Volatility 17
Yibing Xiong, Takashi Yamada, and Takao Terano

Modelling Complex Financial Markets Using Real-Time Human–Agent Trading Experiments 35
John Cartlidge and Dave Cliff

Does High-Frequency Trading Matter? 71
Chia-Hsuan Yeh and Chun-Yi Yang

Modelling Price Discovery in an Agent Based Model for Agriculture in Luxembourg 91
Sameer Rege, Tomás Navarrete Gutiérrez, Antonino Marvuglia, Enrico Benetto, and Didier Stilmant

Heterogeneity, Price Discovery and Inequality in an Agent-Based Scarf Economy 113
Shu-Heng Chen, Bin-Tzong Chie, Ying-Fang Kao, Wolfgang Magerl, and Ragupathy Venkatachalam

Rational Versus Adaptive Expectations in an Agent-Based Model of a Barter Economy 141
Shyam Gouri Suresh

Does Persistent Learning or Limited Information Matter in Forward Premium Puzzle? 161
Ya-Chi Lin

Price Volatility on Investor's Social Network 181
Yangrui Zhang and Honggang Li

The Transition from Brownian Motion to Boom-and-Bust Dynamics in Financial and Economic Systems 193
Harbir Lamba

Product Innovation and Macroeconomic Dynamics 205
Christophre Georges

Part II   New Methodologies and Technologies

Measuring Market Integration: US Stock and REIT Markets 223
Douglas W. Blackburn and N. K. Chidambaran

Supercomputer Technologies in Social Sciences: Existing Experience and Future Perspectives 251
Valery L. Makarov and Albert R. Bakhtizin

Is Risk Quantifiable? 275
Sami Al-Suwailem, Francisco A. Doria, and Mahmoud Kamel

Index 305

About the Editors

Shu-Heng Chen is a Distinguished Professor in the Department of Economics, National Chengchi University (NCCU), Taipei, Taiwan. He serves as the Director of the AI-ECON Research Center at NCCU as well as the editor-in-chief of the Journal of New Mathematics and Natural Computation (World Scientific) and Journal of Economic Interaction and Coordination (Springer), and the associate editor for Computational Economics (Springer) and Evolutionary and Institutional Economics Review (Springer). Prof. Chen holds a Ph.D. in Economics from the University of California at Los Angeles. His research interests include computational intelligence, agent-based computational economics, behavioral and experimental economics, neuroeconomics, computational social sciences, and digital humanities. He has more than 250 referred publications in international journals and edited book volumes. He is the author of the book, Agent-Based Computational Economics: How the Ideas Originated and Where It Is Going (published by Routledge), and Agent-Based Modeling and Network Dynamics (published by Oxford, co-authored with Akira Namatame).

Ying-Fang Kao is a computational social scientist and a research fellow at the AI-Econ Research Center, National Chengchi University, Taiwan. She received her Ph.D. in Economics from the University of Trento, Italy in 2013. Her research focuses on the algorithmic approaches to understanding decision-making in economics and social sciences. Her research interests include Classical Behavioural Economics, Computable Economics, Agent-Based Modelling, and Artificial Intelligence.

Ragupathy Venkatachalam is a Lecturer in Economics at the Institute of Management Studies, Goldsmiths, University of London, UK. He holds a Ph.D. from the University of Trento, Italy. Prior to this, he has held teaching and research positions at the Centre for Development Studies, Trivandrum (India) and AI-Econ Research Center, National Chengchi University, Taipei (Taiwan). His research areas include Computable Economics, Economic Dynamics, Agent-Based Computational Economics, and Methodology and History of Economic Thought.

Ye-Rong Du is an Associate Research Fellow at the Regional Development Research Center, Taiwan Institute of Economic Research, Taiwan. He received his Ph.D. in Economics from the National Chengchi University, Taiwan in 2013. His research focuses on the psychological underpinnings of economic behavior. His research interests include Behavioural Economics, Agent-Based Economics, and Neuroeconomics.

On Complex Economic Dynamics: Agent-Based Computational Modeling and Beyond

Shu-Heng Chen, Ye-Rong Du, Ying-Fang Kao, Ragupathy Venkatachalam, and Tina Yu

Abstract This chapter provides a selective overview of the recent progress in the study of complex adaptive systems. A large part of the review is attributed to agent-based computational economics (ACE). In this chapter, we review the frontier of ACE in light of three issues that have long been grappled with, namely financial markets, market processes, and macroeconomics. Regarding financial markets, we show how the research focus has shifted from trading strategies to trading institutions, and from human traders to robot traders; as to market processes, we empathetically point out the role of learning, information, and social networks in shaping market (trading) processes; finally, in relation to macroeconomics, we demonstrate how the competition among firms in innovation can affect the growth pattern. A minor part of the review is attributed to the recent econometric computing, and methodology-related developments which are pertinent to the study of complex adaptive systems.

Keywords Financial markets · Complexity thinking · Agent-based computational economics · Trading institutions · Market processes

This book is the post-conference publication for *the 21st International Conference on Computing in Economics and Finance* (**CEF 2015**) held on June 20–22, 2015 in Taipei. Despite being the largest conference on computational economics for two decades, CEF has never produced any book volume that documents the path-breaking and exciting developments made in any of its single annual events.

S.-H. Chen (✉) · Y.-F. Kao · T. Yu
AI-ECON Research Center, Department of Economics, National Chengchi University, Taipei, Taiwan

Y.-R. Du
Regional Development Research Center, Taiwan Institute of Economic Research, Taipei, Taiwan

R. Venkatachalam
Institute of Management Studies, Goldsmiths, University of London, London, UK

© Springer Nature Switzerland AG 2018
S.-H. Chen et al. (eds.), *Complex Systems Modeling and Simulation in Economics and Finance*, Springer Proceedings in Complexity,
https://doi.org/10.1007/978-3-319-99624-0_1

For many years, the post-conference publications had always been in the form of journals' special issues, which, unfortunately, have ceased to continue in recent years. Consequently, although the voices of CEF had been loud and clear for many years on many prominent issues, they may have been forgotten as time goes by. Without proper archives, it will be difficult for the new-comers to trace the important contributions that the conferences have made in the field of computational economics.

Two years ago, Springer launched a new series, *Springer Proceedings in Complexity*, to publish proceedings from scholarly meetings on topics related to the interdisciplinary studies of the science of complex systems. The scope of CEF fits the mission of this series perfectly well. Not only does CEF deal with problems which are sufficiently complex to defy an analytical solution from Newtonian Microeconomics [9], but CEF methods also treat economics as a science of complex systems, which requires complexity thinking both in terms of ontology and epistemology [22]. Therefore, when Christopher Coughlin, the publishing editor of the series, invited us contribute a volume, we considered it to be a golden opportunity to archive the works presented at CEF 2015, in a way similar to what we had done previously in the form of journals' special issues.

However, CEF 2015 had a total of 312 presentations, which covered many aspects of CEF. To include all of them in a single volume is doubtlessly impossible. A more practical alternative would be to select an inclusive and involving theme, which can present a sharp focus that is neither too narrow nor too shallow. It is because of this consideration that we have chosen one of the most active areas of CEF, namely *agent-based computational economics (ACE)*, as the main theme of this book and have included ten chapters which contribute to this topic. These ten chapters are further divided into three distinct but related categories: *financial markets*, *market processes* and the *macroeconomy*. Although there are other areas of ACE that have also made important advances, we believe that without tracking the development of these three research areas, the view of ACE will become partial or fragmented. These ten chapters, constituting the first part of the book, will be briefly reviewed in Sect. 1.

In addition to these ten chapters, we include three chapters that present new methodologies and technologies to study the complex economic dynamics. Three chapters are contributions of this kind. The first one is an econometric contribution to the identification of the existence and the extent of financial integration. The second one addresses the role of supercomputers in developing large-scale agent-based models. The last one challenges the capability of formal reasoning in modeling economic and financial uncertainties. It also advocates a reform of the economic methodology of modeling the real-world economy. These three chapters, constituting the second part of the book, will be briefly reviewed in Sect. 2.

1 Agent-Based Computational Economics

One core issue that ACE seeks to address in economics is how well the real economy performs when it is composed of heterogeneous and, to some extent, boundedly rational and highly social agents. In fact, a large body of published ACE works can be connected to this thread. This issue is of particular importance when students of economics are nowadays still largely trained under Newtonian economics using the device of a representative agent who is assumed to be fully rational in seeking to maximize a utility function. As an alternative research paradigm to the mainstream, ACE attempts to see how our understanding of the economy can become different or remain the same when these simplifications are removed.

Part of ACE was originated by a group of researchers, including Brian Arthur, John Holland, Blake LeBaron, Richard Palmer, and Paul Tayler, who developed an agent-based model called the Santa Fe Artificial Stock Market to study financial markets [1, 5]. Quite interestingly, their original focus was not so much on the financial market per se, i.e., the financial market as an institutional parameter, but on the exploration of trading strategies under evolutionary learning, the co-evolution of trading strategies and the emergence of market price dynamics. This focus drove the early ACE research away from the role of trading mechanisms and institutional arrangements in the financial markets, which was later found to be a substantially important subject in computational economics and finance. Section 1.1 will summarize the three ACE works that focus on trading institutions, rather than trading strategies.

Market process theory investigates how a market moves toward a state of general economic equilibrium and how production and consumption become coordinated. Agent-based modeling is a modern tool used to analyze the ideas associated with a theoretical market process. In Sect. 1.2, we will give an overview of six works that investigate the price discovery process, market dynamics under individual, and social learning and market herding behaviors using agent-based simulation.

Macroeconomics studies the performance, structure, behavior, and decision-making of an economy as a whole. ACE is a modern methodology that is applied to examine the macroeconomy. In Sect. 1.3, we introduce the work using ACE models to analyze the macroeconomic dynamics under product innovation.

1.1 Financial Markets

Dark Pools is an alternative trading institution to the regular exchanges that have gained popularity in recent years. In dark pools trading, there is no order book visible to the public; hence the intention of trading is not known until the order is executed. This provides some advantages for the institutional traders who can obtain a better realized price than would be the case if the sale were executed on a regular exchange. However, there are also disadvantages in that the order may not be executed because of the lack of information for the order's counter parties. With the growing usage of dark pools trading, concerns have been raised about its impact on the market quality. In the chapter, entitled "Dark Pool Usage and Equity Market Volatility," Yibing Xiong, Takashi Yamada, and Takao Terano develop a continuous double-auction artificial stock market that has many real-world market features. Using various institutional parametric setups, they conduct market simulations to investigate the market's stability under dark pools trading.

The institutional-level parameters that they investigated are:

- Dark pool usage probability (0, 0.2 or 0.4);
- Market-order proportion (0.3–0.8);
- Dark pool cross-probability (0.1–1.0), which is the probability that the buy orders and sell orders in the dark pool are being crossed at the mid-price of the exchange. A lower cross-probability indicates a relatively longer order execution delay in the dark pool.

Their simulation results indicated that the use of mid-price dark pools decreases market volatility, which makes sense because the transaction is not visible to the public until the order is completed. The transactional impact on the market stock prices is therefore minimized. Moreover, they found that the volatility-suppressing effect is stronger when the dark pool usage is higher and when the market-order proportion submitted to the dark pool is lower.[1] They also reported that the dark pool cross-probability did not have any effects on the market volatility.

Another trend in recent financial markets is the use of computer algorithms to perform *high frequency trading (HFT)*. Since computer programs can execute trades much faster than humans, stocks and other instruments exhibit rapid price fluctuations (fractures) over sub-second time intervals. One infamous example is the *flash crash* on May 6, 2010 when the Dow Jones Industrial Average (DJIA) plunged by around 7% (US$1 trillion) in 5 min, before recovering most of the fall over the following 20 min. To understand the impact of HFT on financial markets, in the chapter, entitled "Modelling Complex Financial Markets Using Real-Time Human-Agent Trading Experiments," John Cartlidge and Dave Cliff used a real-time financial-market simulator (OpEx) to conduct economic trading experiments between humans and automated trading algorithms (robots).

The institutional-level parameters that they investigated included:

- Robots' trading speed, which is controlled by the sleep-wake cycle (t_s) of robots. After each decision (buy, sell, or do nothing) is made, a robot will sleep for t_s milliseconds before waking up to make the next decision. The smaller that t_s is, the faster the robots' trading speed and the higher their trading frequency.
- Cyclical vs. random markets: In each experiment, there are six pre-generated assignment permits, each of which contains a permit number and a limit price—the maximum value at which to buy, or the minimum value at which to sell.

---

[1] See [16] for similar findings using empirical data.

The lower the permit number is, the farther away the limit price is from the equilibrium. In a cyclical market, the permits are issued to humans and robots following the permit numbers. By contrast, the permits are issued in random order in a random market.

Their simulation results showed that, under all robot and human market setups, *robots outperform humans consistently*. In addition, faster robot agents can reduce market efficiency and this can lead to market fragmentation, where humans trade with humans and robots trade with robots more than would be expected by chance. In terms of market type, the cyclical markets gave very different results from those of random markets. Since the demand and supply in the real-world markets do not arrive in neat price-ordered cycles like those in the cyclical markets, the results from cyclical markets cannot be used to explain what happened in the real-world financial markets. The authors used these two types of markets to demonstrate that, if we want to understand complexity in the real-world financial markets, we should move away from the simple experimental economic models first introduced in the 1960s.

In the chapter, entitled "Does High-Frequency Trading Matter?", Chia-Hsuan Yeh and Chun-Yi Yang also investigated the impact of HFT on market stability, price discovery, trading volume, and market efficiency. However, instead of conducting real-time experiments using humans and robots, they developed an agent-based artificial stock market to simulate the interaction between HFT and non-HFT agents. In addition, unlike the robots in the previous chapter that used pre-generated permits to submit buy and sell orders for price matching, the agents in this study are more sophisticated in terms of using heuristics to make trading decisions. Moreover, the agents have learning ability to improve their trading strategies through experiences.

In their agent-based model, the trading speed is implemented as the agents' capability to process market information for decision-making. Although instant market information, such as the best bid and ask, is observable for all traders, only HFT agents have the capability to quickly process all available information and to calculate expected returns for trading decisions. Non-HFT agents, however, only have the capability to process the most recent k periods' information. The smaller that k is, the greater the advantage that the HFT agents have over non-HFT agents.

The institutional-level parameters that they investigated include:

- The number of HFT agents in the market (5, 15, 20);
- The activation frequency of HFT agents, which is specified by the number of non-HFT agents ($m = 40, 20, 10$) that have posted their quotes before an HFT agent can participate in the market. The smaller that m is, the more active the HFT agents are in participating in the trading.

Their simulation results indicated that market volatilities are greater when there are more HFT agents in the market. Moreover, a higher activation frequency of the HFT agents results in greater volatility. In addition, HFT hinders the price discovery process as long as the market is dominated by HFT activities. Finally, the market efficiency is reduced when the number of HFT agents exceeds a threshold, which is similar to that reported in the previous chapter.

1.2 Market Processes

The agriculture market in Luxembourg is thin, in terms of volume turnover, and the number of trades in all commodities is small. While the information on market products can be obtained through an annual survey of the farmers, the market products trading price information is not accessible to the public. In the chapter, entitled "Modelling Price Discovery in an Agent Based Model for Agriculture in Luxembourg," Sameer Rege, Tomás Navarrete Gutiérrez, Antonino Marvuglia, Enrico Benetto, and Didier Stilmant have proposed an agent-based model to simulate the endogenous price discovery process under buyers and sellers who are patient or impatient in submitting their bid/ask quotes.

In this model, agents are farmers whose properties (area, type, crops, etc.) are calibrated using the available survey data. The model is then used to simulate a market that contains 2242 farmers and ten buyers to trade 22 crops for four rounds. In each round, after all buyers and farmers have submitted the quantity and price for a commodity to buy or sell, the buyer who offers the highest price gets to purchase the desired quantity. If only partial quantity is satisfied under the offered price, the unmet quantity is carried over to the remaining rounds. Similarly, the sellers whose products do not get sold under the offered price are carried over to the remaining rounds. Based on the trading price in the initial round, buyers and sellers can adjust their bid/ask prices in the remaining rounds to achieve their trading goals.

Some buyers/sellers are impatient and want to complete the trading in the next round by increasing/decreasing the bid/ask prices to the extreme, while others are more patient and willing to gradually adjust the prices during each of the remaining three rounds. Based on their simulation, they found that the trading quantities and prices produced by patient and by impatient traders have very different distributions, indicating that traders' behaviors in submitting their bids/asks can impact the price discovery process in an economic market.

In the chapter, entitled "Heterogeneity, Price Discovery and Inequality in an Agent-Based Scarf Economy," Shu-Heng Chen, Bin-Tzong Chie, Ying-Fang Kao, Wolfgang Magerl, and Ragupathy Venkatachalam also used an agent-based model to investigate the price discovery process of an economic market. However, their agents are different from those in the previous chapter in that they apply individual and social learning to revise their subjective prices. The focus of this work is to understand how agents' learning behaviors impact the efficacy of price discovery and how prices are coordinated to reach the Walrasian equilibrium.

The model is a pure exchange economy with no market makers. Each agent has its own subjective prices for the commodities and agents are randomly matched for trading. The learning behavior of an agent is influenced by the intensity of choice λ, which specifies the bias toward the better-performing prices in the past. When λ is high, the agent trusts the prices that have done well (the prices can be from self and from other agents) and uses them to *adjust* its prices for the future trades. If λ is low, the agent is more willing to take risk incorporating prices that have not done well in the past for the future trades.

Their simulation results showed that agents with a low λ (0–3) have their subjective prices converging close to the Walrasian equilibrium. This means risk-taking agents are good at discovering prices toward the general equilibrium. Moreover, some agents with a large λ (>4) also have their market prices converging to the general equilibrium. The authors analyzed those high λ (>4) agents in more detail and found those agents to also be imitators who *copied* prices that have done well in the past to conduct most of their trades. This strategy enhanced their price coordination toward the general equilibrium.

In terms of accumulated payoffs, the agents with low λ (0–3) who also mixed innovation and imitation in adjusting their subjective prices have obtained medium or high payoffs. Meanwhile, the agents with high λ (>4) who are also imitators have received very high payoffs. Finally, the high λ (>4) agents who are also reluctant to imitate other agents' prices have received abysmal accumulated payoffs. Based on this emerging inequality of payoffs, the authors suggested that different learning behaviors among individuals may have contributed to the inequality of wealth in an economy.

In the chapter, entitled "Rational Versus Adaptive Expectations in an Agent-Based Model of a Barter Economy," Shyam Gouri Suresh also investigated market dynamics under agents with learning ability in a *pure exchange* or *barter economy*. In this direct exchange market, an agent can apply individual or social learning to predict the productivity level of his next exchange partner. Based on the prediction, the agent then decides his own productivity level. Under the individual learning mode, the prediction is based on the productivity level of the agent's current exchange partner while in the social learning mode, the prediction is based on the productivity level of the entire population.

In this model, the productivity level of an agent can be either high or low and there is a transition table that all agents use to decide their current productivity level according to their previous productivity. Additionally, an agent can incorporate his prediction about the productivity level of his next exchange partner to decide his current productivity level. This prediction can be carried out through either individual or social learning. Finally, to maximize his utility, an agent only adopts high productivity when his transition table indicates high productivity and his next exchange partner is also predicted to have high productivity.

The simulation results showed that the market *per capita outputs* or *average outputs* converged to low productivity under individual learning. This is because each time when an agent trades with another agent with low productivity, the agent will decide to produce low outputs in the next period regardless of the productivity specified by the transition table. This action in turn causes the agent he interacts with in the next period to produce low outputs in the period subsequent to the next. When an agent encounters another agent who has produced a high level of outputs, the agent will only adopt high productivity in the next period if the transition table also specifies high productivity. As a result, the market average outputs converge to low productivity.

By contrast, the market average outputs converge to high productivity under social learning, when the population size is large (100 in their case). This is because, in a large population, the likelihood of the population-wide distribution of productivity level being extreme enough to cause it to fall below the high-productivity threshold is low. Consequently, as all agents started with high productivity, the market average outputs remained high throughout the simulation runs.

In addition to the price discovery process and productivity level prediction, traders' learning behaviors might have impacted the forward premium in the foreign exchange market. In the chapter, entitled "Does Persistent Learning or Limited Information Matter in the Forward Premium Puzzle?, Ya-Chi Lin investigated whether the interactions between adaptive learning and limited market information flows can be used to explain the forward premium puzzle.

The forward premium puzzle in the foreign exchange market refers to the well-documented empirical finding that the domestic currency is expected to appreciate when domestic nominal interest rates exceed foreign interest rates [4, 10, 14]. This is puzzling because economic theory suggests that if all international currencies are equally risky, investors would demand higher interest rates on currencies expected to fall, and not to increase in value. To examine if investors' learning behaviors and their limited accessibility to market information may explain this puzzle, Lin designed a model where each agent can learn to predict the expected exchange rates using either full information (day t and prior) or limited information in the past (day $t-1$ and prior).

In this model, the proportion of agents that have access to full information, n, is an exogenous parameter. In addition, an agent has a learning gain parameter γ that reflects the learning strength. They simulated the model under different values of n, from 0.1 to 1, and γ, from 0.02 to 0.1, and found that the forward premium puzzle exists under small n for all values of γ. Moreover, when agents were allowed to choose between using limited or full information for forecasting, all agents switched to using full information (i.e., $n = 1$) and the puzzle disappeared for all values of γ. This suggests that limited information might play a more important role than learning in explaining the forward premium puzzle. However, regardless of the values of n and γ, the puzzle disappeared when tested in the multi-period mode. This indicates that limited information alone is not sufficient to explain the puzzle. There are other factors involved that will cause the puzzle to occur.

Herding is a well-documented phenomenon in financial markets. For example, using trading data from US brokerages, Barber et al. [3] and Kumar and Lee [13] showed that the trading of individual investors is strongly correlated. Furthermore, based on trading data from an Australian brokerage, Jackson [12] reported that individual investors moved their money in and out of equity markets in a systematic manner. To macroscopically study the effects of herding behavior on the stock return rates and on the price volatility under investors with different interaction patterns, in the chapter, entitled "Price Volatility on the Investor's Social Network," Yangrui Zhang and Honggang Li developed an agent-based artificial stock market model with different network structures.

In their interaction-based herding model, the trading decision of an agent is influenced by three factors: (1) personal belief; (2) public information, and (3) neighbors' opinions. Their work investigated the following institutional-level parameters:

- Agents' interaction structures: regular, small-world, scale-free, and random networks;
- Agents' trust in their neighbors' opinions (1–3);

Their simulation results showed that the market volatility is the lowest when the agents are connected in a regular network structure. The volatility increases when agents are connected under small-world or scale-free structures. The market volatility is the highest when agents are connected under a random network structure. This makes sense as the more irregular the agents' interaction pattern is, the higher the price fluctuations and market volatility. In addition, they found that the more an agent trusts in his neighbors' opinions, the greater the volatility of the stock price. This is also expected, as the more weight an agent attaches to his neighbors' opinions, the more diverse the trading decisions can be, and hence the higher that the price volatility becomes.

In the chapter, entitled "The Transition from Brownian Motion to Boom-and-Bust Dynamics in Financial and Economic Systems," Harbir Lamba also investigated herding behaviors in financial markets. However, instead of using a network model, he proposed a stochastic particle system where each particle is an agent and agents do not interact with each other. Agents' herding behavior is controlled by a herding parameter C, which drives the agents' states toward the market sentiment. Using this system, Lamba demonstrated that even a very low level of herding pressure can cause a financial market to transition to a multi-year boom-and-bust.

At time t, each agent i in the system can be in one of two possible states, owning the asset (+1) or not owning the asset (−1), according to its pricing strategy $[L_i(t), U_i(t)]$. When the asset market price r_t falls outside the interval of $L_i(t)$ and $U_i(t)$, agent i switches its state to the opposite state. In addition, when an agent's state is different from the state of the majority agents, its pricing strategy is updated at a rate of $C|\sigma|$, where σ is the market sentiment, defined as the average state of all agents. Hence, agents have a tendency to evolve toward the state of the majority agents. Finally, the market price r_t is the result of exogenous information and endogenous agent states generated by the agents' evolving pricing strategies.

Using 10,000 agents to simulate the market for 40 years, their results showed that even with a low herding parameter value $C = 20$, which is much lower than the estimated real market herding pressure of $C = 100$, the deviations of market prices away from the equilibrium resemble the characteristics of "boom-and-bust": a multi-year period of low-level endogenous activities that convince equilibrium-believers the system is in an equilibrium state with slowly varying parameters. There then comes a sudden and large reversal involving cascades of agents switching states, triggered by the change in market price.

1.3 Macroeconomy

Product innovation has been shown to play an important role in a firm's performance, growth, and survival in the modern economy. To understand how product innovation drives the growth of the entire economy, causing business cycle fluctuations, in the chapter, entitled "Product Innovation and Macroeconomic Dynamics," Christophre Georges has developed an agent-based macroeconomic model. In this model, a hedonic approach is used, where product characteristics are specified and evaluated against consumer preferences.

The macroeconomic environment consists of a single representative consumer and m firms whose products are described by characteristics that the consumer cares about. To satisfy the consumer's utility function, firms improve their product characteristic values through R&D investment. If the R&D indeed leads to product innovation that also recovers the cost, the firm grows. Otherwise, the firm becomes insolvent and is replaced by a new firm.

A firm can choose to invest or not to invest in R&D activities. The decision is based on the recent profits of other firms engaging in R&D and then tuned by the firm's own intensity parameter γ. When a firm decides to engage in R&D, the probability that the firm will experience successful product innovation increases.

Using 1000 firms and 50 product characteristics to run simulations, the results showed that the evolution of the economy's output (GDP) closely follows the evolution of the R&D investment spending. Meanwhile, the customer's utility grows over time, due to a long-term net improvement in product quality. Moreover, when the R&D intensity parameter γ is increased, the increased R&D spending drives up consumption, output, and utility. Finally, ongoing endogenous product innovation leads to ongoing changes in the relative qualities of the goods and the distribution of product shares. The distribution tends to become skewed, with the degree of skewness depending on the opportunities for niching in the product characteristics space. As the number of firms grows large, the economy's business cycle dynamics tends to become dominated by the product innovation cycle of R&D investment.

2 New Methodologies and Technologies for Complex Economic Dynamics

In addition to the previous ten chapters, this book also includes three chapters, which may not be directly related to agent-based modeling that may provide some useful ideas or tools that can help the modeling, simulation, and analysis of agent-based modeling. We shall also briefly highlight each of them here.

This book is mainly focused on financial markets and market processes. One issue naturally arising is related to how different markets are coupled or connected, and to what degree. In the chapter, entitled "Measuring Market Integration: U.S. Stock and REIT Markets," Douglas Blackburn and N.K. Chidambaran take up the issue of identifying the existence and extent of financial integration. This is an important methodological issue that empirical studies often encounter, given the complex relationships and heterogeneity that underpins financial markets. The authors identify a potential joint hypothesis problem that past studies testing for financial integration may have suffered from. This problem arises when testing for the equality of risk premia across markets for a common (assumed) set of risk factors; nonetheless, there is a possibility that a conclusion claiming a rejection of integration may actually stem from the markets not sharing a common factor.

Overcoming the joint hypothesis problem means disentangling the two issues and examining them separately. They present an approach based on factor analysis and canonical correlation analysis. This approach can be summarized in two steps. First, one should determine the correct factor model in each market and determine whether the markets share a common factor. Second, one should develop economic proxies for the shared common factor and test for the equality of risk premia conditional on a common factor being present. The equality of risk premia is tested *only if* common factors exist. The authors argue that this procedure in fact gives more power to the tests. They test their method on US REIT and stock markets for 1985–2013.

When one attempts to understand social systems as complex systems, for instance, through agent-based models, computers and simulations play a very important role. As the scale and scope of these studies increase, simulations can be highly demanding in terms of data-storage and performance. This is likely to motivate more and more researchers to use highly powerful, supercomputers for their studies as the field matures. In the chapter, entitled "Supercomputer Technologies in Social Sciences: Existing Experience and Future Perspectives," Valery Makarov and Albert Bakhtizin document several forays into supercomputing in the social science literature.

The authors introduce some open-source platforms that already exist in the scientific community to perform large-scale, parallel computations. They discuss their hands-on experience in transforming a pre-existing agent-based model into a structure that can be executed on supercomputers. They also present their own valuable experiences and lessons in applying their models to supercomputers. From their experiences, C++ appears to be more efficient than Java for developing softwares running on supercomputers. The processes and issues related to translating a Java-based system into a C++ based system are also explained in the chapter.

Social sciences are distinct from natural sciences in terms of the potential of their theories to have an impact, for better or worse, on the actual lives of people. The great financial crisis of 2008, as some have argued, is a result of over reliance on unrealistic models with a narrow world-view, ignoring the complexities of the financial markets. Should more complex, sophisticated mathematical models be the solution? In the chapter, entitled "Is Risk Quantifiable?", Sami Al-Suwailem, Francisco Doria, and Mahmoud Kamel take up this issue and examine the methodological issues related to the use of or over-reliance on "formal" models in the social sciences, in particular in economics and finance.

The authors question whether the indeterminacy associated with future economic losses or failures can be accurately modeled and systematically quantified using *formal* mathematical systems. Using insights from metamathematics—in particular, Kurt Gödel's famous theorems on incompleteness from the 1930s—they point to the inherent epistemological limits that exist while using formal models. Consequently, they argue that a systematic evaluation or quantification of risk using formal models may remain an unachievable dream. They draw several examples and applications from real-world financial markets to strengthen their argument and the chapter serves as a cautionary message.

3 Conclusion and Outlook

Computational economics is a growing field [6]. With the advancement of technologies, modern economies exhibit complex dynamics that demand sophisticated methods to understand. As manifested in this book, agent-based modeling has been used to investigate contemporary financial institutions of dark pools and high-frequency trading (chapters "Dark Pool Usage and Equity Market Volatility", "Modelling Complex Financial Markets Using Real-Time Human-Agent Trading Experiments", and "Does High-Frequency Trading Matter?"). Meanwhile, agent-based modeling is also used to shed light on the market processes or the price discovery processes by examining the roles of traders' characteristics (chapter "Modelling Price Discovery in an Agent Based Model for Agriculture in Luxembourg"), learning schemes (chapters "Heterogeneity, Price Discovery and Inequality in an Agent-Based Scarf Economy" and "Rational Versus Adaptive Expectations in an Agent-Based Model of a Barter Economy"), information exposure (chapter "Does Persistent Learning or Limited Information Matter in Forward Premium Puzzle?"), social networks (chapter "Price Volatility on Investor's Social Network"), and herding pressure (chapter "The Transition from Brownian Motion to Boom-and-Bust Dynamics in Financial and Economic Systems"). Each of these efforts made is a contribution to enhancing our understanding and awareness of market complexity. Given this extent of complexity, markets may not perform well for many reasons, not just economic ones, but also psychological, behavioral, sociological, cultural, and even humanistic ones. Indeed, market phenomena have constituted an interdisciplinary subject for decades [11, 15, 17–19]. What agent-based modeling can offer is a framework that can integrate these interdisciplinary elements into a coherent body of knowledge.

Furthermore, agent-based modeling can also help modern economies that have been greatly influenced by the big data phenomenon [7]. By applying computational methods to big data, economists have addressed microeconomic issues in the internet marketplaces, such as pricing and product design. For example, Michael Dinerstein and his co-authors [8] ranked products in response to a consumer's search to decide which sellers get more business as well as the extent of price competition. Susan Athey and Denis Nekipelov [2] modeled advertiser behavior and looked at the impact of algorithm changes on welfare. To work with big data, Google chief economist Hal Varian proposed machine learning tools as new computational methods for econometrics [20]. What will the impact of machine learning be on economics? "Enormous" answered Susan Athey, Economics of Technology Professor at Stanford Graduate School of Business. "Econometricians will modify the methods and tailor them so that they meet the needs of social scientists primarily interested in conducting inference about causal effects and estimating the impact of counterfactual policies," explained Athey [21]. We also expect the collaborations between computer scientists and econometricians to be productive in the future.

Acknowledgements The authors are grateful for the research support in the form of the Taiwan Ministry of Science and Technology grants, MOST 104-2916-I-004-001-Al, 103-2410-H-004-009-MY3, and MOST 104-2811-H-004-003.

References

1. Arthur, W. B., Holland, J., LeBaron, B., Palmer, R., & Tayler, P. (1997). Asset pricing under endogenous expectations in an artificial stock market. In W. B. Arthur, S. Durlauf, & D. Lane (Eds.), *The economy as an evolving complex system II* (pp. 15–44). Reading, MA: Addison-Wesley.
2. Athey, S., & Nekipelov, D. (2010). *A Structural Model of Sponsored Search Advertising Auctions*. Sixth ad auctions workshop (Vol. 15).
3. Barber, B. M., Odean, T., & Zhu, N. (2009). Systematic noise. *Journal of Financial Markets, 12*(4), 547–569. Amsterdam: Elsevier.
4. Bilson, J. F. O. (1981), The 'speculative efficiency' hypothesis. *Journal of Business, 54*(3), 435–451.
5. Chen, S. H. (2017). Agent-based computational economics: How the idea originated and where it is going. Routledge.
6. Chen, S. H., Kaboudan, M., & Du, Y. R. (Eds.), (2018). *The Oxford handbook of computational economics and finance*. Oxford: Oxford University Press.
7. Chen, S. H., & Venkatachalam, R. (2017). Agent-based modelling as a foundation for big data. *Journal of Economic Methodology, 24*(4), 362–383.
8. Dinerstein, M., Einav, L., Levin, J., & Sundaresan, N. (2018), Consumer price search and platform design in internet commerce. *American Economic Review, 108*(7), 1820–59.
9. Estola, M. (2017). Newtonian microeconomics: A dynamic extension to neoclassical micro theory. Berlin: Springer.
10. Fama, E. (1984). Forward and spot exchange rates. *Journal of Monetary Economics, 14*(3), 319–338.
11. Halteman, J., & Noell, E. S. (2012). *Reckoning with markets: The role of moral reflection in economics*. Oxford: Oxford University Press.
12. Jackson, A. (2004). The aggregate behaviour of individual investors, working paper. http://ssrn.com/abstract=536942
13. Kumar, A., & Lee, C. M. C. (2006). Retail investor sentiment and return comovements. *Journal of Finance, LXI*(5). https://doi.org/10.1111/j.1540-6261.2006.01063.x
14. Longworth, D. (1981). Testing the efficiency of the Canadian-U.S. exchange market under the assumption of no risk premium. *The Journal of Finance, 36*(1), 43–49.
15. Lonkila, M. (2011). *Networks in the Russian market economy*. Basingstoke: Palgrave Macmillan.
16. Petrescu, M., Wedow, M., & Lari, N. (2017). Do dark pools amplify volatility in times of stress? *Applied Economics Letters, 24*(1), 25–29.
17. Rauch, J. E., & Casella, A. (Eds.), (2001). *Networks and markets*. New York: Russell Sage Foundation.

18. Staddon, J. (2012). *The malign hand of the markets: The insidious forces on wall street that are destroying financial markets—and what we can do about it*. New York: McGraw Hill Professional.
19. Tuckett, D. (2011). *Minding the markets: An emotional finance view of financial instability*. Berlin: Springer.
20. Varian, H. R. (2014, Spring). Big data: New tricks for econometrics. *Journal of Economic Perspectives, 28*(2), 3–28.
21. What Will The Impact Of Machine Learning Be On Economics? https://www.forbes.com/.../what-will-the-impact-of-machine-learning-be-on-economics/), Forbes, Jan 27, 2016.
22. Wolfram, S. (2002). *A new kind of science* (Vol. 5). Champaign: Wolfram Media.

Part I
Agent-Based Computational Economics

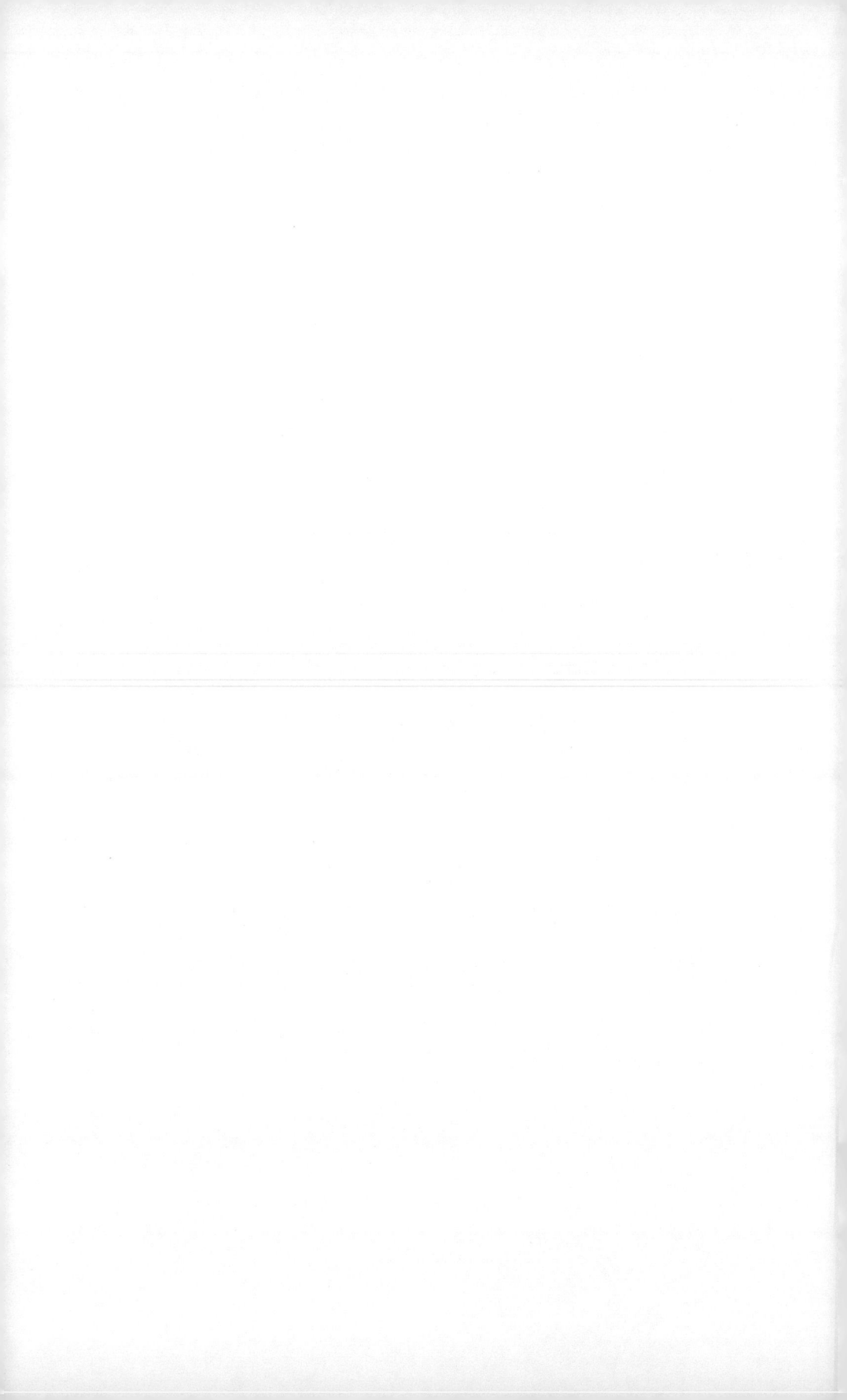

Dark Pool Usage and Equity Market Volatility

Yibing Xiong, Takashi Yamada, and Takao Terano

Abstract An agent-based simulation is conducted to explore the relationship between dark pool usage and equity market volatility. We model an order-driven stock market populated by liquidity traders who have different, but fixed, degrees of dark pool usage. The deviation between the order execution prices of different traders and the volume weighted average price of the market is calculated in an attempt to measure the effect of dark pool usage on price volatility. By simulating the stock market under different conditions, we find that the use of the dark pool enhances market stability. This volatility-decreasing effect is shown to become stronger as the usage of the dark pool increases, when the proportion of market orders is lower, and when market volatility is lower.

Keywords Dark pool · Market volatility · Agent-based model · Behavioral economics · Order-driven market

1 Introduction

In recent years, equity markets have become decentralized electronic networks, with increasingly fragmented liquidity over multiple venues. Take the US equity market as an example: between January 2009 and April 2014, the market shares of NYSE Euronext and NASDAQ OMX declined by approximately one-third and one-quarter, respectively [13], whereas off-exchange trading volume increased from one-quarter to more than one-third of the market.

Over the same period, the off-exchange market trading volume in dark pools (alternative trading venues where orders placed are not visible to other market participants) increased from 9% to 15%. Trading in dark pools has advantages and challenges. When an institutional investor uses a dark pool to sell a block of one

---

Y. Xiong (✉) · T. Yamada · T. Terano
Tokyo Institute of Technology, Yokohama, Kanagawa, Japan
e-mail: ybxiong@trn.dis.titech.ac.jp; tyamada@trn.dis.titech.ac.jp; terano@dis.titech.ac.jp

million shares, the lack of transparency actually works in the institutional investors favor, as it may result in a better realized price than if the sale was executed on an exchange. This is because dark pool participants do not disclose their trading intention to the exchange prior to execution—as there is no order book visible to the public, large orders have a much smaller market impact. In addition, there is the possibility of price improvement if the midpoint of the quoted bid and ask price is used for the transaction. This can help the investor save half of the spread. However, the lack of information about counterparties can lead to the uncertain execution of orders in the dark pool. In general, dark pools offer potential price improvements, but do not guarantee execution.

The increasing usage of the dark pool raises concerns about the impact of dark trading on market quality [18]. Although previous studies offered consistent conclusions on a variety of issues, i.e., market impact is significantly reduced for large orders, the relationship between dark pool usage and market volatility remains unclear. Some studies [5, 20] suggest that a higher dark pool market share is associated with higher market volatility, whereas others draw the opposite [1, 10] or more complex conclusions [19]. One potential explanation for such contrasting results is that these studies are conducted based on different dark pools and market conditions.

Previous studies concerning dark pool usage and market volatility can be classified into three categories according to their methodologies: (1) using empirical data from the market; (2) using an equilibrium model to predict the behavior of market participants; and (3) using an agent-based model to simulate the market dynamics. In the first case, because different traders have different dark trading participation rates and trading strategies, conclusions drawn from various markets are likely to be inconsistent. For example, using transaction data from 2005to 2007 covering three block-crossing dark pools (Liquidnet, Posit, and Pipeline), Ready [16] showed that dark pool usage is lower for stocks with the lowest spreads per share. However, using daily data collected by SIFMA (Securities Industry and Financial Markets Association) from 11 anonymous dark pools in 2009, Buti et al. [2] found that the market share of dark pools is higher for lower-spread and higher-volume stocks [2]. In the second category, the traders are considered to be fully rational in seeking to maximize a utility function [1, 20]. However, the equilibrium methods of these models are too abstract to be observed in financial markets [6], especially for dark trading. In addition, because traders are unlikely to exist as monopolists, their models are only applicable to short trading periods (single-period optimization model). Unlike such approaches, we investigate the relationship between dark pool usage and market volatility through zero-intelligence and repeatedly transacting traders. The transaction scenarios considered by previous agent-based models are relatively simple: in each trading session, only one agent submits one share order that will never be repeated [10]. To model the real market more accurately, we build a continuous double-auction market that allows multiple traders to trade continuously in environments with different order sizes, and explore how the use of the dark pool affects market volatility under different market conditions.

As of April 2014, there were 45 dark pools of three types in the USA: agency broker or exchange-owned, broker–dealer owned, and electronic market makers [4]. Agency broker and exchange-owned dark pools adopt a classical pricing mechanism that matches customer orders at prices derived from lit venues, such as the midpoint of the National Best Bid and Offer (NBBO). Examples include ITG Posit and Liquidnet, as well as midpoint dark order types offered by NASDAQ, BATS, and Direct Edge. The simplicity of this kind of dark pool means it is frequently used for model building [1, 10, 11, 20]. In this paper, we mainly focus on this type of dark pool.

Without loss of generality, traders can be considered to use dark pools in a simple way: each one has a fixed probability of using the dark pool during a transaction. This zero-intelligence agent design, which relies not on utility functions but an empirically generated distribution that characterizes the aggregate market participant behavior, has been widely used to represent agent actions. It was first introduced by Gode and Sunder [8] in a double-auction market, and its modifications have come to dominate the agent-based model (ABM) limit order book literature because of the ability to replicate dynamics such as spread variance and price diffusion rates [3, 7, 9, 14]. With zero-intelligence agents, we can focus on the overall influence of dark pool usage in the absence of possible nonlinear interactions arising from diverse individual trading strategies. Thus, we overcome the problems encountered when using empirical market data.

After describing a model that represents the usage of dark pools in equity markets, we answer the following two questions:

- How does the usage of dark pool affect market volatility?
- How does this influence vary according to different market conditions?

By analyzing the simulation results, we find that the use of dark pools decreases market volatility. Higher usage of dark pools strengthens this effect, as does a lower proportion of market orders and lower market volatility. However, changes in the cross-probability of dark pool use do not affect market volatility.

The remainder of this paper proceeds as follows: The next section briefly reviews the relevant literature considering dark pools and market quality. Section 3 describes the design of an artificial stock market with a dark pool. In Sect. 4, we describe simulations carried out to explore the relationship between dark pool usage and market volatility. Section 5 analyzes the results, and Sect. 6 gives our conclusions.

2 Literature Review

Many empirical studies have examined the relationship between dark trading and market quality. Ready [16] investigated the determinants of trading volume for NASDAQ stocks in three dark pools, and his results suggest that dark pool usage is lower for stocks with lower spreads, but higher for higher market volatility. Similarly, Ray [15] modeled the decision of whether to use a crossing network

(CN) or a traditional quoting exchange, and derived hypotheses regarding the factors that affect this decision. He then tested these hypotheses on realized CN volumes, and found that the likelihood of using CNs increases and then decreases as the relative bid–ask spread and other measures of market liquidity increase. Nimalendran and Ray [12] analyzed a proprietary high-frequency dataset, and found that the introduction of a CN resulted in increased bid–ask spreads and provided short-term technical information for informed traders. In contrast, Ye [19] modeled the market outcome when an informed trader can split trades between an exchange and a CN (dark pool). He found that the CN reduces price discovery and volatility, and that the dark pool share decreases with increasing volatility and spread.

Although the results of empirical studies largely depend on the data available, equilibrium models can examine the influence exerted by the dark pool on market quality, regardless of data limitations. Buti et al. [1] modeled a dynamic financial market in which traders submit orders either to a limit order book (LOB) or to a dark pool. They demonstrated that dark pool market share is higher when the LOB is deeper and has a narrower spread, when the tick size is large, and when traders seek protection from price impacts. Further, though the inside quoted depth of the LOB always decreases when a dark pool is introduced, the quoted spreads can narrow for liquid stocks and widen for illiquid ones. Zhu [20] formed a different opinion. In his model, the dark pool share is likely to increase and then decrease with increasing volatility and spread, and higher usage of the dark pool results in a higher spread and has a greater market impact.

Agent-based simulations can also be applied to test hypotheses regarding the influence of dark trading. Mo et al. [11] described the costs and benefits of trading small orders in dark pool markets through agent-based modeling. The simulated trading of 78 selected stocks demonstrated that dark pool market traders can obtain better execution rates when the dark pool market has more uninformed traders relative to informed traders. In addition, trading stocks with larger market capitalization yields better price improvement in dark pool markets. In another study, Mizuta et al. [10] built an artificial market model and found that as the dark pool was used more often, the markets stabilized. In addition, higher usage of the dark pool reduced the market impacts.

Table 1 summarizes these contrasting findings regarding dark pool usage and market quality, grouped by different analytical methods.

To develop a more realistic model representing the trading activity in real equity markets, we introduce a number of improvements on the basis of previous work. For example, in previous studies, traders only submit their orders once [10] or within a few trading periods [20], but our model allows the continuous submission of orders toward the exchange or dark pool. In addition, many studies limit orders to one share (Buti et al. [2]), but our model handles different order sizes obeying some statistical property and simulates the process of splitting orders into smaller chunks. Other important characteristics of real markets are also considered, such as the cancelation of orders, partially executed orders, and the change of order urgencies according to price movement. With these new features, we are able to explore various factors that affect the influence of the dark pool on market volatility.

Table 1 Selected studies and findings on dark pool usage and market quality (DPS: dark pool share)

Method	Paper	Findings	
		Market condition → DPS	DPS → Market quality
Empirical study	Ready [16]	spread↓ → DPS↓ volatility↑ → DPS↑	
	Ray [15]	volatility↑ → DPS(↑)↘	DPS↑ → spread↑ market impact↑
	Ye [19]	spread↓ → DPS↑ volatility↑ → DPS↓	
Equilibrium model	Buti ([2])	depth↑ tick size↑ → DPS↑	DPS↑ → volume↑
		spread↓ → DPS↑	DPS↑ → inside quoted depth↓
		stock liquidity↑ → DPS↑	DPS↑ → (liquidity) spread↓
		large orders↑ → DPS↑	→ (illiquidity)spread↑
	Zhu [20]	volatility↑ spread↑ → DPS↑(↘)	DPS↑ → spread↑ market impact↑
		uniformed traders↑ → DPS↑	DPS↑ → adverse selection↓
Agent-based model	Mizuta [10]	tick size↑ → DPS↑	DPS↑ → volatility↓ DPS↑ → market impact↓

3 Model

Our model is based on a simple limit order-driven market model developed by Maslov [9]. This agent-based model is chosen because many realistic features of the financial market are generated by only a few parameters. In the original model, a new trader appears and attempts to make a transaction at each time step. There is an equal probability of this new trader being a seller or a buyer, and fixed probabilities of trading one unit of stock at the market price or placing a limit order. Our model employs the idea of zero-intelligence traders who make their decisions based on exogenous distributions. To make this model more realistic, we allow multiple traders to continuously transact in one trading session. The order sizes follow a log-normal distribution, and large orders are split into small pieces. Furthermore, order urgency is introduced to determine order price, and this is updated according to market conditions. Finally, a midpoint dark pool is added as an alternative trading venue.

Thus, in our model, the market consists of an exchange and a dark pool, and one single stock is traded by liquidity traders. Their order types, order sizes, and order urgencies are assigned exogenously, but will be updated based on price movements in the market and the execution situations of individual orders. After receiving a trading task, the trader splits his/her order into smaller ones and submits them

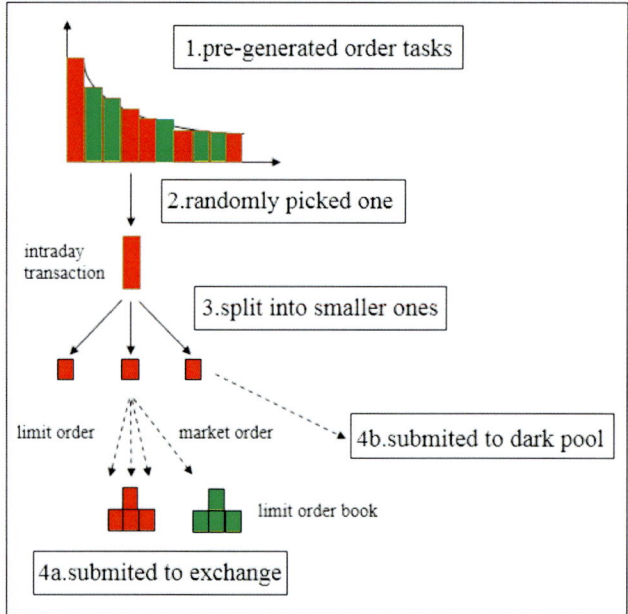

Fig. 1 Intraday trading process

successively. For each submission, the order will be submitted to either the exchange or the dark pool. The submission prices of orders to the exchange are determined by both the midprice of the exchange and the order urgency, and the latter will be adjusted according to price movements and order execution. Orders submitted to the dark pool do not have a specified price, as they will be executed at the midprice of the exchange when the dark pool is crossing and when there are enough orders on the other side of the dark pool. The intraday trading process of a trader is illustrated in Fig. 1.

3.1 Trading Sessions and Traders

In the model, intraday trading sessions are set as S, a total of D trading days are considered in the scenario.

The total number of liquidity traders is T, and they are evenly divided into M groups. Each group is distinguished from others with different but fixed levels of dark pool usage, denoted as U_1 to U_M.

3.2 Order Type, Order Size, and Order Urgency

Trading tasks are generated before simulation. In the task pool, there is an equal number of buy and sell order types. The order size follows a log-normal distribution. The order urgency is divided into three levels (low = −1, middle = 0, high = 1). In the initial stage, the total number of orders of each urgency level follows an $a : b : c$ distribution. The order urgency is updated after each intraday transaction. Each trader randomly picks one task from the pool as their trading task at the beginning of the day.

3.3 Order Submission and Cancelation

When submitting orders, traders use a time-weighted average price with randomization (TWAP)-like order placement strategy to mitigate the market impact. Each trader first splits his/her task order into F equal fractions, then randomly selects F trading sessions in the day and submits one piece of the order in each chosen session.

If a prior order has been submitted to the exchange but not fully executed by the time a new order is about to be submitted to the exchange, the old order will be canceled and the remaining amount will be added to the new one. However, orders submitted to the dark pool will not be canceled. This is because orders submitted to the dark pool follow the time priority mechanism.

3.4 Order Submitted Price

The order submitted price is mainly determined by order urgency. Take buy orders as an example, and suppose the current best ask is A_0 and best bid is B_0. If the urgency is low (−1), the trader will place it at Δ ticks away from the best bid in the LOB ($B_0 - \Delta \times tick$), where $\Delta \sim U[0, \lambda]$ (λ is the maximum value for the "offset") is an integer that follows an even distribution. At the medium urgency level (0), the trader will place the order at the midquote ($0.5 \times (A_0 + B_0)$). This is also the price executed in the dark pool. If the urgency is high (1), the trader will aggressively take liquidity up to a limit price impact, denoted as PI_{max} (PI_{max} refers to the highest price impact a trader can suffer), so the submitting price P_s will be

$$P_s = 0.5 \times (A_0 + B_0) \times (1 + PI_{max}). \tag{1}$$

This is a market order that attempts to absorb sell orders with prices lower than the submitting price. The relationship between order urgency and order price is illustrated in Table 2.

Table 2 Order urgency and submitting price

Urgency	Buy order		Sell order	
	Submit price	Probability	Submit price	Probability
−1	B_0	$1/(\lambda + 1)$	A_0	$1/(\lambda + 1)$
	...	$1/(\lambda + 1)$	...	$1/(\lambda + 1)$
	$B_0 - tick \times \lambda$	$1/(\lambda + 1)$	$A_0 + tick \times \lambda$	$1/(\lambda + 1)$
0	$0.5 \times (A_0 + B_0)$		$0.5 \times (A_0 + B_0)$	
1	$0.5 \times (A_0 + B_0) \times (1 + PI_{max})$		$0.5 \times (A_0 + B_0) \times (1 - PI_{max})$	

3.5 Order Execution in Exchange and Dark Pool

Order execution in the exchange follows the traditional price-then-time mechanism, and unexecuted orders are stored in the LOB. Order execution in the dark pool follows the time priority mechanism. Each trader is assumed to have a fixed probability of using the dark pool, denoted as U_i (the probability of using the dark pool is exogenously specified). Hence, when a trader decides to submit an order, he/she has a probability of U_i of placing it on the dark pool. At the end of each trading session, buy orders and sell orders in the dark pool have a probability of cp of being crossed at the midprice ($0.5 \times (A_0 + B_0)$) of the exchange. Unexecuted orders in the dark pool are left for the next crossing.

There are two types of midpoint dark pools in the market. For example, in Liquidnet, the matching acts continuously, whereas ITG only performs order matching a few times a day [20]. The probability cp is introduced to consider both situations.

3.6 Order Urgency Updated After Intraday Transaction

The order urgency will be updated by price movements and the order execution condition after intraday transactions. For example, if the stock price has increased by a significant level in one trading day, then the sell order urgency will increase and the buy order urgency will decrease for the next day. Another adjustment considers the order execution condition. Suppose that one trader has a task order size for a day of S_t, and the order executed that day is S_{exe}. At the end of the day, if there are still a considerable number of orders that have not been fully executed ($(S_t - S_{exe})/S_t > UT$, where UT is some threshold for unexecuted orders), this trader may continue to execute the order the next day, but with higher urgency. Suppose that the open price and close price for a trading day are P_{open} and P_{close}. At the beginning of the next day, the order urgency will be adjusted according to the following rules:

- if $P_{close} > P_{open} \times (1 + \theta)$ (θ is the threshold for urgency adjustment), traders will pick a new task. If it is a buy order, its urgency has a 50% probability to

minus 1 (50% chance of maintaining the original urgency); if it is a sell order, its urgency has a 50% probability to plus 1.
- if $P_{close} < P_{open} \times (1-\theta)$, traders will pick up a new task. If it is a buy order, its urgency has a 50% probability to plus 1 (50% chance of maintaining the original urgency); if it is a sell order, its urgency has a 50% probability to minus 1.
- else if $(S_t - S_{exe})/S_t > UT$, traders have an equal probability (50%) of continuing to execute the previous remaining order and increase its urgency with 1, or dropping it and picking a new one.

4 Experiment

To test how different extents of dark pool usage affect market volatility, T traders are divided into three groups, each distinguished from the others by different but fixed levels of dark pool usage, set as U_1, U_2, and U_3, respectively. The probabilities of dark order submission of these three groups are denoted as $[U_1:U_2:U_3]$. U_1 is assigned as the benchmark probability of 0. U_2 and U_3 refer to low and high usage of the dark pool, respectively.

In addition to dark pool usage, there are two other factors that may have a significant influence on the order executed price in the dark pool, and thus reflect market volatility. The first is the dark pool cross-probability. A lower cross-probability indicates a relatively longer order execution delay in the dark pool. In this case, there may be a significant difference between the midprice in the submission session and the execution price in the dark pool, especially when the market is volatile. The second factor is the proportion of market orders. Although an increase in market orders makes the market more volatile and implies a higher price improvement in dark trading, this increase also reflects a lack of liquidity in the main exchange, and implies the same situation in the dark pool. Thus, there is an increased execution risk for the dark orders. During the simulations, different values are assigned to these two parameters to examine their influence on market volatility. Table 3 lists the parameter values used in the simulation.

These parameters fall into three categories. Some are effectively insensitive. For example, the experiments show that there is no significant influence on the results for 100–3000 agents. The experiments described in this paper consider a relatively small simulation time. The model parameters evaluated by this method include T, D, S, F, and P_0. Some of the parameters are based on the values used in the previous models, like λ, θ, and UT [9, 10], whereas others are taken from empirical studies, such as $tick$ and PI_{max} [16, 19].

Before conducting the experiments, we first confirm that the model can account for the main characteristics of financial markets, which reflect empirical statistical moments of a market pricing series. In validating that markets are efficient, it is common practice to show there is no predictability in the price returns of assets. To demonstrate this, the autocorrelation of price returns should show that there is no

Table 3 Parameters in simulation

Description	Symbol	Value
Number of traders	T	100
Number of groups	M	3
Trading days	D	10
Intraday trading sessions	S	100
Order split fractions	F	10
Stock initial price	P_0	10
Tick size	$tick$	0.01
Dark pool cross-probability	cp	[0.1,1]
Order urgency distribution	[a:b:c]	[0.9-MO:0.1:MO]
Market order proportion	MO	[0.3,0.8]
Dark order submission probability	$[U_1:U_2:U_3]$	[0:0.2:0.4]
Order placement depth	λ	4
Max price impact	PI_{max}	0.003
Urgency adjust threshold	θ	0.05
Unexecuted order threshold	UT	0.2

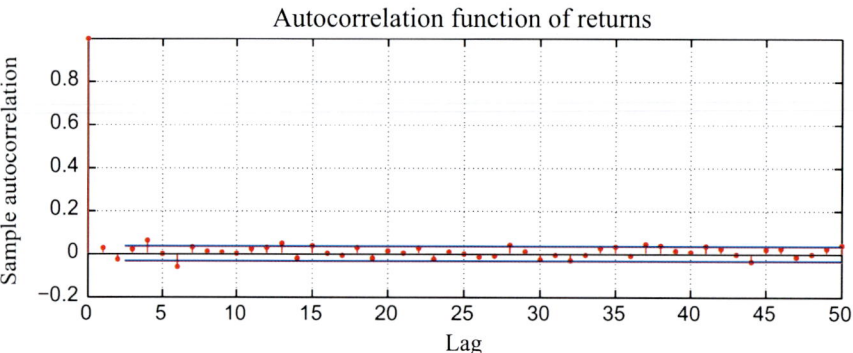

Fig. 2 Autocorrelation of returns for dark pool trading model

correlation in the time series of returns. Figure 2 shows that the price movements generated by the model are in line with the empirical evidence in terms of an absence of autocorrelation.

The characteristic of volatility clustering is seen in the squared price returns for securities that have a slowly decaying autocorrelation in variance. In our model, the autocorrelation function of squared returns displays a slow decaying pattern, as shown in Fig. 3.

The autocorrelation values are listed in Table 4.

In addition, it has been widely observed that the empirical distribution of financial returns has a fat tail. The distribution of returns is shown in Fig. 4. The kurtosis of this distribution is greater than 3, indicating the existence of a fat tail.

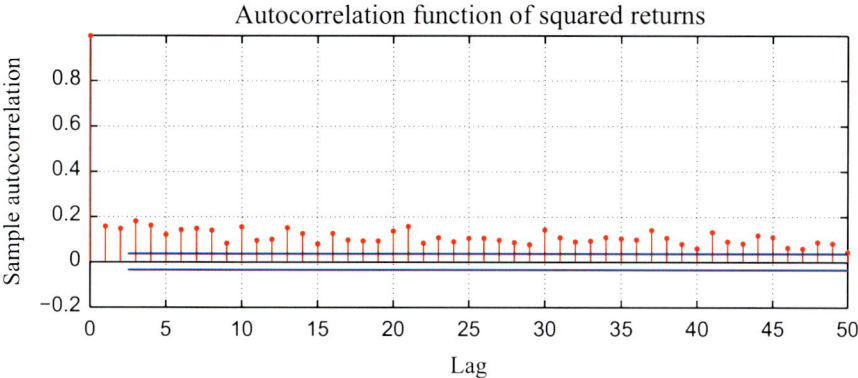

Fig. 3 Autocorrelation of squared returns for dark pool trading model

Table 4 Values of autocorrelation returns and squared returns

Returns	1	0.0282	−0.0253	0.0217	0.0628	0.0007	−0.0591	0.0323	0.0078 ...
Squared returns	1	0.1578	0.1481	0.1808	0.162	0.1216	0.1426	0.1401	0.0836 ...

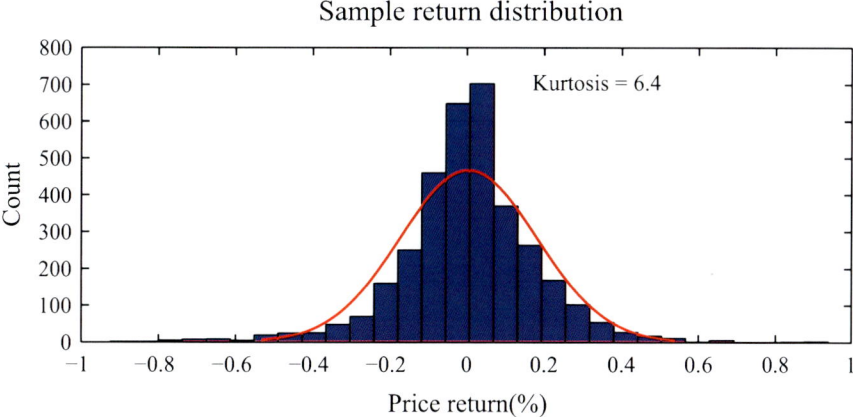

Fig. 4 Return distribution for dark pool trading model

In each simulation, there are $TS = D \times S = 1000$ trading sessions. Assuming the stock price at session t is P_t, the return at session t is R_t, which is calculated as:

$$R_t = Ln(P_t/P_{t-1}) \tag{2}$$

The daily volatility (Vol) of the market is calculated as:

$$\text{Vol} = \sqrt{S\left(\frac{1}{TS-1}\sum_{i=1}^{n} R_t^2 - \frac{1}{TS(TS-1)}\left(\sum_{i=1}^{n} R_t\right)^2\right)} \quad (3)$$

Assuming that the trading volume at session t is V_t, the volume weighted average price (VWAP) of the market is calculated as:

$$\text{VWAP(market)} = \frac{\sum_{t=1}^{TS}(P_t \times V_t)}{\sum_{t=1}^{TS} V_t} \quad (4)$$

In the same way, we can calculate the VWAP of trader k, denoted as VWAP(k). The price slippage of trader k ($PS(k)$) is calculated as:

$$PS(k) = \begin{cases} (\text{VWAP}(k) - \text{VWAP(market)})/\text{VWAP(market)}, & \text{sell order} \\ (\text{VWAP(market)} - \text{VWAP}(k))/\text{VWAP(market)}, & \text{buy order} \end{cases} \quad (5)$$

Let group(A) refer to the group of n traders ($k_1, k_2, \ldots, k_n$) in which the dark pool submission probability is A. $V(k_i)$ refers to the trading volume of trader k_i. The VWAP slippage of group A (VWAPS(A)) is then calculated as:

$$\text{VWAPS}(A) = \frac{\sum_{i=1}^{n} PS(k_i) \times V(k_i)}{\sum_{i=1}^{n} V(k_i)} \quad (6)$$

The variance of executed prices (with respect to VWAP) of group A (VEP(A)) is calculated as the weighted square of the price slippage:

$$\text{VEP}(A) = \frac{\sum_{i=1}^{n} PS(k_i)^2 \times V(k_i)}{\sum_{i=1}^{n} V(k_i)} \quad (7)$$

Inference

Price volatility measures the extent of price changes with respect to an average price. For different types of traders, the average prices set here are the same (VWAP of the market). Higher variance in the executed price of a group indicates that the execution of orders in this group tends to make the stock price deviate more from the previous price, thus making the market more volatile [17]. Take VEP(0) as the benchmark. The impact of group(A) on market volatility with respect to the benchmark (RVEP(A)) is calculated as:

$$\text{RVEP}(A) = \frac{\text{VEP}(A)}{\text{VEP}(0)} \quad (8)$$

As the market volatility is the combination of the price volatilities of these three groups, higher price volatility in one group will increase the price volatility of the whole market. This leads to:

1. If RVEP(A) > 1, usage of the dark pool makes the market more volatile;
2. If RVEP(A) < 1, usage of the dark pool makes the market less volatile.

The price slippage is a derived measurement based on price volatility. The volatility level is mainly determined by the liquidity situation of the market, and the price slippage shows how different traders contribute to this volatility level. The advantage in using price slippage is its ability to compare differences in the order execution situations of traders (with different dark pool usages) within one simulation, rather than measuring volatility differences among different simulations. Because the order-balance, order-quantity, and order-urgency in different simulations may slightly affect market volatility levels, it is more accurate to compare the price slippages of different dark pool users within one simulation.

5 Result

During each simulation, cp was assigned a value from the set $\{0.1, 0.2, \ldots, 0.9, 1.0\}$ and MO took a value from $\{0.3, 0.4, \ldots, 0.7, 0.8\}$. The simulation was repeated 50 times for each setting, giving a total of 3000 simulations. For each simulation, the market volatility, dark pool usage, and the variance of executed prices of each group were recorded. The analysis of variance (ANOVA) was used to test whether different dark pool usage, dark pool cross-probability, and market order proportion had a significant effect on the values of VEP and RVEP.

The means of VEP(0), VEP(0.2), and VEP(0.4) over the 3000 simulations were compared by one-way ANOVA to see whether different dark pool usages led to different values of VEP. The results in Table 5 indicate that VEP changes with the dark pool usage.

In addition, two-way ANOVA was applied to test the effects of dark pool cross-probability and market order proportion on the mean of RVEP(0.2) and RVEP(0.4). Table 6 presents the analysis results for RVEP(0.2).

Table 5 One-way ANOVA to analyze the effect of dark pool usage on VEP

Source of variation	ANOVA table					
	SS	df	MS	F	P-value	F crit
Between groups	12,534	2	6267	9.62	<0.01	3.0
Within groups	5,862,478	8997	652			
Total	5,875,012	8999				

SS: sum of the squared errors; df: degree of freedom, MS: mean squared error

Table 6 Two-way ANOVA to analyze the effect of dark pool cross-probability and market order proportion on RVEP(0.2)

Source of variation	SS	df	MS	F	P-value	F crit
ANOVA table						
Cross-probability	0.51	9	0.057	0.78	0.63	1.88
Market order	5.48	5	1.10	15.2	<0.01	2.22
Interaction	3.20	45	0.07	0.99	0.50	1.37
Within	211	2940	0.07			
Total	221	2999				

SS: sum of the squared errors; df: degree of freedom; MS: mean squared error

Table 7 Linear regression (1) (Objective variable: RVEP. Explanation variables: dark pool usage, cross-probability, market order proportion, volatility)

Regression statistics	
Multiple R	0.92
R square	0.84
Adjusted R square	0.84
Standard error	0.02
Observation	6000

Table 8 Linear regression (2) (Objective variable: RVEP. Explanation variables: dark pool usage, cross-probability, market order proportion, volatility. SE: standard error)

	Coefficients	SE	t Stat	P-value	Lower 95%	Upper 95%
Intercept	0.73	0.003	247	0	0.72	0.73
DP usage	−0.24	0.003	−68	<0.01	−0.24	−0.23
Cross-prob.	0.001	0.0012	0.88	0.38	−0.001	0.003
Market order	0.3	0.006	50	<0.01	0.29	0.32
Volatility	1.33	0.12	11	<0.01	1.10	1.57

Table 6 illustrates that the value of RVEP is affected by the proportion of market orders but is not affected by the cross-probability of the dark pool.

Next, we analyzed how RVEP changes according to these factors. Taking RVEP as the objective variable, we conducted a linear regression to observe the effect of dark pool usage, cross-probability, market order proportion, and volatility on RVEP. Among these factors, market volatility is a statistic. RVEP and the market volatility were calculated after each simulation, and the analysis investigates the relationship between the two. The following tables present the linear regression results.

Tables 7 and 8 indicate that the market volatility and market order proportion are important and have a positive relationship with RVEP. Moreover, the dark pool usage exhibits a negative relationship with RVEP, which suggests that higher usage of the dark pool leads to smaller RVEP values and tends to stabilize the market. The dark pool cross-probability does not have a significant effect on RVEP.

Figures 5, 6, 7 plot RVEP(0.2) and RVEP(0.4) according to different dark pool cross-probabilities, market order proportions, and market volatilities, respectively.

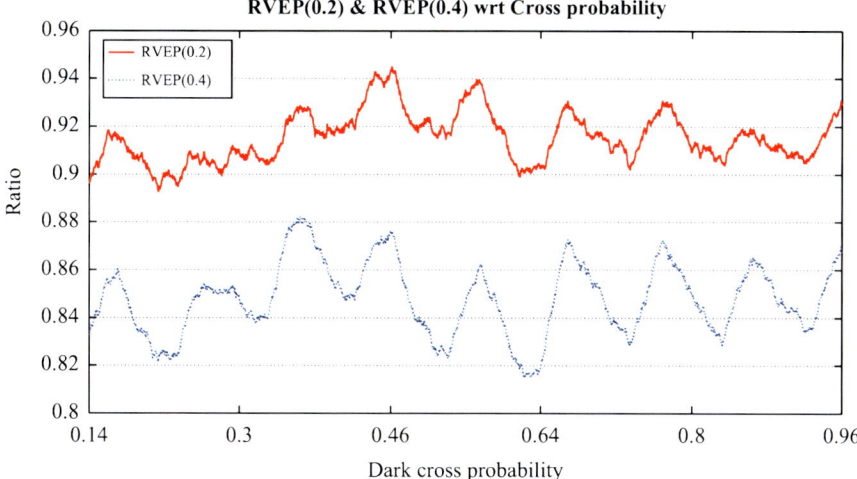

Fig. 5 RVEP with respect to dark pool cross-probability

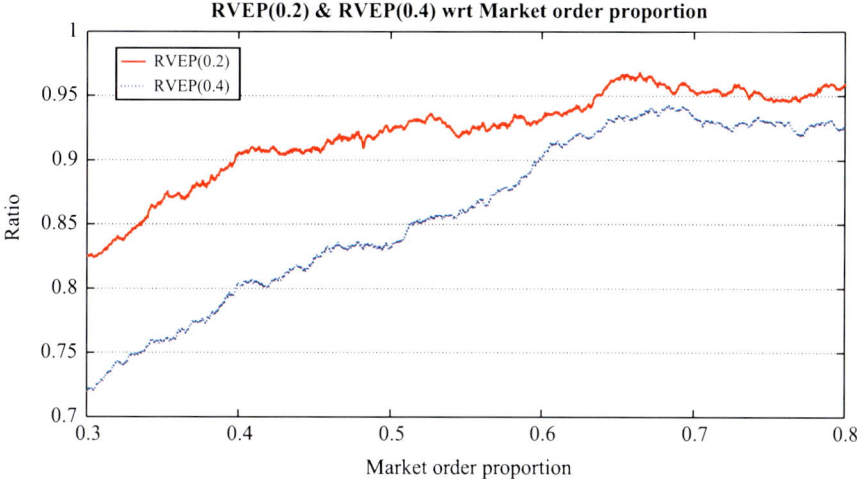

Fig. 6 RVEP with respect to market order proportion

In Fig. 5, the results are ordered by the cross-probability of the 3000 simulations and the RVEP is shown as a moving average (over intervals of 500). The red solid line and blue dashed line are the RVEP values given by low and high usage of the dark pool, respectively. Both RVEP(0.2) and RVEP(0.4) are less than 1 for all cross-probabilities, which indicates that using the dark pool decreases market volatility. In addition, the value of RVEP(0.4) is consistently lower than that of RVEP(0.2), suggesting that higher usage of the dark pool makes the market more stable. Both

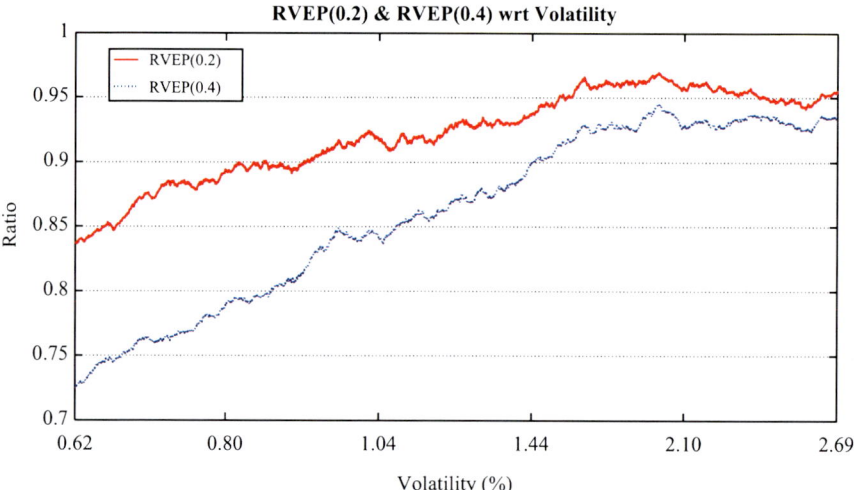

Fig. 7 RVEP with respect to market volatility

curves fluctuate within small ranges, indicating that the dark pool cross-probability does not affect market volatility. These fluctuations are caused by different market order proportions.

The market tends to be stable when there is a balance between buy orders and sell orders. In this case, the price will move up and down around the midprice. Such order-balance guarantees the executions in the dark pool, and these executions at the midprice of the exchange stabilize the price movements. However, the market tends to be more volatile when there exists an extreme order-imbalance. When these imbalanced orders are submitted to the exchange, they cause rapid price changes and larger spreads. However, if such an imbalance occurs in the dark pool, only a few executions are made until a new balance is formed between buy orders and sell orders. In this sense, orders submitted to the dark pool tend to inhibit the trading volume when the price is changing rapidly, but enhance the trading volume when the price is relatively stable.

According to the above analysis, increased usage of the midpoint dark pool leads to a less volatile market. Although different midpoint dark pools have different crossing mechanisms, the results indicate that the crossing time may not have a significant influence on market volatility.

Figure 6 shows an upward trends in RVEP with an increase in market order proportion. Moreover, the difference between RVEP(0.2) and RVEP(0.4) decreases as the market order proportion rises. This indicates that when the proportion of market orders is small, higher usage of the dark pool decreases market volatility (corresponding to lower RVEP values and a larger difference between RVEP(0.2) and RVEP(0.4)). For larger numbers of market orders, higher usage of the dark pool has less of an effect on decreasing the market volatility (corresponding to higher RVEP values and a smaller difference between RVEP(0.2) and RVEP(0.4)).

Similar to Figs. 6, 7 reveals the relationship between RVEP and market volatility. This graph shows that when market volatility is low, higher usage of the dark pool has a definite suppressing effect. However, when market volatility is high, higher usage of the dark pool causes a less noticeable suppression.

When market volatility is low, there is a balance between buy orders and sell orders, and executions in the midpoint dark pool tend to make the market stable. When market volatility is high, only a small number of executions will happen in the dark pool, so this volatility-decreasing effect is weakened. The difference between RVEP(0.2) and RVEP(0.4) decreases under high market volatility, possibly because the higher usage of the dark pool will not lead to many more executions when the market is very volatile and there is an extreme order-imbalance.

This volatility-decreasing effect of the dark pool is consistent with the findings of Buti et al. [1, 2] and Mizuta et al. [10]. Buti et al. [1, 2] stated that increased dark pool trading activity tends to be associated with lower spreads and lower return volatilities. Similarly, Mizuta et al. [10] found that as dark pool use increases, the markets become more stable. In contrast, Ray [15] reported that following dark pool transactions, the bid–ask spreads tend to widen and price impacts tend to increase. This may be because the CN in his sample was relatively small: the average dark pool usage in Ray's sample was 1.5%, whereas we considered a usage of 20% (which is closer to the present market rate of 15%). In another model, Zhu [20] suggested that a higher dark pool market share is associated with wider spreads and higher price impacts. This difference may be because Zhu's conclusion is based on the assumption that all traders are fully rational. Our model, however, used zero-intelligence agents to investigate the relationship between dark pool usage and market volatility, and did not consider the impact of specific trading strategies.

6 Conclusion

We used an agent-based model to analyze the relationship between dark pool usage and market volatility. We focused on one typical type of dark pool in which customer orders are matched at the midpoint of the exchange's bid and ask prices. Simulation results showed that the use of this midpoint dark pool decreases market volatility. This volatility-suppressing effect becomes stronger when the usage of the dark pool is higher, when the proportion of market orders is lower, and when market volatility is lower. However, changes in the cross-probability of the dark pool were not found to have any effects.

References

1. Buti, S., Rindi, B., & Werner, I. M. (2011). *Dark Pool Trading Strategies*. Charles A Dice Center Working Paper (2010-6).

2. Buti S, Rindi, B., & Werner, I. M. (2011). *Diving into Dark Pools*. Charles A Dice Center Working Paper (2010-10).
3. Challet, D., & Stinchcombe, R. (2001). Analyzing and modeling 1+ 1D markets. *Physica A: Statistical Mechanics and its Applications, 300*(1), 285–299.
4. Cheridito, P., & Sepin, T. (2014). Optimal trade execution with a dark pool and adverse selection. Available at SSRN 2490234.
5. Degryse, H., De Jong, F., Van Kervel, V. (2011). The impact of dark and visible fragmentation on market quality. SSRN eLibrary.
6. Farmer, J. D., & Foley, D. (2009). The economy needs agent-based modelling. *Nature, 460*(7256), 685–686.
7. Farmer, J. D., Patelli, P., & Zovko, I. I. (2005). The predictive power of zero intelligence in financial markets. *Proceedings of the National Academy of Sciences of the United States of America, 102*(6), 2254–2259.
8. Gode, D. K., & Sunder, S. (1993). Allocative efficiency of markets with zero-intelligence traders: Market as a partial substitute for individual rationality. *Journal of Political Economy, 101*, 119–137.
9. Maslov, S. (2000). Simple model of a limit order-driven market. *Physica A: Statistical Mechanics and Its Applications, 278*(3), 571–578.
10. Mizuta, T., Matsumoto, W., Kosugi, S., Izumi, K., Kusumoto, T., & Yoshimura, S. (2014). Do dark pools stabilize markets and reduce market impacts? Investigations using multi-agent simulations. In *2104 IEEE Conference on Computational Intelligence for Financial Engineering & Economics (CIFEr)* (pp. 71–76). New York: IEEE.
11. Mo, S. Y. K., Paddrik, M., & Yang, S. Y. (2013). A study of dark pool trading using an agent-based model. In *2013 IEEE Conference on Computational Intelligence for Financial Engineering & Economics (CIFEr)* (pp. 19–26). New York: IEEE.
12. Nimalendran, M., & Ray, S. (2011), Informed trading in dark pools. SSRN eLibrary.
13. Preece, R., & Rosov, S. (2014). Dark trading and equity market quality. *Financial Analysts Journal, 70*(6), 33–48.
14. Preis, T., Golke, S., Paul, W., & Schneider, J. J. (2006). Multi-agent-based order book model of financial markets. *Europhysics Letters, 75*(3), 510.
15. Ray, S. (2010). A match in the dark: understanding crossing network liquidity. Available at SSRN 1535331.
16. Ready, M. J. (2010). *Determinants of volume in dark pools*. Working Paper, University of Wisconsin-Madison.
17. Satchell, S., & Knight, J. (2011). *Forecasting volatility in the financial markets*. Oxford: Butterworth-Heinemann.
18. Securities and Exchange Commission. (2010). Concept release on equity market structure. *Federal Register, 75*(13), 3594–3614.
19. Ye, M. (2011). A glimpse into the dark: Price formation, transaction cost and market share of the crossing network. Transaction Cost and Market Share of the Crossing Network June 9 (2011).
20. Zhu, H. (2013). Do dark pools harm price discovery? *Review of Financial Studies. 27*(3), 747–789.

Modelling Complex Financial Markets Using Real-Time Human–Agent Trading Experiments

John Cartlidge and Dave Cliff

Abstract To understand the impact of high-frequency trading (HFT) systems on financial-market dynamics, a series of controlled real-time experiments involving humans and automated trading agents were performed. These experiments fall at the interdisciplinary boundary between the more traditional fields of behavioural economics (human-only experiments) and agent-based computational economics (agent-only simulations). Experimental results demonstrate that: (a) faster financial trading agents can reduce market efficiency—a worrying result given the race towards zero-latency (ever faster trading) observed in real markets; and (b) faster agents can lead to market fragmentation, such that markets transition from a regime where humans and agents freely interact to a regime where agents are more likely to trade between themselves—a result that has also been observed in real financial markets. It is also shown that (c) realism in experimental design can significantly alter market dynamics—suggesting that, if we want to understand complexity in real financial markets, it is finally time to move away from the simple experimental economics models first introduced in the 1960s.

Keywords Agent-based computational economics · Automated trading · Continuous double auction · Experimental economics · High-frequency trading · Human–agent experiments · Robot phase transition · Trading agents

1 Introduction

In recent years, the financial markets have undergone a rapid and profound transformation from a highly regulated human-centred system to a less-regulated and more fragmented computerised system containing a mixture of humans and automated trading systems (ATS)—computerised systems that automatically select and execute

J. Cartlidge (✉) · D. Cliff
Department of Computer Science, University of Bristol, Bristol, UK
e-mail: john.cartlidge@bristol.ac.uk; dc@cs.bris.ac.uk

a trade with no human guidance or interference. For hundreds of years, financial trading was conducted by humans, for humans, via face-to-face (or latterly telephone) interactions. Today, the vast majority of trades are executed electronically and anonymously at computerised trading venues where human traders and ATS interact. Homogeneous human-only markets have become heterogeneous human-ATS markets, with recent estimates suggesting that ATS now initiate between 30% and 70% of all trades in the major US and European equity markets [25].

As computerisation has altered the structure of financial markets, so too the dynamics (and systemic risk) have changed. In particular, trading velocity (the number of trades that occur in unit time) has dramatically increased [25]; stocks and other instruments exhibit rapid price fluctuations (*fractures*) over subsecond time-intervals [36]; and widespread system crashes occur at astonishingly high speed. Most infamously, the *flash crash* of 6th May 2010 saw the Dow Jones Industrial Average (DJIA) plunge around 7% ($1 trillion) in 5 min, before recovering most of the fall over the following 20 min [21, 37]. Alarmingly, during the crash, some major company stocks (e.g., Accenture) fell to just one cent, while others (e.g., Hewlett-Packard) increased in value to over $100,000. These dynamics were unprecedented, but are not unique. Although unwanted, flash crashes are now an accepted feature of modern financial markets.[1]

To accurately model financial systems, it is now no longer sufficient to consider human traders only; it is also necessary to model ATS. To this end, we take a bottom-up, agent-based experimental economics approach to modelling financial systems. Using purpose built financial trading platforms, we present a series of controlled real-time experiments between human traders and automated trading agents, designed to observe and understand the impact of ATS on market dynamics. Conducted at the University of Bristol, UK, these experiments fall at the interdisciplinary boundary between the more traditional fields of experimental economics (all human participants) and agent-based computational economics (all agent simulation models) and offer a new insight into the effects that agent strategy, agent speed, human experience, and experiment design have on the dynamics of heterogeneous human–agent markets.

Results demonstrate that: (a) the speed of financial agents has an impact on market efficiency—in particular, it is shown that faster financial trading agents can lead to less efficient markets, a worrying result given the race towards zero-latency (ever faster trading) observed in real markets; (b) faster agents can lead to market fragmentation, such that markets transition from a regime where humans and agents freely interact to a regime where agents are more likely to trade between themselves—a result that has also been observed in real financial markets; (c) experiment design, such as discrete-time (where participants strictly act in turns) versus real-time systems (where participants can act simultaneously and at any

---

[1]Flash crashes are now so commonplace that during the writing of this chapter, a flash crash occurred in the FX rate of the British Pound (GBP). On 7 Oct 2016, GBP experienced a 6% drop in 2 min, before recovering most of the losses [53]—a typical flash crash characteristic.

time), can dramatically affect results, leading to the conclusion that, where possible, a more *realistic* experiment design should be chosen—a result that suggests it is finally time to move away from Vernon Smith's traditional discrete time models of experimental economics, first introduced in the 1960s.

This chapter is organised as follows. Section 2 introduces the argument that financial systems are inherently complex ecosystems that are best modelled using agent-based approaches rather than neoclassical economic models. Understanding the causes and consequences of transient non-linear dynamics—e.g., *fractures* and *flash crashes* that exacerbate systemic risk for the entire global financial system—provides the primary motivation for this research. In Sect. 3, the agent-based experimental economics approach—i.e., human–agent financial trading experiments—is introduced and contextualised with a chronological literature review. Section 4 introduces the trading platform used for experiments, and details experiment design and configuration. Empirical results presented in Sect. 5 demonstrate market fragmentation—a significantly higher proportion of agent-only and human-only trading in markets containing super-humanly fast agents. Since the experimental market we construct is too constrained to exhibit fractures directly, in Sect. 6 we interpret this result as proxy evidence for the robot phase transition associated with fractures in real markets. In Sect. 7, conclusions are drawn, and some avenues for future research are outlined.

2 Motivation

There exists a fundamental problem facing financial-market regulators—current understanding of the dynamics of financial systems is woefully inadequate; there is simply no sound theoretical way of knowing what the systemic effect of a structural change will be [7]. Therefore, when policy makers introduce new market regulation, they are effectively trial and error testing in the live markets. This is a concerning state of affairs that has negative ramifications for us all, and it provides adequate motivation for the research presented here.

In this section, it is argued that our lack of understanding is a symptom of the dominant neoclassical economic paradigms of rational expectations and oversimplified equilibrium models. However, a solution is proposed. It has been compellingly argued elsewhere that economic systems are best considered through the paradigm of complexity [41]. Agent-based models—dynamic systems of heterogeneous interacting agents—present a way to model the financial economy as a complex system [22] and can naturally be extended to incorporate human (as living agent) interactions. In addition, the converging traditions of behavioural [38] and experimental [48] economics can address non-rational human behaviours such as overconfidence and fear using controlled laboratory experiments. Here, we present a hybrid approach—mixed human–agent financial trading experiments—that we believe offers a path to enlightenment.

2.1 Complex Economic Systems

Neoclassical economics relies on assumptions such as market efficiency, simple equilibrium, agent rationality, and Adam Smith's invisible hand. These concepts have become so ingrained that they tend to supersede empirical evidence, with many economists subliminally nurturing an implicit Platonic idealism about market behaviour that is divorced from reality. As Robert Nelson argued in his book, *Economics as Religion*, it is almost "as if the marketplace has been deified" [42]. Consequently, no neoclassical framework exists to understand and mitigate *wild* market dynamics such as flash crashes and fractures. It is necessary, therefore, to develop "a more pragmatic and realistic representation of what is going on in financial markets, and to focus on data, which should always supersede perfect equations and aesthetic axioms" [7]. Disturbingly, despite global capitalism's existential reliance on well-functioning financial markets, there exist no mature models to understand and predict issues of systemic risk [14]. Policy makers, therefore, are essentially acting in the dark, with each new regulatory iteration perturbing the market in unanticipated ways.

Fuelled by disillusionment with orthodox models, and a desire to address the inadequacies of naïve policy making, there is a trend towards alternative economic modelling paradigms: (1) *Non-equilibrium economics* focuses on non-equilibrium processes that transform the economy from within, and include the related and significantly overlapping fields of *evolutionary economics* (the study of processes that transform the economy through the actions of diverse agents from experience and interactions, using an evolutionary methodology, e.g., [43]), *complexity economics* (seeing the economy not as a system in equilibrium, but as one in motion, perpetually constructing itself anew, e.g., [2]), *circular and cumulative causation* (CCC) (understanding the real dynamic and self-reinforcing aspects of economic phenomena, e.g., [5]), and *network effects and cascading effects* (modelling the economy as a network of entities connected by inter-relationships) [3, 6, 20, 46]; (2) *Agent-based models* potentially present a way to model the financial economy as a complex system, while taking human adaptation and learning into account [7, 22, 23, 41]; (3) *Behavioural economics* addresses the effects of social, cognitive, and emotional factors on the economic decisions of individuals [3, 38]; (4) *Experimental economics* is the application of experimental methods to study economic questions. Data collected in experiments are used to test the validity of economic theories, quantify the effects, and illuminate market mechanisms [48].

Following this movement away from traditional economic models, in the research presented here, markets are modelled using an approach that straddles the interdisciplinary boundary between experimental economics and agent-based computational economics. Controlled real-time experiments between human traders and automated financial trading agents (henceforth, referred to simply as *agents*, or alternatively as *robots*) provide a novel perspective on real-world markets, through which we can hope to better understand, and better regulate for, their complex dynamics.

2.2 Broken Markets: Flash Crashes and Subsecond Fractures

As algorithmic trading has become common over the past decade, automated trading systems (ATS) have been developed with truly super-human performance, assimilating and processing huge quantities of data, making trading decisions, and executing them, on subsecond timescales. This has enabled what is known as *high-frequency trading* (HFT), where ATS take positions in the market (e.g., by buying a block of shares) for a very short period of perhaps 1 or 2 s or less, before reversing the position (e.g., selling the block of shares); each such transaction may generate relatively small profit measured in cents, but by doing this constantly and repeatedly throughout the day, steady streams of significant profit can be generated. For accounts of recent technology developments in the financial markets, see [1, 29, 40, 44].

In February 2012, Johnson et al. [35] published a working paper—later revised for publication in Nature Scientific Reports [36]—that immediately received widespread media attention, including coverage in New Scientist [26], Wired [39], and Financial News [45]. Having analysed millisecond-by-millisecond stock-price movements over a 5 year period between 2006 and 2011, Johnson et al. argued that there is evidence for a step-change or *phase transition* in the behaviour of financial markets at the subsecond timescale. At the point of this transition—approximately equal to human response times—the market dynamics switch from a domain where humans and automated *robot* (i.e., *agent*) trading systems freely interact with one another to a domain newly identified by Johnson et al. in which humans cannot participate and where all transactions result from robots interacting only among themselves, with no human traders involved.[2] Here, we refer to this abrupt system-wide transition from mixed human-algorithm phase to a new all-algorithm phase, the *robot phase transition* (RPT).

At subsecond timescales, below the robot transition, the robot-only market exhibits *fractures*—ultrafast extreme events (UEEs) in Johnson et al.'s parlance,

---

[2]The primary reason for no human involvement on these timescales is not because of granularity in decision making—i.e., limitations in human abilities to process information, e.g., [12]—but rather that humans are simply too slow to react to events happening, quite literally, in the blink of an eye.

akin to mini flash crashes—that are undesirable, little understood, and intriguingly appear to be linked to longer-term instability of the market as a whole. In Johnson et al.'s words, "[w]e find 18,520 crashes and spikes with durations less than 1500 ms in our dataset... We define a crash (or spike) as an occurrence of the stock price ticking down (or up) at least ten times before ticking up (or down) and the price change exceeding 0.8% of the initial price... Their rapid subsecond speed and recovery... suggests [UEEs are] unlikely to be driven by exogenous news arrival" [36].

In other words, while fractures are relatively rare events at human time scales—those above the RPT—at time scales below the RPT, fractures are commonplace, occurring many thousands of times over a 5 year period (equivalent to more than ten per day when averaged uniformly). This is interesting. The price discovery mechanism of markets is generally assumed to be driven by the actions of buyers and sellers acting on external information, or news. For instance, the announcement of poor quarterly profits, a new takeover bid, or civil unrest in an oil producing region will each affect the sentiment of buyers and sellers, leading to a shift in price of financial instruments. The prevalence of ATS means that markets can now absorb new information rapidly, so it is not unusual for prices to shift within (milli)seconds of a news announcement. However, fractures are characterised by a shift in price followed by an immediate recovery, or inverse shift (e.g., a spike from $100 to $101; returning to $100). To be driven by news, therefore, fractures would require multiple news stories to be announced in quick succession, with opposing sentiment (positive/negative) of roughly equal net weighting. The speed and frequency of fractures makes this highly unlikely. Therefore, fractures must be driven by an endogenous process resulting from the interaction dynamics of traders in the market. Since fractures tend to occur only below the RPT, when trading is dominated by robots, it is reasonable to conclude that they are a direct result of the interaction dynamics of HFT robot strategies.

What Johnson et al. have identified is a phase transition in the behaviour of markets in the temporal domain caused by fragmentation of market participants—i.e., at time scales below the RPT, the only active market participants are HFT robots, and the interactions between these robots directly result in fractures that are not observed over longer time scales above the RPT. Intriguingly, however, Johnson et al. also observe a correlation between the frequency of fractures and global instability of markets over much longer time scales. This suggests that there may be a causal link between subsecond fractures and market crashes. "[Further, data] suggests that there may indeed be a degree of causality between propagating cascades of UEEs and subsequent global instability, despite the huge difference in their respective timescales ... [Analysis] demonstrates a coupling between extreme market behaviours below the human response time and slower global instabilities above it, and shows how machine and human worlds can become entwined across timescales from milliseconds to months ... Our findings are consistent with an emerging ecology of competitive machines featuring 'crowds' of predatory algorithms, and highlight the need for a new scientific theory of subsecond financial phenomena" [36].

This discovery has the potential for significant impact in the global financial markets. If short-term micro-effects (fractures) can indeed give some indication of longer-term macro-scale behaviour (e.g., market crashes), then it is perhaps possible that new methods for monitoring the stability of markets could be developed—e.g., using fractures as early-warning systems for impending market crashes. Further, if we can better understand the causes of fractures and develop methods to avoid their occurrence, then long-term market instability will also be reduced. This provides motivation for our research. To understand fractures, the first step is to model the RPT.

Here, we report on using a complementary approach to the historical data analysis employed by Johnson et al. [35, 36]. We conduct laboratory-style experiments where human traders interact with algorithmic trading agents (i.e., robots) in minimal experimental models of electronic financial markets using Marco De Luca's *OpEx* artificial financial exchange (for technical platform details, see [19, pp. 26–33]). Our aim is to see whether correlates of the two regimes suggested by Johnson et al. can occur under controlled laboratory conditions—i.e., we attempt to *synthesise* the RPT, such that we hope to observe the market transition from a regime of mixed human–robot trading to a regime of robot-only trading.

3 Background

Experimental human-only markets have a rich history dating back to Vernon Smith's seminal 1960s research [48]. "Before Smith's experiments, it was widely believed that the competitive predictions of supply/demand intersections required very large numbers of well-informed traders. Smith showed that competitive efficient outcomes could be observed with surprisingly small numbers of traders, each with no direct knowledge of the others' costs or values" [32]. This was a significant finding, and it has spawned the entire field of experimental economics; whereby markets are studied by allowing the market equilibration process to *emerge* from the interacting population of actors (humans and/or agents), rather than assuming an *ideal* market that is trading at the theoretical equilibrium. By measuring the distance between the experimental equilibrium and the theoretical equilibrium, one can quantify the *performance* of the market. Further, by altering the rules of interaction (the market mechanism) and varying the market participants (human or agent), one can begin to understand and quantify the relative effects of each. This is a powerful approach and it is one that we adopt for our experimental research.[3]

The following sections present a detailed background. Section 3.1 introduces the continuous double auction mechanism used for experiments; Sect. 3.2 provides metrics for evaluating the performance of markets; and Sect. 3.3 presents a review of previous human–agent experimental studies.

---

[3] For a more thorough background and literature review, refer to [19, pp. 6–25].

3.1 The Continuous Double Auction

An auction is a mechanism whereby sellers and buyers come together and agree on a transaction price. Many auction mechanisms exist, each governed by a different set of rules. In this chapter, we focus on the *Continuous Double Auction* (CDA), the most widely used auction mechanism and the one used to control all the world's major financial exchanges. The CDA enables buyers and sellers to freely and independently exchange quotes at any time. Transactions occur when a seller accepts a buyer's *bid* (an offer to buy), or when a buyer accepts a seller's *ask* (an offer to sell). Although it is possible for any seller to accept any buyer's bid, and *vice-versa*, it is in both of their interests to get the best deal possible at any point in time. Thus, transactions execute with a counter party that offers the most competitive quote.

Vernon Smith explored the dynamics of CDA markets in a series of Nobel Prize winning experiments using small groups of human participants [47]. Splitting participants evenly into a group of buyers and a group of sellers, Smith handed out a single card (an *assignment*) to each buyer and seller with a single *limit price* written on each, known only to that individual. The limit price on the card for buyers (sellers) represented the maximum (minimum) price they were willing to pay (accept) for a fictitious commodity. Participants were given strict instructions to not bid (ask) a price higher (lower) than that shown on their card, and were encouraged to bid lower (ask higher) than this price, regarding any difference between the price on the card and the price achieved in the market as profit.

Experiments were split into a number of *trading days*, each typically lasting a few minutes. At any point during the trading day, a buyer or seller could raise their hand and announce a quote. When a seller and a buyer agreed on a quote, a transaction was made. At the end of each trading day, all stock (sellers assignment cards) and money (buyer assignment cards) was recalled, and then reallocated anew at the start of the next trading day. By controlling the limit prices allocated to participants, Smith was able to control the market's supply and demand schedules. Smith found that, typically after a couple of trading days, human traders achieved very close to 100% allocative efficiency, a measure of the percentage of profit in relation to the maximum theoretical profit available (see Sect. 3.2). This was a significant result: few people had believed that a very small number of inexperienced, self-interested participants could effectively self-equilibrate.

3.2 Measuring Market Performance

An *ideal* market can be perfectly described by the aggregate quantity supplied by sellers and the aggregate quantity demanded by buyers at every price point (i.e., the market's supply and demand schedules; see Fig. 1). As prices increase, in general there is a tendency for supply to increase, with increased potential revenues from

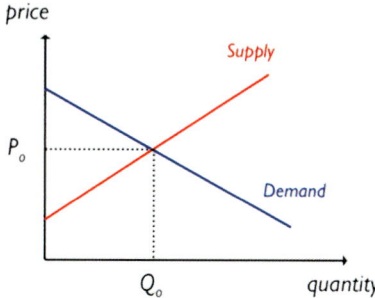

Fig. 1 Supply and demand curves (here illustrated as straight lines) show the quantities supplied by sellers and demanded by buyers at every price point. In general, as price increases, the quantity supplied increases and the quantity demanded falls. The point at which the two curves intersect is the theoretical equilibrium point, where Q_0 is the equilibrium quantity and P_0 is the equilibrium price

sales encouraging more sellers to enter the market; while, at the same time, there is a tendency for demand to decrease as buyers look to spend their money elsewhere. At some price point, the quantity demanded will equal the quantity supplied. This is the theoretical market equilibrium. An idealised theoretical market has a *market equilibrium* price and quantity (P_0, Q_0) determined by the intersection between the supply and demand schedules. The dynamics of competition in the market will tend to drive transactions towards this partial equilibrium point.[4] For all prices above P_0, supply will exceed demand, forcing suppliers to reduce their prices to make a trade; whereas for all prices below P_0, demand exceeds supply, forcing buyers to increase their price to make a trade. Any quantity demanded or supplied below Q_0 is called *intra-marginal*; all quantity demanded or supplied in excess of Q_0 is called *extra-marginal*. In an ideal market, all intra-marginal units and no extra-marginal units are expected to trade.

In the real world, markets are not ideal. They will always trade away from equilibrium at least some of the time. We can use metrics to calculate the *performance* of a market by how far from ideal equilibrium it trades. In this chapter, we make use of the following metrics:

Smith's Alpha

Following Vernon Smith [47], we measure the equilibration (equilibrium-finding) behaviour of markets using the coefficient of convergence, α, defined as the root mean square difference between each of n transaction prices, p_i (for $i = 1 \ldots n$) over some period, and the P_0 value for that period, expressed as a percentage of the equilibrium price:

---

[4]The micro-economic supply and demand model presented only considers a single commodity, *ceteris paribus*, and is therefore a partial equilibrium model. The market is considered independently from other markets, so this is not a general equilibrium model.

$$\alpha = \frac{100}{P_0} \sqrt{\frac{1}{n} \sum_{i=1}^{n} (p_i - P_0)^2} \qquad (1)$$

In essence, α captures the standard deviation of trade prices about the theoretical equilibrium. A low value of α is desirable, indicating trading close to P_0.

Allocative Efficiency

For each trader, i, the maximum theoretical profit available, π_i^*, is the difference between the price they are prepared to pay (their *limit price*) and the theoretical market equilibrium price, P_0. Efficiency, E, is used to calculate the performance of a group of n traders as the mean ratio of realised profit, π_i, to theoretical profit, π_i^*:

$$E = \frac{1}{n} \sum_{i=1}^{n} \frac{\pi_i}{\pi_i^*} \qquad (2)$$

As profit values cannot go below zero (traders in these experiments are not allowed to enter into loss-making deals, although that constraint can easily be relaxed), a value of 1.0 indicates that the group has earned the maximum theoretical profit available, π_i^*, on all trades. A value below 1.0 indicates that some opportunities have been missed. Finally, a value above 1.0 means that additional profit has been made by taking advantage of a trading counterparty's willingness to trade away from P_0. So, for example, a group of sellers might record an allocative efficiency of, say, 1.2 if their counterparties (a group of buyers) consistently enter into transactions at prices greater than P_0; in such a situation, the buyers' allocative efficiency would not be more than 0.8.

Profit Dispersion

Profit dispersion is a measure of the extent to which the profit/utility generated by a group of traders in the market differs from the profit that would be expected of them if all transactions took place at the equilibrium price, P_0. For a group of n traders, profit dispersion is calculated as the root mean square difference between the profits achieved, π_i, by each trader, i, and the maximum theoretical profit available, π_i^*:

$$\pi_{disp} = \sqrt{\frac{1}{n} \sum_{i=1}^{n} (\pi_i - \pi_i^*)^2} \qquad (3)$$

Low values of π_{disp} indicate that traders are extracting actual profits close to profits available when all trades take place at the equilibrium price P_0. In contrast, higher values of π_{disp} indicate that traders' profits differ from those expected at equilibrium. Since zero-sum effects between buyers and sellers do not mask profit dispersion, this statistic is attractive [28].

Delta Profit

Delta profit is used to calculate the difference in profit maximising performance between two groups, x and y, as a percentage difference relative to the mean profit of the two groups, π_x, π_y:

$$\Delta P(x - y) = \frac{2(\pi_x - \pi_y)}{\pi_x + \pi_y} \qquad (4)$$

Delta profit directly measures the difference in profit gained by two groups. In a perfect market, we expect $\Delta P(x - y) = 0$, with both groups trading at the equilibrium price P_0. A positive (negative) value indicates that group x secures more (less) profit than group y. Using this measure enables us to determine which, if either, of the two groups competitively outperforms the other.

3.3 Human vs. Agent Experimental Economics

In 1993, after three decades of human-only experimental economics, a landmark paper involving a mix of traditional human experimental economics and software-agent market studies was published in the *Journal of Political Economy* by Gode and Sunder (G&S) [28]. G&S were interested in understanding how much of the efficiency of the CDA is due to the intelligence of traders, and how much is due to the organisation of the market. To test this, G&S introduced a very simple *Zero Intelligence Constrained* (ZIC) trading agent that generate random bid or ask prices drawn from a uniform distribution, subject to the constraint that prices generated cannot be loss-making—i.e., sell prices are equal or above limit price, buy prices are equal or below limit price. G&S performed a series of ZIC-human experiments, with results demonstrating that the simple ZIC agents produced convergence towards the theoretical equilibrium and had human-like scores for allocative efficiency (Eq. (2)); suggesting that market convergence towards theoretical equilibrium is an emergent property of the CDA market mechanism and not the intelligence of the traders. Indeed, G&S found that the only way to differentiate the performance of humans and ZIC traders was by using their profit dispersion statistics (Eq. (3)). These results were striking and attracted considerable attention.

In 1997, Dave Cliff [13] presented the first detailed mathematical analysis and replication of G&S's results. Results demonstrated that the ability of ZIC traders to converge on equilibrium was dependent on the shape of the market's demand and supply curves. In particular, ZIC traders were unable to equilibrate when acting in markets with demand and supply curves very different to those used by G&S. To address this issue, Cliff developed the *Zero Intelligence Plus* (ZIP) trading algorithm. Rather than issuing randomly generated bid and ask prices in the manner of ZIC, Cliff's ZIP agents contain an internal profit margin from which bid and ask prices are calculated. When a buyer (seller) sees transactions happen at a price below (above) the trader's current bid (ask) price, profit margin is raised, thus resulting in a lower (higher) bid (ask) price. Conversely, a buyer's (seller's) profit margin is lowered when order and transaction prices indicate that the buyer (seller) will need to raise (lower) bid (ask) price in order to transact [13, p.43]. The size of ZIP's profit margin update is determined using a well-established machine learning mechanism (derived from the Widrow–Hoff *Delta rule* [56]). Cliff's autonomous and adaptive ZIP agents were shown to display human-like efficiency and equilibration behaviours in all markets, irrespective of the shape of demand and supply.

Around the same time that ZIP was introduced, economists Steve Gjerstad and his former PhD supervisor John Dickhaut independently developed a trading algorithm that was later named GD after the inventors [27]. Using observed market activity—frequencies of bids, asks, accepted bids, and accepted asks—resulting in the most recent L transactions (where $L = 5$ in the original study), GD traders calculate a private, subjective "belief" of the probability that a counterparty will accept each quote price. The belief function is extended over all prices by applying cubic-spline interpolation between observed prices (although it has previously been suggested that using *any* smooth interpolation method is likely to suffice [19, p.17]). To trade, GD quotes a price to buy or sell that maximises expected surplus, calculated as price multiplied by the belief function's probability of a quote being accepted at that price. Simulated markets containing GD agents were shown to converge to the competitive equilibrium price and allocation in a fashion that closely resembled human equilibration in symmetric markets, but with greater efficiency than human traders achieved [27]. A modified GD (MGD) algorithm, where the belief function of bid (ask) prices below (above) the previous lowest (highest) transaction price was set to probability zero, was later introduced to counter unwanted price volatility.

In 2001, a series of experiments were performed to compare ZIP and MGD in real-time heterogeneous markets [52]. MGD was shown to outperform ZIP. Also in 2001, the first ever human–agent experiments—with MGD and ZIP competing in the same market as human traders—were performed by Das et al., a team from IBM [15]. Results had two major conclusions: (a) firstly, mixed human–agent markets were off-equilibrium—somehow the mixture of humans and agents in the market reduces the ability of the CDA to equilibrate; (b) secondly, in all experiments reported, the efficiency scores of humans were lower than the efficiency scores of agents (both MGD and ZIP). In Das et al.'s own words, "...the successful demonstration of machine superiority in the CDA and other common auctions could have a much more direct and powerful impact—one that might be measured in billions of dollars annually" [15]. This result, demonstrating for the first time in human-algorithmic markets that agents can outperform humans, implied a future financial-market system where ATS replace humans at the point of execution.

Despite the growing industry in ATS in real financial markets, in academia there was a surprising lack of further human–agent market experiments over the following decade. In 2003 and 2006, Grossklags & Schmidt [30, 31] performed human–agent market experiments to study the effect that human behaviours are altered by their knowledge of whether or not agent traders are present in the market. In

2011, De Luca & Cliff successfully replicated Das et al.'s results, demonstrating that GDX (an extension of MGD, see [51]) outperforms ZIP in agent–agent and agent–human markets [17]. They further showed that *Adaptive Aggressive* (AA) agents—a trading agent developed by Vytelingum in 2006 that is loosely based on ZIP, with significant novel extensions including short-term and long-term adaptive components [54, 55]—dominate GDX and ZIP, outperforming both in agent–agent and agent–human markets [18]. This work confirmed AA as the dominant trading-agent algorithm. (For a detailed review of how ZIP and AA have been modified over time, see [49, 50].) More recent human–agent experiments have focused on emotional arousal level of humans, monitoring heart rate over time [57] and monitoring human emotions via EEG brain data [8].

Complementary research comparing markets containing only humans against markets containing only agents—i.e., human-only or agent-only markets rather than markets in which agents and humans interact—can also shed light on market dynamics. For instance, Huber, Shubik, and Sunder (2010) compare dynamics of three market mechanisms (sell-all, buy-all, and double auction) in markets containing all humans against markets containing all agents. "The results suggest that abstracting away from all institutional details does not help understand dynamic aspects of market behaviour and that inclusion of mechanism differences into theory may enhance our understanding of important aspects of markets and money, and help link conventional analysis with dynamics" [33]. This research stream reinforces the necessity of including market design in our understanding of market dynamics. However, it does not offer the rich interactions between humans and ATS that we observe in real markets, and that only human–agent interaction studies can offer.

4 Methodology

In this section, the experimental methodology and experimental trading platform (OpEx) are presented. Open Exchange (OpEx) is a real-time financial-market simulator specifically designed to enable economic trading experiments between humans and automated trading algorithms (robots). OpEx was designed and developed by Marco De Luca between 2009 and 2010 while he was a PhD student at the University of Bristol, and since Feb. 2012 is freely available for open-source download from SourceForge, under the terms of the Creative Commons Public License.[5] Figure 2 shows the *Lab-in-a-box* hardware arranged ready for a human–agent trading experiment. For a detailed technical description of the OpEx platform, refer to [19, pp. 26–33].

At the start of each experiment, 6 human participants were seated at a terminal around a rectangular table—with three buyers on one side and three sellers opposite—and given a brief introduction and tutorial to the system (explaining

---

[5]OpEx download available at: www.sourceforge.net/projects/open-exchange.

Fig. 2 The *Lab-in-a-box* hardware ready to run an Open Exchange (OpEx) human versus agent trading experiment. Six small netbook computers run human trader Sales GUIs, with three buyers (near-side) sitting opposite three sellers (far-side). Netbook clients are networked via Ethernet cable to a network switch for buyers and a network switch for sellers, which in turn are connected to a router. The central exchange and robots servers run on the dedicated hardware server (standing vertically, top-left), which is also networked to the router. Finally, an *Administrator* laptop (top table, centre) is used to configure and run experiments. Photograph: © J. Cartlidge, 2012

the human trading GUI illustrated in Fig. 3), during which time they were able to make test trades among themselves while no robots were present in the market. Participants were told that their aim during the experiment was to maximise profit by trading client orders (assignments or alternatively named *permits* to distinguish that traders will simultaneously have multiple client orders to work, whereas in the traditional literature, a new assignment would only be received once the previous assignment had been completed) that arrive over time. For further details on the experimental method, refer to [9, pp. 9–11].

Trading Agents (Robots)

Agent-robots are independent software processes running on the multi-core hardware server that also hosts the central exchange server. Since agents can act at any time—there is no central controller coordinating when, or in which order, an agent can act—and since the trading logic of agents does not explicitly include temporal information, in order to stop agents from issuing a rapid stream of quotes, a sleep timer is introduced into the agent architecture. After each action, or decision to not act, an agent will *sleep* for t_s milliseconds before *waking* and deciding upon the next action. We name this the *sleep-wake* cycle of agents. For instance, if $t_s = 100$, the sleep-wake cycle is 0.1 s. To ensure agents do not miss important events during sleep, agents are also set to wake (i.e., sleep is interrupted) when a new assignment permit is received and/or when an agent is notified about a new trade execution. The parameter t_s is used to configure the "speed" of agents for each experiment.

Modelling Financial Markets Using Human–Agent Experiments

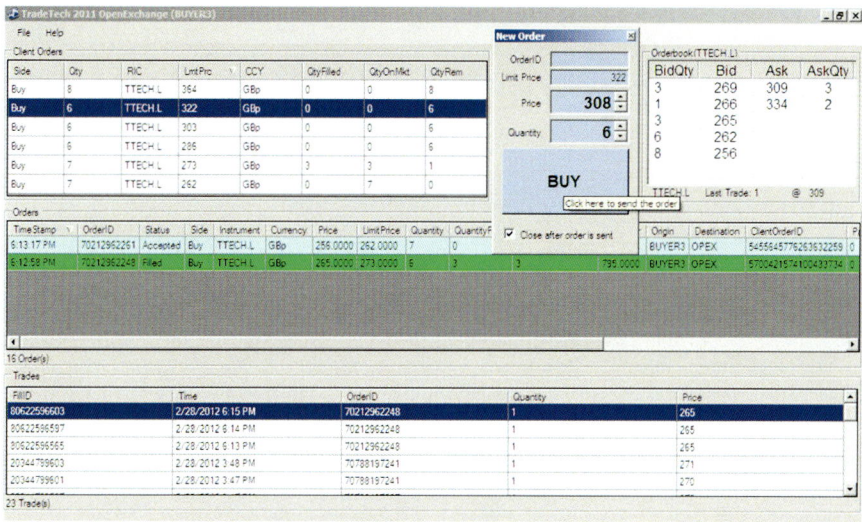

Fig. 3 Trading GUI for a human buyer. New order assignments (or *permits*) arrive over time in the *Client Orders* panel (top-left) and listed in descending order by potential profit. Assignments are selected by double-clicking. This opens a *New Order* dialogue pop-up (top-centre) where bid price and quantity are set before entering the new bid into the market by pressing button *BUY*. The market *Order Book* is displayed top-right, with all bids and asks displayed. Bid orders that the trader currently has live in the market are listed in the *Orders* panel (middle), and can be amended from here by double-clicking. When an order executes it is removed from the orders panel and listed in the *Trades* history panel (bottom). For further GUI screen shots, refer to [9, Appendix C]

Trading agents are configured to use the *Adaptive Aggressive* (AA) strategy logic [54, 55], previously shown to be the dominant trading agent in the literature (see Sect. 3.3). AA agents have short-term and long-term adaptive components. In the short term, agents use learning parameters β_1 and λ to adapt their order aggressiveness. Over a longer time frame, agents use the moving average of the previous N market transactions and a learning parameter β_2 to estimate the market equilibrium price, $\hat{p}_0$. The *aggressiveness* of AA represents the tendency to accept lower profit for a greater chance of transacting. To achieve this, an agent with high (low) aggression will submit orders better (worse) than the estimated equilibrium price $\hat{p}_0$. For example, a buyer (seller) with high aggression and estimated equilibrium value $\hat{p}_0 = 100$ will submit bids (asks) with price $p > 100$ (price $p < 100$). Aggressiveness of buyers (sellers) increases when transaction prices are higher (lower) than $\hat{p}_0$, and decreases when transaction prices are lower (higher) than $\hat{p}_0$. The Widrow–Hoff mechanism [56] is used by AA to update aggressiveness in a similar way that it is used by ZIP to update profit margin (see Sect. 3.3). For all experiments reported here, we set parameter values $\beta_1 = 0.5$, $\lambda = 0.05$, $N = 30$, and $\beta_2 = 0.5$. The convergence rate of bids/asks to transaction price is set to $\eta = 3.0$.

Table 1 Permit schedule for market efficiency experiments.

	1	2[a]	3	4	5	6
Buyer 1	350 (0)	250 (4)	220 (7)	190 (09)	150 (14)	140 (16)
Buyer 2	340 (1)	270 (3)	210 (8)	180 (10)	170 (12)	130 (17)
Buyer 3	330 (2)	260 (4)	230 (6)	170 (11)	160 (13)	150 (15)
Seller 1	50 (0)	150 (4)	180 (7)	210 (09)	250 (14)	260 (16)
Seller 2	60 (1)	130 (3)	190 (8)	220 (10)	230 (12)	270 (17)
Seller 3	70 (2)	140 (4)	170 (6)	230 (11)	240 (13)	250 (15)

Six permit types are issued to each market participant, depending on their role. For each role (e.g., *Buyer 1*), there are two traders: one human (*Human Buyer 1*) and one robot (*Robot Buyer 1*). Thus, there are 12 traders in the market. Permit values show *limit price*—the maximum value at which to buy, or minimum value at which to sell—and the time-step they are issued (in parentheses). The length of each time-step is 10 s, making one full permit cycle duration 170 s. During a 20-min experiment there are seven full cycles

[a] Type 2 permits were accidentally issued to Buyer1/Seller1 at time-step 4 rather than time-step 5

Exploring the Effects of Agent Speed on Market Efficiency: April–June 2011

All experiments were run at the University of Bristol between April and July 2011 using postgraduate students in non-financial but analytical subjects (i.e., students with skills suitable for a professional career in finance, but with no specific trading knowledge or experience). Participants were paid £20 for participating and a further £40 bonus for making the most profit, and £20 bonus for making the second highest profit. Moving away from the artificial constraint of regular simultaneous replenishments of currency and stock historically used, assignment permits were issued at regular intervals. AA agents had varying sleep-wake cycle: $t_s = 100$, and $t_s = 10,000$. We respectively label these agents AA-0.1 to signify a sleep-wake cycle of 0.1 s, and AA-10 to signify a sleep-wake cycle of 10 s. A total of 7 experiments were performed, using the assignment permit schedules presented in Table 1. The supply and demand curves generated by these permits are shown in Fig. 4. We can see that for all experiments, $P_0 = 200$ and $Q_0 = 126$. Since each human only participates in one experiment, and since trading agents are reset at the beginning of each run, traders have no opportunity to learn the fixed value of P_0 over repeated runs. For further details of experimental procedure, see [11].

Exploring the Robot Phase Transition (RPT): March 2012

Twenty-four experiments were run on 21st March 2012, at Park House Business Centre, Park Street, Bristol, UK. Participants were selected on a first-come basis from the group of students that responded to adverts broadcast to two groups: (1) students enrolled in final year undergraduate and postgraduate module in computer science that includes coverage of the design of automated trading agents; (2) members of the Bristol Investment Society, a body of students interested in pursuing a career in finance. We assume that these students have the knowledge and skills to embark on a career as a trader in a financial institution. Volunteers were paid £25 for participating, and the two participants making the greatest profit received an iPad valued at £400. To reduce the total number of participants required, each group were

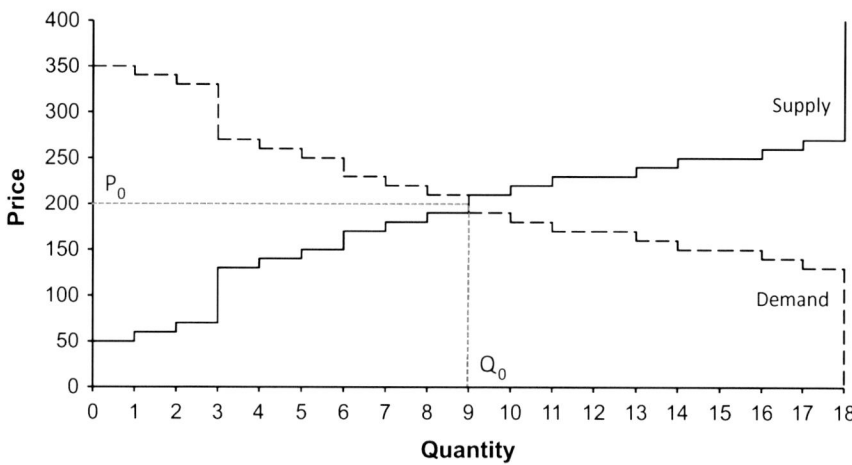

Fig. 4 Stepped supply and demand curves for permit schedule defined in Table 1. Curves show the aggregate quantity that participants are prepared to buy (demand) and sell (supply) at every price point. The point at which the two curves intersect is the theoretical equilibrium point for the market: $P_0 = 200$ is the equilibrium price, and Q_0 is the equilibrium quantity. As there are two traders in each role—one human and one robot—each permit cycle $Q_0 = 2 \times 9 = 18$, and over the seven permit cycles of one full experiment, $Q_0 = 18 \times 7 = 126$. The market is symmetric about P_0

used in a session of six separate experiments. Therefore, 24 experiments were run using the 24 participants. Between experiments, human participants rotated seats, so each played every role exactly once during the session of 6 experiments. Human roles were purposely mixed between experiment rounds to reduce the opportunity for collusion and counteract any bias in market role. Once again, agents used the AA algorithm with varying sleep-wake cycle, and assignment orders were released into the market at regular intervals.

Table 2 presents the assignment permit schedules used for each experiment, and the full supply and demand curves generated by these permits are plotted in Fig. 5. At each price point—i.e., at each *step* in the *permit* schedule—two assignment permits are sent simultaneously to a human trader and to a robot trader, once every replenishment cycle. For all experiments, permits are allocated in pairs symmetric about P_0 such that the equilibrium is not altered, and the inter-arrival time of permits is 4 s. Cycles last 72 s and are repeated eight times during a 10 min experiment. Therefore, over a full experiment there are $2 \times 8 = 16$ permits issued at each price point. The expected equilibrium number of trades for the market, Q_0, is 144 intra-marginal units. Each experiment, P_0 is varied in the range 209–272 to stop humans from learning the equilibration properties of the market between experiments. Agents are reset each time and have no access to data from previous experiments. In *cyclical* markets, permits are allocated in strict sequence that is unaltered between cycles. In *random* markets, the permit sequence across the entire run is randomised. For further details on experimental procedure, see [9, 10].

Table 2 Permit schedule for RPT experiments

	1	2	3	4	5	6
Buyer 1	77 (1)	27 (4)	12 (7)	−9 (10)	−14 (13)	−29 (16)
Buyer 2	73 (2)	35 (5)	8 (8)	−5 (11)	−22 (14)	−25 (17)
Buyer 3	69 (3)	31 (6)	16 (9)	−1 (12)	−18 (15)	−33 (18)
Seller 1	−77 (1)	−27 (4)	−12 (7)	9 (10)	14 (13)	29 (16)
Seller 2	−73 (2)	−35 (5)	−8 (8)	5 (11)	22 (14)	25 (17)
Seller 3	−69 (3)	−31 (6)	−16 (9)	1 (12)	18 (15)	33 (18)

Six permit types are issued to each market participant, depending on their role. For each role, there is one human and one robot participant. Permit values show $limitprice - P_0$. Thus, e.g., if $P_0 = 100$, a permit of type 4 to Buyer1 would have a limit price of 91. For buyers, limit prices are the maximum value to bid; and for sellers, limit prices are the minimum value to ask. Numbers in brackets show the time-step sequence in which permits are allocated. Thus, after 11 time-steps, Buyer2 and Seller2 each receive a permit of type 4. For all experiments, the inter-arrival time-step between permits is 4 s. Permits are always allocated in pairs, symmetric about P_0. In cyclical markets, the sequence is repeated eight times: the last permits are issued to Buyer3 and Seller3 at time 576 s, and the experiment ends 24 s later. In non-cyclical or "random" markets, the time-step of permits is randomised across the run. Participants receive the same set of permits in both cyclical and random markets, but in a different order

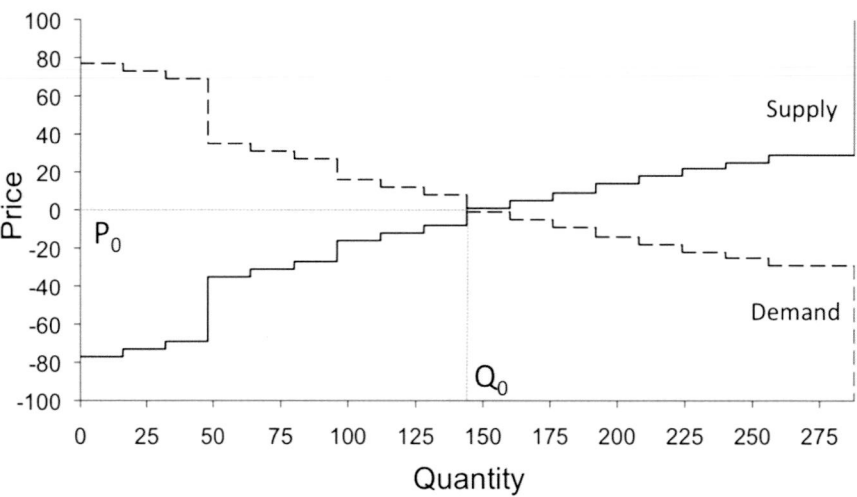

Fig. 5 Stepped supply and demand curves for an entire run of the RPT experiments, defined by the permit schedules shown in Table 2. Curves show the aggregate quantity that participants are prepared to buy (demand) and sell (supply) at every price point. The point at which the two curves intersect is the theoretical equilibrium point for the market: $Q_0 = 144$ is the equilibrium quantity, and P_0 is the equilibrium price. Each experiment the value of P_0 is varied in the range 209–272 to avoid humans learning a fixed value of P_0 over repeated trials. The market is symmetric about P_0.

5 Results

Here, we present empirical results from the two sets of experiments: (a) exploring the robot phase transition, performed in March 2012, and (b) exploring the effects of agent speed on market efficiency, performed in April–June 2011. Throughout this section, for detecting significant differences in location between two samples we use the nonparametric Robust Rank-Order (RRO) test and critical values reported by Feltovich [24]. RRO is particularly useful for small sample statistics of the kind we present here, and is less sensitive to changes in distributional assumptions than the more commonly known Wilcoxon–Mann–Whitney test [24].

5.1 Exploring the Robot Phase Transition: March 2012

Experiments were run using AA agents with sleep-wake cycle times (in seconds) $t_s = 0.1$ (AA-0.1), $t_s = 1$ (AA-1), $t_s = 5$ (AA-5), and $t_s = 10$ (AA-10). Of the 24 runs, one experienced partial system failure, so results were omitted. Runs with agent sleep time 5 s (AA-5) are also omitted from analysis where no significant effects are found. For further detail of results, see [9, 10].

5.1.1 Market Data

OpEx records time-stamped data for every exchange event. This produces rich datasets containing every quote (orders submitted to the exchange) and trade (orders that execute in the exchange) in a market. In total, we gathered 4 h of trading data across the four one-hour sessions, but for brevity we explore only a small set of indicative results here; however, for completeness, further datasets are presented in [9, Appendix A]. Figure 6 plots time series of quotes and trades for a cyclical market containing AA-0.1 agents. The dotted horizontal line represents the theoretical market equilibrium, P_0, and vertical dotted lines indicate the start of each new permit replenishment cycle (every 72 s). We see the majority of trading activity (denoted by filled markers) is largely clustered in the first half of each permit replenishment cycle; this correlates with the phase in which intra-marginal units are allocated and trades are easiest to execute. After the initial *exploratory* period, execution prices tend towards P_0 in subsequent cycles. In the initial period, robots (diamonds for sellers; inverted triangle for buyers) explore the space of prices. In subsequent periods, robots quote much closer to equilibrium. Agent quotes are densely clustered near to the start of each period, during the phase that intra-marginal units are allocated. In contrast, humans (squares for sellers; triangles for buyers) tend to enter exploratory quotes throughout the market's open period.

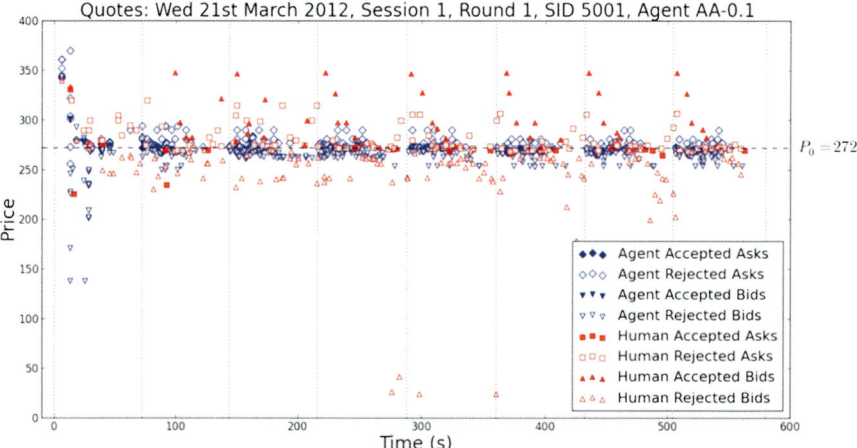

Fig. 6 Time series of quote and trade prices from a cyclical market containing AA-0.1 agents. The dotted horizontal line represents the theoretical market equilibrium, P_0. Vertical dotted lines indicate the start of each new permit replenishment cycle

5.1.2 Smith's α

We can see the equilibration behaviour of the markets more clearly by plotting Smith's α for each cycle period. In Fig. 7 we see mean α ($\pm 95\%$ confidence interval) plotted for cyclical and random markets. Under both conditions, α follows a similar pattern, tending to approx 1% by market close. However, in the first period, cyclical markets produce significantly greater α than random markets (RRO, $p < 0.0005$). This is due to the sequential order allocation of permits in cyclical markets, where limit prices farthest from equilibrium are allocated first. This enables exploratory shouts and trades to occur far from equilibrium. In comparison, in random markets, permits are not ordered by limit price, thus making it likely that limit prices of early orders are closer to equilibrium than they are in cyclical markets.

5.1.3 Allocative Efficiency

Tables 3 and 4 display the mean allocative efficiency of agents, humans, and the whole market grouped by agent type and market type, respectively. Across all groupings, $E(agents) > E(humans)$. However, when grouped by robot type (Table 3), the difference is only significant for AA-0.1 and AA-5 (RRO, $0.051 < p < 0.104$). When grouped by market type (Table 4), $E(agents) > E(humans)$ is significant in cyclical markets (RRO, $0.05 < p < 0.1$), random markets (RRO, $0.05 < p < 0.1$), and across all 23 runs (RRO, $0.01 < p < 0.025$). These results suggest that agents outperform humans.

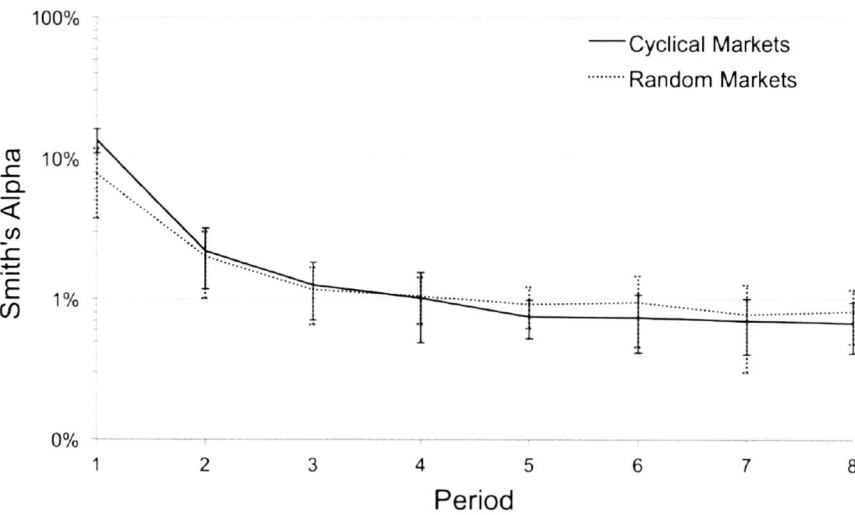

Fig. 7 Mean α ($\pm 95\%$ confidence interval) plotted using log scale for results grouped by market type. In cyclical markets, α values are significantly higher than in random markets during the initial period (RRO, $p < 0.0005$). In subsequent periods all markets equilibrate to $\alpha < 1\%$ with no statistical difference between groups

Table 3 Efficiency and profit for runs grouped by robot type

Robot Type	Trials	$E(agents)$	$E(humans)$	$E(market)$	$\Delta P(agents - humans)$
AA-0.1	6	0.992	0.975	0.984	1.8%
AA-1	5	0.991	0.977	0.984	1.4%
AA-5	6	0.990	0.972	0.981	1.8%
AA-10	6	0.985	0.981	0.983	0.4%
All	23	0.989	0.976	0.983	1.34%

Agents achieve greater efficiency $E(agents) > E(humans)$, and greater profit $\Delta P(agents - humans) > 0$, under all conditions

Table 4 Efficiency and profit for runs grouped by market type

Market type	Trials	$E(agents)$	$E(humans)$	$E(market)$	$\Delta P(agents - humans)$
Cyclical	12	0.991	0.978	0.985	1.32%
Random	11	0.987	0.974	0.981	1.36%
All	23	0.989	0.976	0.983	1.34%

Agents achieve greater efficiency $E(agents) > E(humans)$, and greater profit $\Delta P(agents - humans) > 0$, under all conditions

In Table 3, it can be seen that as sleep time increases the efficiency of agents decreases (column 3, top-to-bottom). Conversely, the efficiency of humans tends to increase as sleep time increases (column 4, top-to-bottom). However, none of these differences are statistically significant (RRO, $p > 0.104$). In Table 4, efficiency of agents, humans, and the market as a whole are all higher when permit schedules are issued cyclically rather than randomly, suggesting that cyclical markets lead to greater efficiency. However, these differences are also not statistically significant (RRO, $p > 0.104$). Finally, when comparing $E(agents)$ grouped by robot type using only data from cyclical markets (data not shown), AA-0.1 robots attain a significantly higher efficiency than AA-1 (RRO, $p = 0.05$), AA-5 (RRO, $p = 0.05$), and AA-10 (RRO $p = 0.1$), suggesting that the very fastest robots are most efficient in cyclical markets.

5.1.4 Delta Profit

From the right-hand columns of Tables 3 and 4, it can be seen that agents achieve greater profit than humans under all conditions, i.e., $\Delta P(agents - humans) > 0$. Using data across all 23 runs, the null hypothesis $H_0 : \Delta P(agents - humans) \leq 0$ is rejected (t-test, $p = 0.0137$). Therefore, the profit of agents is significantly greater than the profit of humans, i.e., agents outperform humans across all runs. Differences in $\Delta P(agents - humans)$ between robot groupings and market groupings are not significant (RRO, $p > 0.104$).

5.1.5 Profit Dispersion

Table 5 shows the profit dispersion of agents $\pi_{disp}(agents)$, humans $\pi_{disp}(humans)$, and the whole market $\pi_{disp}(market)$, for runs grouped by market type. It is clear that varying between cyclical and random permit schedules has a significant effect on profit dispersion, with random markets having significantly lower profit dispersion of agents (RRO, $0.001 < p < 0.005$), significantly lower profit dispersion of humans (RRO, $0.025 < p < 0.05$), and significantly lower profit dispersion of the market as a whole (RRO, $0.005 < p < 0.01$). These results indicate that traders in random markets are extracting actual profits closer to profits available when all trades take place at the equilibrium price, P_0; i.e., random markets are trading closer to equilibrium, likely due to the significant difference in α during the initial trading period (see Sect. 5.1.2). When grouping data by robot type (not shown), there is no significant difference in profit dispersion of agents, humans, or markets (RRO, $p > 0.104$).

Table 5 Profit dispersion for runs grouped by market type

Market type	Trials	$\pi_{disp}(agents)$	$\pi_{disp}(humans)$	$\pi_{disp}(market)$
Cyclical	12	89.6	85.4	88.6
Random	11	50.2	57.2	55.6
All	23	70.0	71.9	72.8

Profit dispersion in random markets is significantly lower than in cyclical markets for agents $\pi_{disp}(agents)$, humans $\pi_{disp}(humans)$, and the whole market $\pi_{disp}(market)$

5.1.6 Execution Counterparties

Let *aa* denote a trade between agent buyer and agent seller, *hh* a trade between human buyer and human seller, *ah* a trade between agent buyer and human seller, and *ha* a trade between human buyer and agent seller. Then, assuming a fully mixed market where any buyer (seller) can independently and anonymously trade with any seller (buyer), we generate null hypothesis, H_0: the proportion of trades with homogeneous counterparties—*aa* trades or *hh* trades—should be 50%. More formally:

$$H_0 : \frac{\Sigma aa + \Sigma hh}{\Sigma aa + \Sigma hh + \Sigma ah + \Sigma ha} = 0.5$$

In Fig. 8, box-plots present the proportion of homogeneous counterparty trades for markets grouped by robot type (AA-0.1, AA-1, and AA-10); the horizontal dotted line represents the H_0 value of 50%. It can clearly be seen that the proportion of homogeneous counterparty trades for markets containing AA-0.1 robots is significantly greater than 50%; and H_0 is rejected (t-test, $p < 0.0005$). In contrast, for markets containing AA-1 and AA-10 robots, H_0 is not rejected at the 10% level of significance. This suggests that for the fastest agents (AA-0.1) the market tends to fragment, with humans trading with humans and robots trading with robots more than would be expected by chance. There also appears to be an inverse relationship

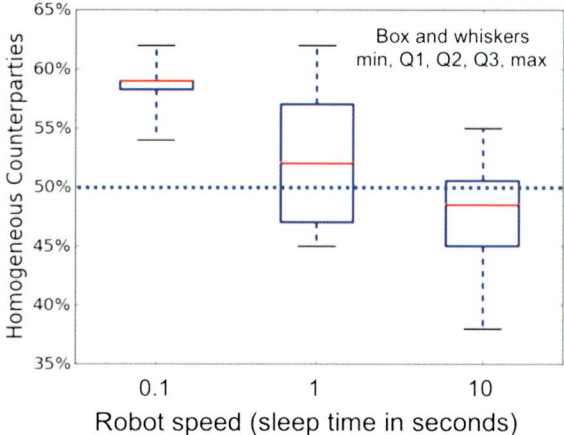

Fig. 8 Box-plot showing the percentage of homogeneous counterparty executions (i.e., trades between two humans, or between two agents). In a fully mixed market, there is an equal chance that a counterparty will be agent or human, denoted by the horizontal dotted line, H_0. When agents act and react at time scales equivalent to humans (i.e., when sleep time is 1 s or 10 s), counterparties are selected randomly—i.e., there is a mixed market and H_0 is not rejected ($p > 0.1$). However, when agents act and react at super-human timescales (i.e., when sleep time is 0.1 s), counterparties are more likely to be homogeneous—H_0 is rejected ($p < 0.0005$). This result suggests that, even under simple laboratory conditions, when agents act at super-human speeds the market fragments.

between robot sleep time and proportion of homogeneous counterparty trades. RRO tests show that the proportion of homogeneous counterparty trades in AA-0.1 markets is significantly higher than AA-1 markets ($p < 0.051$) and AA-10 markets ($p = 0.0011$); and for AA-1 markets the proportion is significantly higher than AA-10 ($p < 0.104$). For full detail of RRO analysis of execution counterparties, see [9, Appendix A.2.1].

5.2 Effect of Agent Speed on Market Efficiency: April–June 2011

Experiments were run using AA agents with sleep-wake cycle times (in seconds) $t_s = 0.1$ (AA-0.1) and $t_s = 10$ (AA-10). A total of 8 experiments were performed. However, during one experiment, a human participant began feeling unwell and could no longer take part, so results for this trial are omitted. Here, we present results of the remaining 7 experiments. For further detail of results, see [11].

5.2.1 Smith's α

Figure 9 plots mean α ($\pm 95\%$ confidence interval) for each permit replenishment cycle, grouped by robot type: AA-0.1 and AA-10. Under both conditions, $\alpha > 10\%$ in the initial period, and then equilibrates to a value $\alpha < 10\%$. For every period, i,

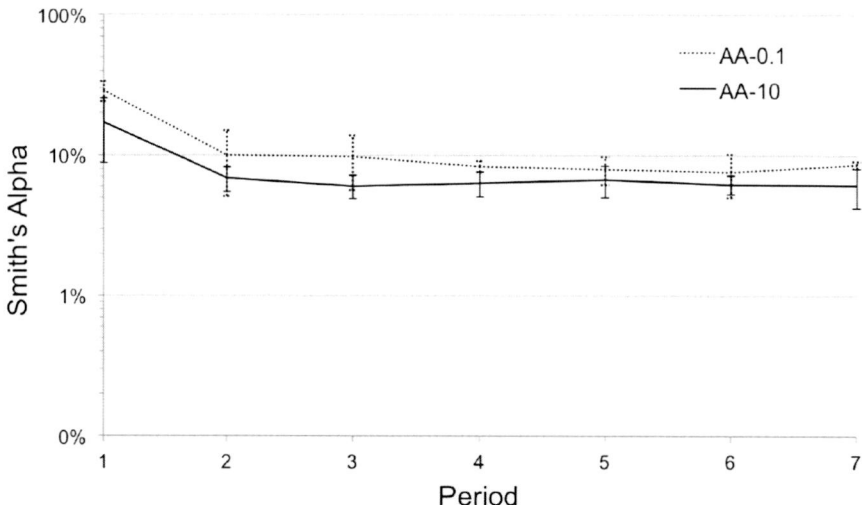

Fig. 9 Mean Smith's α ($\pm 95\%$ confidence interval) plotted using log scale for results grouped by robot type. In markets containing fast AA-0.1 robots, α values are significantly higher than in markets containing slow AA-10 robots. After the initial period, all markets equilibrate to $\alpha < 10\%$.

Table 6 Efficiency and profit for runs grouped by robot type

	Trials	$E(agents)$	$E(humans)$	$E(market)$	$\Delta P(agents - humans)$
AA-0.1	3	0.966	0.906	0.936	3.2%
AA-10	4	0.957	0.963	0.960	−0.3%

Agents achieve greater efficiency $E(agents) > E(humans)$, and greater profit $\Delta P(agents - humans) > 0$, when robots are fast AA-0.1. In contrast, humans achieve greater efficiency $E(humans) > E(agents)$, and greater profit $\Delta P(agents - humans) < 0$, when robots are slow AA-10

mean value α_i is lower for markets containing AA-10 robots than it is for markets containing AA-0.1 robots. Using RRO, this difference is significant at every period: α_1 ($p < 0.029$), α_2 ($p < 0.029$), α_3 ($p < 0.029$), α_4 ($p < 0.029$), α_5 ($p < 0.114$), α_6 ($p < 0.057$), α_7 ($p < 0.029$). This suggests that markets with slower agents (AA-10) are able to equilibrate better than markets with faster agents (AA-0.1).

5.2.2 Allocative Efficiency and Delta Profit

Table 6 presents mean allocative efficiency and delta profit for runs grouped by robot type. The efficiency of agents is similar under both conditions, with no statistical difference (RRO, $p > 0.114$). However, runs with slow AA-10 robots result in significantly higher efficiency of humans (RRO, $0.114 < p < 0.029$), and significantly higher efficiency of the whole market (RRO, $0.114 < p < 0.029$). In markets containing slower AA-10 robots, humans are able to secure greater profit than agents $\Delta P(agents - humans) < 0$; whereas in markets containing fast AA-0.1 robots, agents secure more profit than humans $\Delta P(agents - humans) > 0$. However, the difference in delta profit between the two groups is not significant (RRO, $p > 0.114$).

These data provide evidence that markets containing fast AA-0.1 robots are less efficient than markets containing slow AA-10 robots. However, this does not imply that AA-10 outperform AA-0.1, as their efficiency shows no significant difference. Rather, we see that humans perform more poorly when competing in markets containing faster trader-agents, resulting in lower efficiency for the market as a whole.

5.2.3 Profit Dispersion

Table 7 presents profit dispersion for runs grouped by robot type. In markets containing fast AA-0.1 robots, profit dispersion is significantly higher for agents (RRO, $p < 0.114$), humans (RRO, $p < 0.114$), and the market as a whole (RRO, $0.029 < p < 0.057$). These data provide evidence that fast AA-0.1 agents result in higher profit dispersion than slow AA-10 agents, an undesirable result.

Table 7 Profit dispersion for runs grouped by market type

	Trials	$\pi_{disp}(agents)$	$\pi_{disp}(humans)$	$\pi_{disp}(market)$
AA-0.1	3	105	236	185
AA-10	4	100	164	139

In markets with slow AA-10 robots, profit dispersion of agents $\pi_{disp}(agents)$, humans $\pi_{disp}(humans)$, and the whole market $\pi_{disp}(market)$ is significantly lower than in markets with fast AA-0.1 robots

6 Discussion

Here we discuss results presented in the previous section. First, in Sect. 6.1 we summarise the main results that hold across all our market experiments presented in Sect. 5.1. Subsequently, in Sect. 6.2 we discuss potentially conflicting results from experiments presented in Sect. 5.2. Finally, in Sect. 6.3 we discuss results that demonstrate significant differences between cyclical and random markets.

6.1 Evidence for the Robot Phase Transition (RPT)

Results in Sect. 5.1.2 show that, across all markets, α values start relatively high ($\alpha \approx 10\%$) as traders *explore* the space of prices, and then quickly reduce, with markets tending to an equilibration level of $\alpha \approx 1\%$. This suggests that the market's price discovery is readily finding values close to P_0. Further, in Sects. 5.1.3 and 5.1.4, agents are shown to consistently outperform humans, securing greater allocative efficiency $E(agents) > E(humans)$, and gaining greater profit $\Delta P(agents - humans) > 0$. These results demonstrate a well-functioning robot-human market trading near equilibrium, with robots out-competing humans. This is an interesting result, but for our purpose of exploring the RPT described by [35, 36] it only serves as demonstrative proof that our experimental markets are performing as we would expect. The real interest lies in whether we can observe a phase transition between two regimes: one dominated by robot–robot interactions, and one dominated by human–robot interactions. We seek evidence of this by observing the proportion of homogeneous counterparties within a market; that is, the number of trade executions that occur between a pair of humans or a pair of robots, as a proportion of all market trades. Since traders interact anonymously via the exchange, there can be no preferential selection of counterparties. Therefore, every buyer (seller) has an equal opportunity to trade with every seller (buyer), as long as both have a pending assignment. The experimental market is configured to have an equal number of robot traders and human traders, and an equal number of identical assignments are issued to both groups. Hence, in the limit, we should expect 50% of trade counterparties to be homogeneous (both robot, or both human), and 50% to be heterogeneous (one robot and one human), as traders execute with counterparties drawn at random from the population.

From Sect. 5.1.6, our results demonstrate that for markets containing AA-0.1 robots (with sleep-wake cycle $t_s = 100$ ms; faster than human response time), the proportion of homogeneous counterparties is significantly higher than we would expect in a mixed market; whereas for markets containing robots AA-1 ($t_s = 1000$ ms; a similar magnitude to human response time) and AA-10 ($t_s = 10{,}000$ ms; slower than human response time), the proportion of homogeneous counterparties cannot be significantly differentiated from 50%. We present this as tentative first evidence for a robot-phase transition in experimental markets with a boundary between 100 ms and 1 s; although, in our experiments the effects of increasing robot speed appear to give a progressive response rather than a step-change. However, we feel obliged to caveat this result as non-conclusive proof until further experiments have been run, and until our results have been independently replicated.

The careful reader may have noticed that the results presented have not demonstrated *fractures*—ultrafast series of multiple sequential up-tick or down-tick trades that cause market price to deviate rapidly from equilibrium and then just as quickly return—phenomena that [35, 36] revealed in real market data. Since we are constraining market participants to one role (as buyer, or seller) and strictly controlling the flow of orders into the market and limit prices of trades, the simple markets we have constructed do not have the capacity to demonstrate such fractures. For this reason, we use the proportion of homogeneous counterparties as proxy evidence for the robot phase transition.

6.2 Fast Agents and Market Efficiency

Results presented in Sect. 5.2 compare markets containing fast AA-0.1 robots to markets containing slower AA-10 robots. It is shown that markets containing fast AA-0.1 robots have higher α (Sect. 5.2.1), lower allocative efficiency (Sect. 5.2.2), and higher profit dispersion (Sect. 5.2.3). Together, these facts suggest that when agents act at super-human speeds, human performance suffers, causing an overall reduction in the efficiency of the market. The reason for this could be, perhaps, that the presence of very fast acting agents causes confusion in humans, resulting in poorer efficiency. If an analogous effect occurs in real financial markets, it may imply that high-frequency trading (HFT) can reduce market efficiency.

However, these findings largely contradict the findings presented in Sect. 5.1 and discussed in Sect. 6.1; where market equilibration α (Sect. 5.1.2), market efficiency (Sect. 5.1.3), and profit dispersion (Sect. 5.1.5) are shown to be unaffected by robot speed. The reason for this disparity is primarily due to an unanticipated feature (a bug) in the behaviour of AA agents used in the experiments of Sect. 5.2 that was not discovered at the time (for details, see [9, pp. 25–26] and [50, p. 8]). These AA agents included a *spread jumping* rule such that agents will execute against a counterparty in the order-book if the relative spread width (the difference in price between the highest bid and the lowest ask, divided by the mean of the highest bid and lowest ask) is below a relative threshold of $MaxSpread = 15\%$. This is a large, unrealistic threshold, and it was reduced to $MaxSpread = 1\%$ for experiments presented in Sect. 5.1.

It is reasonable to infer that the spread jumping behaviour of AA agents is directly responsible for the higher α values presented in Sect. 5.2.1, compared with results for non-spread jumping agents shown in Sect. 5.1.2.[6] However, when considering market efficiency (Sect. 5.2.2), the explanation is not quite so simple. Despite the *bug* existing in the agent, efficiency for agents is largely unaffected when agent speed is increased; whereas human efficiency drops from a level comparable with agents when sleep-wake cycle time is 10 s to 6% lower than agents when agents act with 0.1 s sleep-wake cycle time. Therefore, the effect of the bug on efficiency only affects humans, and does so only when agents act at super-human speeds. We also see a similar affect on the magnitude of profit dispersion (Sect. 5.2.3), such that $\pi_{disp}(humans)$ is 76% higher in AA-0.1 markets compared with AA-10 markets, whereas $\pi_{disp}(agents)$ is only 5% higher.

This slightly counter-intuitive result is perhaps again further evidence for the RPT. When agents act at super-human speeds, fragmentation in the market means that agents are more likely to trade with other agents. While agents that execute a trade due to the spread jumping bug will lose out, the agent counterparty to the trade will gain, thus cancelling out the negative efficiency effects for agents overall. Human efficiency, however, is negatively affected by the resulting market dynamics, as the market trades away from equilibrium. A similar phenomena has also been observed in a recent pilot study performed at the University of Nottingham Ningbo China (UNNC) in July 2016, using a different agent (ZIP) and performed on a different experimental platform (ExPo2—details forthcoming in future publication).[7] In the pilot study, agents were allowed to submit loss-making orders into the market (i.e., agent quote price was not constrained by assignment limit price). Interestingly, when agents acted at human speeds (10 s sleep-wake cycle), markets equilibrated as expected. However, when agents acted at super-human speeds (0.1 s sleep-wake cycle), the market did not equilibrate to P_0. This demonstrates that when agents act on human timescales, i.e., above the RPT, the equilibration behaviour of humans can dampen idiosyncratic agent behaviour. However, at super-human timescales (i.e., below the RPT), the cumulative effects of agent behaviour dominate the market. Therefore, as we move from timescales above the RPT to below the RPT, the market transitions from a more efficient to a less efficient domain.

We see similar effects occur in real markets, for example, Knight Capital's fiasco, previously dubbed elsewhere as the *Knightmare on Wall Street*. On 1st August 2012, Knight Capital—formerly the largest US equities trader by volume,

---

[6]Some of the variation in α between results presented in Sects. 5.1.2 and 5.2.1 may be explained by the different permit schedules used for the two experiments (compare Tables 1 and 2). However, previous results from a direct comparison using an identical permit schedule to Table 2 show that $MaxSpread = 15\%$ results in higher α than $MaxSpread = 1\%$ [9, Appendix B]. Although, a more recent study [16] suggests the opposite result, so there is some uncertainty around this effect.

[7]ExPo: the exchange portal: www.theexchangeportal.org.

trading an average of 128,000 shares per second—started live trading their new Retail Liquidity Provider (RLP) market making software on NYSE. Within 45 min, RLP executed four million trades across 154 stocks, generating a pre-tax loss of $440 million. The following day, Knight's share price collapsed over 70%. Knight subsequently went into administration, before being acquired by Getco, a smaller rival, forming KCG Holdings (for further details, see [4]). It is widely accepted that Knight's failure was due to repurposing, and inadvertently releasing, deprecated test code that began executing trades deliberately designed to move the market price. In the live markets, and at high frequencies well above the RPT, this resulted in Knight's RLP effectively trading with itself, but at a loss on either side of the trade. The parallel here with spread-jumping AA agents is clear; if RLP acted at much lower frequencies, below the RPT, it is likely, perhaps, that the market could have dampened the instability caused. Of further interest is that the market perturbance caused by RLP percolated across a large number of stocks as other automated trading systems reacted to the behaviour. This demonstrates how single stock fractures below the RPT can have wider market impact over longer timescales.

6.3 Realism in Market Experiments: Artefacts or Evidence?

The cyclical-replenishment permit schedules presented in Sect. 4 approximate real-world markets more poorly than random-replenishment permit schedules. In real markets, demand and supply does not arrive in neat price-ordered cycles. For that reason, where results from cyclical markets (presented in Sect. 5.1) show a significant effect that is not also present in random markets, we interpret it as an indication that introducing artificial constraints into experimental markets for ease of analysis runs the risk of also introducing artefacts that, because they are statistically significant, can be misleading.

The following relationships were all observed to be statistically significant in cyclical markets and not statistically significant in random markets, providing further support for the argument for *realism* in artificial-market experiment design, previously advanced at length in [19]:

1. Cyclical-replenishment markets have significantly greater α in the first period of trade (see Sect. 5.1.2). This is a direct consequence of cyclical-replenishment allocating orders in a monotonically decreasing sequence from most profitable to least profitable. As such, the first orders allocated into the market have limit prices far from equilibrium. Since the market is empty, there is no mechanism for price discovery other than trial-and-error exploration, leading to large α. In random-replenishment markets, the initial orders entering the market are drawn at random from the demand and supply schedules. This leads to lower bounds on limit prices and hence lower α. Subsequently, price discovery is led by the order book, resulting in lower α that is statistically similar in both cyclical and random markets.

2. In cyclical-replenishment markets, the efficiency of AA-0.1 robots is significantly higher than the efficiency of the other robot types (see Sect. 5.1.3). While there is some evidence of an inverse relationship between robot sleep time and robot efficiency across all markets, we infer that this difference is an artefact of cyclical replenishment until further experimental trials can confirm otherwise.
3. In cyclical-replenishment markets, profit dispersion is significantly higher for agents, humans, and the market as a whole (see Sect. 5.1.5). Since lower profit dispersion is a desirable property of a market, this suggests that the relatively high profit dispersion observed in previous cyclical-replenishment experiments [11, 19] is an artefact of the experimental design.

7 Conclusion

We have presented a series of laboratory experiments between agent traders and human traders in a controlled financial market. Results demonstrate that, despite the simplicity of the market, when agents act on super-human timescales—i.e., when the sleep-wake cycle of agents is 0.1 s—the market starts to fragment, such that agents are more likely to trade with agents, and humans are more likely to trade with humans. In contrast, when agents act on human timescales—i.e., when the sleep-wake cycle of agents is 1 s, or above—the markets are well mixed, with agents and humans equally likely to trade between themselves and between each other. This transition to a fragmented market from a mixed market intriguingly appears to be linked to market inefficiency, such that below the threshold of human reaction times (i.e., at 0.1 s timescale) any idiosyncratic agent behaviours can adversely perturb the market; whereas above the threshold (i.e., at timescales of 1 s and above) human interactions help to dampen market perturbations, ensuring better equilibration and efficiency.

This behaviour has parallels with the real financial markets, and in particular, we present this as tantalising evidence for the robot phase transition (RPT), discovered by Johnson et al. [35, 36]. In Johnson et al.'s words, "a remarkable new study by Cliff and Cartlidge provides some additional support for our findings. In controlled lab experiments, they found when machines operate on similar timescales to humans (longer than 1 s), the 'lab market' exhibited an efficient phase (c.f. few extreme price-change events in our case). By contrast, when machines operated on a timescale faster than the human response time (100 ms) then the market exhibited an inefficient phase (c.f. many extreme price-change events in our case)" [36].

In the final quarter of 2016, a new exchange node containing the first ever intentional delay was introduced in the USA. To achieve a delay of 350 µs in signal transmission, the exchange embedded a 38-mile coil of fibre optic cable. The desired intention is to "level out highly asymmetric advantages available to faster participants" in the market [34]. However, the impact this might have at the system level is unknown. To address this, Johnson declares that more academic studies need to focus on subsecond resolution data, and he identifies the work we have reported here as one of the few exceptions in the literature that attempts to understand subsecond behaviours [34].

This work is presented as a demonstration of the utility of using experimental human–agent laboratory controlled markets: (a) to better understand real-world complex financial markets; and (b) to test novel market policies and structures before implementing them in the real world. We hope that we are able to encourage the wider scientific community to pursue more research endeavour using this methodology.

Future Work

For results presented here, we used De Luca's OpEx experimental trading software, running on the *Lab-in-a-box* hardware, a self-contained wired-LAN containing networked exchange server, netbooks for human participants, and an administrator's laptop. This platform is ideally suited for controlled real-time trading experiments, but is designed for relatively small-scale, synchronous markets where participants are physically co-located. If experiments are to be scaled up, to run for much longer periods and to support large-scale human participation, an alternative platform architecture is required. To this end, development began on ExPo—the *Exchange Portal*—in 2011. ExPo has a Web service architecture, with humans participating via interaction through a Web browser (see [50]). This enables users to connect to the exchange via the Internet, and participate remotely. Immediately, ExPo negates the requirement for specific hardware, and enables long-term and many-participant experimentation, with users able to leave and return to a market via individual account log-in. Currently, an updated version—ExPo2—is under development at UNNC, in collaboration with Paul Dempster. As with OpEx and ExPo, ExPo2 will be released open-source to encourage replication studies and engagement in the wider scientific community.

In [8] a detailed proposal for future research studies is presented. In particular, future work will concentrate on relaxing some experimental constraints, such as enabling agents to trade on their own account, independent of permit schedules. This relaxation—effectively changing the function of agents from an agency trader (or "broker") design to a proprietary "prop" trader design—should enable the emergence of more realistic dynamics, such as Johnson et al.'s UEE price swing fractures. If we are able to reproduce these dynamics in the lab, this will provide compelling evidence for the RPT. Further, market structures and regulatory mechanisms such as financial circuit breakers, intentional network delays, and periodic (rather than real-time) order matching at the exchange will be tested to understand the impact these have on market dynamics. In addition, preliminary studies to monitor human emotional responses to market shocks, using EEG brain data, are underway. Hopefully these studies can help us better understand how emotional reactions can exacerbate market swings, and how regulatory mechanisms, or trading interface designs, can be used to dampen such adverse dynamics.

Acknowledgements The experimental research presented in this chapter was conducted in 2011–2012 at the University of Bristol, UK, in collaboration with colleagues Marco De Luca (the developer of OpEx) and Charlotte Szostek. They both deserve a special thanks. Thanks also to all the undergraduate students and summer interns (now graduated) that helped support related work, in particular Steve Stotter and Tomas Gražys for work on the original ExPo platform. Finally, thanks to Paul Dempster and the summer interns at UNNC for work on developing the ExPo2 platform, and the pilot studies run during July 2016. Financial support for the studies at Bristol was provided by EPSRC grants EP/H042644/1 and EP/F001096/1, and funding from the UK Government Office for Science (Go-Science) Foresight Project on *The Future of Computer Trading in Financial Markets*. Financial support for ExPo2 development at UNNC was provided by FoSE summer internship funding.

References

1. Angel, J., Harris, L., & Spratt, C. (2010). Equity trading in the 21st century. Working Paper FBE-09-10, Marshall School of Business, University of Southern California, February 2010. Available via SSRN https://ssrn.com/abstract=1584026 Accessed 22.03.2017
2. Arthur, W. B. (2014). *Complexity and the economy*. Oxford: Oxford University Press.
3. Battiston, S., Farmer, J. D., Flache, A., Garlaschelli, D., Haldane, A. G., Heesterbeek, H., et al. (2016). Complexity theory and financial regulation: Economic policy needs interdisciplinary network analysis and behavioral modeling. *Science, 351*(6275), 818–819
4. Baxter, G., & Cartlidge, J. (2013). Flying by the seat of their pants: What can high frequency trading learn from aviation? In G. Brat, E. Garcia, A. Moccia, P. Palanque, A. Pasquini, F. J. Saez, & M. Winckler (Eds.), *Proceedings of 3rd International Conference on Applied and Theory of Automation in Command and Control System (ATACCS)*, Naples (pp. 64–73). New York: ACM/IRIT Press, May 2013.
5. Berger, S. (Ed.), (2009). *The foundations of non-equilibrium economics*. New York: Routledge.
6. Bisias, D., Flood, M., Lo, A. W., & Valavanis, S. (2012). A survey of systemic risk analytics. *Annual Review of Financial Economics, 4*, 255–296.
7. Bouchaud, J. P. (2008). Economics needs a scientific revolution. *Nature, 455*(7217), 1181.
8. Cartlidge, J. (2016). Towards adaptive ex ante circuit breakers in financial markets using human-algorithmic market studies. In *Proceedings of 28th International Conference on Artificial Intelligence* (ICAI), Las Vegas (pp. 77–80). CSREA Press, Athens, GA, USA. July 2016.
9. Cartlidge, J., & Cliff, D. (2012). Exploring the 'robot phase transition' in experimental human-algorithmic markets. In Future of computer trading. Government Office for Science, London, UK (October 2012) DR25. Available via GOV.UK https://www.gov.uk/government/publications/computer-trading-robot-phase-transition-in-experimental-human-algorithmic-markets Accessed 22.03.2017.
10. Cartlidge, J., & Cliff, D. (2013). Evidencing the robot phase transition in human-agent experimental financial markets. In J. Filipe & A. Fred (Eds.), *Proceedings of 5th International Conference on Agents and Artificial Intelligence (ICAART)*, Barcelona (Vol. 1, pp. 345–352). Setubal: SciTePress, February 2013.
11. Cartlidge, J., De Luca, M., Szostek, C., & Cliff, D. (2012). Too fast too furious: Faster financial-market trading agents can give less efficient markets. In J. Filipe & A. Fred (Eds.), *Proceedings of 4th International Conference on Agents and Artificial Intelligent (ICAART)*, Vilamoura (Vol. 2, pp. 126–135). Setubal: SciTePress, February 2012.
12. Chen, S. H., & Du, Y. R. (2015). Granularity in economic decision making: An interdisciplinary review. In W. Pedrycz & S. M. Chen (Eds.), *Granular computing and decision-making: Interactive and iterative approaches* (pp. 47–72). Berlin: Springer (2015)
13. Cliff, D., & Bruten, J. (1997). *Minimal-Intelligence Agents for Bargaining Behaviours in Market-Based Environments*. Technical Report HPL-97-91, Hewlett-Packard Labs., Bristol, August 1997.

14. Cliff, D., & Northrop, L. (2017). The global financial markets: An ultra-large-scale systems perspective. In: Future of computer trading. Government Office for Science, London, UK (September 2011) DR4. Available via GOV.UK https://www.gov.uk/government/publications/computer-trading-global-financial-markets Accessed 22.03.2017
15. Das, R., Hanson, J., Kephart, J., & Tesauro, G. (2001) Agent-human interactions in the continuous double auction. In Nebel, B. (Ed.), *Proceedings of 17th International Conference on Artificial Intelligence (IJCAI)*, Seattle (pp. 1169–1176). San Francisco: Morgan Kaufmann, August 2001
16. De Luca, M. (2015). Why robots failed: Demonstrating the superiority of multiple-order trading agents in experimental human-agent financial markets. In S. Loiseau, J. Filipe, B. Duval, & J. van den Herik, (Eds.), *Proceedings of 7th International Conference on Agents and Artificial Intelligence (ICAART)*, Lisbon (Vol. 1, pp. 44–53). Setubal: SciTePress, January 2015.
17. De Luca, M., & Cliff, D. (2011). Agent-human interactions in the continuous double auction, redux: Using the OpEx lab-in-a-box to explore ZIP and GDX. In J. Filipe, & A. Fred (Eds.), *Proceedings of 3rd International Conference on Agents and Artificial Intelligents (ICAART)* (Vol. 2, pp. 351–358) Setubal: SciTePress, January 2011.
18. De Luca, M., & Cliff, D. (2011). Human-agent auction interactions: Adaptive-aggressive agents dominate. In Walsh, T. (Ed.), *Proceedings of 22nd International Joint Conference on Artificial Intelligence (IJCAI)* (pp. 178–185). Menlo Park: AAAI Press, July 2011.
19. De Luca, M., Szostek, C., Cartlidge, J., & Cliff, D. (2011). Studies of interactions between human traders and algorithmic trading systems. In: Future of Computer Trading. Government Office for Science, London, September 2011, DR13. Available via GOV.UK https://www.gov.uk/government/publications/computer-trading-interactions-between-human-traders-and-algorithmic-trading-systems Accessed 22.03.17.
20. Easley, D., & Kleinberg, J. (2010). *Networks, crowds, and markets: Reasoning about a highly connected world.* Cambridge: Cambridge University Press
21. Easley, D., Lopez de Prado, M., & O'Hara, M. (Winter 2011). The microstructure of the 'flash crash': Flow toxicity, liquidity crashes and the probability of informed trading. *Journal of Portfolio Management, 37*(2), 118–128
22. Farmer, J. D., & Foley, D. (2009). The economy needs agent-based modelling. *Nature, 460*(7256), 685–686
23. Farmer, J. D., & Skouras, S. (2011). An ecological perspective on the future of computer trading. In: Future of Computer Trading. Government Office for Science, London, September 2011, DR6. Available via GOV.UK https://www.gov.uk/government/publications/computer-trading-an-ecological-perspective Accessed 22.03.2017.
24. Feltovich, N. (2003). Nonparametric tests of differences in medians: Comparison of the Wilcoxon-Mann-Whitney and Robust Rank-Order tests. *Experimental Economics, 6*, 273–297.
25. Foresight. (2012). *The Future of Computer Trading in Financial Markets.* Final project report, The Government Office for Science, London, UK (October 2012). Available via GOV.UK http://www.cftc.gov/idc/groups/public/@aboutcftc/documents/file/tacfuturecomputertrading1012.pdf Accessed 22.03.17
26. Giles, J. (2012). Stock trading 'fractures' may warn of next crash. New Scientist (2852) (February 2012). Available Online: http://www.newscientist.com/article/mg21328525.700-stock-trading-fractures-may-warn-of-next-crash.html Accessed 22.03.17.
27. Gjerstad, S., & Dickhaut, J. (1998). Price formation in double auctions. *Games and Economic Behavior, 22*(1), 1–29
28. Gode, D., & Sunder, S. (1993). Allocative efficiency of markets with zero-intelligence traders: Markets as a partial substitute for individual rationality. *Journal of Political Economy, 101*(1), 119–137.
29. Gomber, P., Arndt, B., Lutat, M., & Uhle, T. (2011). *High Frequency Trading.* Technical report, Goethe Universität, Frankfurt Am Main (2011). Commissioned by Deutsche Börse Group.

30. Grossklags, J., & Schmidt, C. (2003). Artificial software agents on thin double auction markets: A human trader experiment. In J. Liu, B. Faltings, N. Zhong, R. Lu, & T. Nishida (Eds.), *Proceedings of IEEE/WIC Conference on Intelligent Agent and Technology (IAT)*, Halifax (pp. 400–407). New York: IEEE Press.
31. Grossklags, J., & Schmidt, C. (2006). Software agents and market (in)efficiency: A human trader experiment. *IEEE Transactions on Systems, Man and Cybernetics, Part C (Applications and Review) 36*(1), 56–67.
32. Holt, C. A., & Roth, A. E. (2004). The Nash equilibrium: A perspective. *Proceedings of the National Academy of Sciences of the United States of America, 101*(12), 3999–4002
33. Huber, J., Shubik, M., & Sunder, S. (2010). Three minimal market institutions with human and algorithmic agents: Theory and experimental evidence. *Games and Economic Behavior, 70*(2), 403–424
34. Johnson, N. (2017). To slow or not? Challenges in subsecond networks. *Science, 355*(6327), 801–802.
35. Johnson, N., Zhao, G., Hunsader, E., Meng, J., Ravindar, A., Carran, S., et al. (2012). *Financial Black Swans Driven by Ultrafast Machine Ecology*. Working paper published on arXiv repository, Feb 2012.
36. Johnson, N., Zhao, G., Hunsader, E., Qi, H., Johnson, N., Meng, J., et al. (2013). Abrupt rise of new machine ecology beyond human response time. *Scientific Reports, 3*(2627), 1–7 (2013)
37. Joint CFTC-SEC Advisory Committee on Emerging Regulatory Issues. (2010). *Findings Regarding the Market Events of May 6, 2010*. Report, CTFC-SEC, Washington, DC, September 2010. Available via SEC https://www.sec.gov/news/studies/2010/marketevents-report.pdf Accessed 22.03.2017.
38. Kahneman, D. (2011). *Thinking, fast and slow*. New York, NY: Farrar, Straus and Giroux.
39. Keim, B. (2012). Nanosecond trading could make markets go haywire. Wired (February 2012). Available Online: http://www.wired.com/wiredscience/2012/02/high-speed-trading Accessed 22.03.2017.
40. Leinweber, D. (2009). *Nerds on wall street*. New York: Wiley.
41. May, R. M., Levin, S. A., & Sugihara, G. (2008) Complex systems: Ecology for bankers. *Nature, 451*, 893–895
42. Nelson, R. H. (2001). *Economics as religion: From Samuelson to Chicago and beyond*. University Park, PA: Penn State University Press.
43. Nelson, R. R., & Winter, S. G. (1982). *An evolutionary theory of economic change*. Harvard: Harvard University Press.
44. Perez, E. (2011). *The speed traders*. New York: McGraw-Hill.
45. Price, M. (2012). New reports highlight HFT research divide. Financial News (February 2012). Available Online: https://www.fnlondon.com/articles/hft-reports-highlight-research-divide-cornell-20120221 Accessed 22.03.2017.
46. Schweitzer, F., Fagiolo, G., Sornette, D., Vega-Redondo, F., & White, D. R. (2009). Economic networks: what do we know and what do we need to know? *Advances in Complex Systems, 12*(04n05), 407–422
47. Smith, V. (1962). An experimental study of comparative market behavior. *Journal of Political Economy, 70*, 111–137
48. Smith, V. (2006). *Papers in experimental economics*. Cambridge: Cambridge University Press.
49. Stotter, S., Cartlidge, J., & Cliff, D. (2013). Exploring assignment-adaptive (ASAD) trading agents in financial market experiments. In J. Filipe & A. Fred (Eds.), *Proceedings of 5th International Conference on Agents and Artificial Intelligence (ICAART)*, Barcelona. Setubal: SciTePress, February 2013.
50. Stotter, S., Cartlidge, J., & Cliff, D. (2014). Behavioural investigations of financial trading agents using Exchange Portal (ExPo). In N. T. Nguyen, R. Kowalczyk, A. Fred, & F. Joaquim (Eds.), *Transactions on computational collective intelligence XVII*. Lecture notes in computer science (Vol. 8790, pp. 22–45). Berlin: Springer.

51. Tesauro, G., & Bredin, J. (2002). Strategic sequential bidding in auctions using dynamic programming. In C. Castelfranchi & W. L. Johnson (Eds.), *Proceedings of 1st International Conference on Autonomous Agents and Multiagent Systems (AAMAS)*, Bologna (pp. 591–598). New York: ACM.
52. Tesauro, G., & Das, R. (2001). High-performance bidding agents for the continuous double auction. In *Proceedings of the ACM Conference on Electronic Commerce (EC)*, Tampa, FL (pp. 206–209), October 2001.
53. Treanor, J. (2017). Pound's flash crash 'was amplified by inexperienced traders'. The Guardian, January 2017. Available Online https://www.theguardian.com/business/2017/jan/13/pound-flash-crash-traders-sterling-dollar Accessed 22.03.2017.
54. Vytelingum, P. (2006). *The Structure and Behaviour of the Continuous Double Auction*. PhD thesis, School of Electronics and Computer Science, University of Southampton.
55. Vytelingum, P., Cliff, D., & Jennings, N. (2008). Strategic bidding in continuous double auctions. *Artificial Intelligence, 172*, 1700–1729
56. Widrow, B., & Hoff, M. E. (1960). Adaptive switching circuits. Institute of Radio Engineers, Western Electronic Show and Convention, Convention Record, Part 4 (pp. 96–104).
57. Zhang, S. S. (2013). *High Frequency Trading in Financial Markets*. PhD thesis, Karlsruher Institut für Technologie (KIT).

Does High-Frequency Trading Matter?

Chia-Hsuan Yeh and Chun-Yi Yang

Abstract Over the past few decades, financial markets have undergone remarkable reforms as a result of developments in computer technology and changing regulations, which have dramatically altered the structures and the properties of financial markets. The advances in technology have largely increased the speed of communication and trading. This has given birth to the development of algorithmic trading (AT) and high-frequency trading (HFT). The proliferation of AT and HFT has raised many issues regarding their impacts on the market. This paper proposes a framework characterized by an agent-based artificial stock market where market phenomena result from the interaction between many heterogeneous non-HFTs and HFTs. In comparison with the existing literature on the agent-based modeling of HFT, the traders in our model adopt a genetic programming (GP) learning algorithm. Since they are more adaptive and heuristic, they can form quite diverse trading strategies, rather than zero-intelligence strategies or pre-specified fundamentalist or chartist strategies. Based on this framework, this paper examines the effects of HFT on price discovery, market stability, volume, and allocative efficiency loss.

Keywords High-frequency trading · Agent-based modeling · Artificial stock market · Continuous double action · Genetic programming

C.-H. Yeh (✉)
Department of Information Management, Yuan Ze University, Taoyuan, Chungli, Taiwan
e-mail: imcyeh@saturn.yzu.edu.tw

C.-Y. Yang
Department of Computational and Data Sciences, College of Science, Krasnow Institute for Advanced Study, George Mason University, Fairfax, VA, USA

1 Introduction

Over the past few decades, financial markets around the world have undergone a remarkable transformation as a result of developments in computer technology and regulatory changes, which have dramatically altered the structures as well as the properties of financial markets. For example, in 1976, an electronic system was introduced to the New York Stock Exchange for scheduling the sequence of market orders and limit orders. In the 1980s, the order system of the NYSE was upgraded to the SuperDot system, which made traders capable of trading whole portfolios of stocks simultaneously, i.e. program trading. In 1998, the US Securities and Exchange Commission (SEC) authorized the adoption of electronic communication networks (ECNs), which was referred to the Regulation of Alternative Trading Systems (Reg. ATS). In such a system, traders are allowed to trade assets outside of stock exchanges and their orders are matched automatically without the intermediation of dealers or brokers. In 2005, the SEC employed the Regulation National Market System (Reg. NMS) which was dedicated to the promotion of automatic and immediate access to intermarket quotations. In addition, the advances in communications technology and trading greatly increased the speed of information dissemination and aggregation. The high-speed processing and communication of data as well as computationally intensive mathematical models became more important for the management of financial portfolios. Therefore, the competition for transactions became increasingly fierce and the duration of arbitrage opportunities correspondingly shorter.

All of these institutional and informational transitions have motivated financial professionals to develop automated and intelligent trading algorithms to execute orders via a computer network, giving birth to the development of *algorithmic trading* (AT) and high-frequency trading (HFT). AT represents the implementation of a set of trading algorithms implemented via computers, while HFT, a subset of AT, refers to the execution of proprietary trading strategies through employing advanced computing technology to automatically submit and manage orders at quite high frequencies. Nowadays, HFT accounts for a large portion of transactions in financial markets. According to [28], over 70% of dollar trading volume in the US capital market is transacted via HFT. There is no doubt that the development of computer technology has persistently had a profound influence on the market at many levels and scales, and has affected the connectivity between market participants. Therefore, the importance of HFT will monotonically increase as computer technology unceasingly fosters innovation and change in financial markets which consist of human traders and program traders (software agents). The emergence of AT and HFT trading has had a significant impact on the market and has permanently changed the market's financial structure and other properties. It has also increased market complexity and resulted in new market dynamics, so that it will be harder for regulators to understand and regulate the market.

The proliferation of AT and HFT raises many issues regarding their impacts on the market, such as price discovery, market efficiency, market stability, liquidity, risk, and financial regulations. In actual fact, the role of HFT is quite controversial. The extent to which HFT may improve or undermine the functions of financial markets as well as its impact on market volatility and risk are still being debated. A consensus has not yet been reached. The proponents of HFT argue that HFT provides liquidity to the market, improves price discovery, reduces transaction costs, and lowers spreads and market volatility. By contrast, its opponents claim that HFT increases the systematic risk associated with both financial instability and management, and raises rather than lowers volatility. In addition, HFT may induce market manipulation. In this regard, the Flash Crash of 6 May 2010 that occurred in the Dow Jones stock index is arguably attributable to HFT. Therefore, researchers in the past few years have paid much attention to its impact from both a theoretical and empirical perspective. They have consistently investigated this uncharted hybrid financial ecosystem of human and program traders, in the hope of preempting any deceitful practice and providing adaptive guidelines for effective regulations that ensure the efficiency and stability of financial markets and the welfare of market participants.

To understand the channels through which and the extent to which HFT may affect the market, a number of theoretical models as well as empirical studies have been proposed to examine these issues, e.g. [5, 9, 11, 13, 17], and [4]. Although the literature regarding this topic has experienced precipitous growth in the past few years, many challenges remain, as mentioned in [21]:

> From a regulatory perspective, HFT presents difficult and partially unresolved questions. The difficulties stem partly from the fact that HFT encompasses a wide range of trading strategies, and partly from a dearth of unambiguous empirical findings about HFT's effects on markets. ...Overall, the empirical research does not demonstrate that HFT has substantial social benefits justifying its clear risks. Regulatory measures including stronger monitoring, order cancellation taxes, and resting rules deserve more urgent attention. (p. 1)

Subject to mathematical tractability, in order to derive analytical results, theoretical approaches tend to restrict their analyses to some quite simplified frameworks where traders' heterogeneity as well as their interactions are represented in reduced forms. However, heterogeneity is the main driving force in financial systems, which fuels the market dynamics with emergent properties. Without taking into account this characteristic, the results obtained in theoretical models could be biased. Empirical studies also suffer from serious problems in that the analyses depend on the availability of information regarding the quotes submitted by high-frequency traders (HFTs) as well as the trading strategies employed by them. However, such information are mostly proprietary. Without such information, empirical studies can only be conducted based on the inconclusive evidence regarding the behavior and order submissions of HFTs. As mentioned in [5]:

> Research in this area is limited by the fact that the contents of the algorithms that define HF trading are closely guarded secrets, and hence unavailable to academics. Without the possibility of looking into the black boxes, empirical research is left with the task of studying HF trading indirectly, via observed trading behavior. Some studies work with pure trade data, others have been given tags that distinguish trades generated by one group of firms, labeled HFTs, from those generated by the rest of the markets. (p. 1250014-2)

The conclusions reached through such an indirect channel could be problematic. Therefore, under theoretical frameworks or empirical studies, it is rather difficult to understand and analyze how and why HFT may affect the market when isolating and tracing the effects of a specific trading strategy in a highly interactive system may be unrealistic. In addition, its impact may also be affected by financial regulations such as price limit rules or tick size rules, and the trading mechanism employed to determine transaction prices. Moreover, as [22] point out:

> The systematic danger lays in the possibility of cross-excitations to other markets causing additional herding of low latency and/or fundamental traders. We believe that a complex systems approach to future research is crucial in the attempt to capture this inter-dependent and out-of-equilibrium nature of financial markets. Particularly relevant in assessing HFT risk is to understand their behaviour in conjunction with other market participants across various asset classes. Agent-based modeling (ABM hereafter) offers a key method to improve our understanding of the systems' dynamics. (p. 5)

According to research funded by the U.K. government, "Foresight: The Future of Computer Trading in Financial Markets" of the Final Project Report of the Government Office for Science (2012), any regulation however effective it is should be based on solid evidence and reliable analysis. Nevertheless, the regulators encounter two particular challenges in the investigation of HFT. First, the development and application of new technology together with the complexity of financial trading has advanced rapidly. Such a situation makes understanding the nature of HFT, developing associated policies, and executing regulations more difficult. Second, the research regarding the effects of HFT often fails to maintain the same pace as that of the advances in technology, and we are short of data that are accessible, comprehensive, and consistent. These restrictions call for novel approaches, in particular a computational one like agent-based modeling (ABM). The method of heterogeneous agent models (HAM) is a promising endeavor that applies ABM to the study of the forces driving the dynamics in financial markets. In addition to serving as an attempt to uncover the origin of stylized facts observed in real financial markets, it further serves as a tool to examine the effectiveness of financial policies. The developments in this field are still a work in progress.[1]

However, the study of HFT by means of ABM remains in its infancy. To the best of our knowledge, [8] is the first paper that adopts this technique to examine the effects of HFT. So far, papers studying this issue have been few in number. Gsell [8] examines the impacts of AT on price formation and volatility. Two types of traders are employed in this model. One comprises the stylized traders who are assumed to know the fundamental value of the asset. They are simulated along with a special combination of informed traders, momentum traders and noise traders. The other consists of algorithmic trading traders. The preliminary results show that large volumes traded by algorithmic traders generate an increasingly large impact on prices, while lower latency results in lower volatility. Hanson [10] adopts an

---

[1]In [15] and [19], they provide a rather detailed review and description regarding a number of different types of HAMs and show the importance of HAMs to the study of financial markets.

agent-based model where a continuous double auction (CDA) market is employed to investigate the impacts of a simple HFT strategy on a group of zero-intelligence (ZI) traders. Each of them is assigned a random reservation price. The HF traders do not learn from their experience, but act as liquidity buyers at the beginning of the simulation. However, when they engage in trading, they become potential sellers based on new reservation prices calculated by adding a markup to the transaction price. The results indicate that the population size of HFTs generates significant impacts on market liquidity, efficiency, and the overall surplus. Market volatility increases with the number of HFTs, and the profits of HFTs may be realized at the cost of long-term investors. Walsh et al. [24] design a hybrid experiment consisting of human traders and algorithmic traders. These traders adopt either fundamental strategies or technical strategies. The difference between human traders and algorithmic traders is that the latter behave like HFTs. Their findings indicate that liquidity increases with the share of algorithmic traders, while there exist no significant patterns regarding volatility. In a simple model of latency arbitrage, [23] examine how latency arbitrage may affect allocative efficiency and liquidity in fragmented financial markets. In a two-market model with a CDA mechanism, two types of traders, i.e. the background traders and the latency arbitrageur (LA), buy and sell a single security. The former adopt ZI strategies and each of them is confined to a randomly assigned market without direct access to the information outside its own market except, subject to some degree of latency, via an information intermediating channel of the best price quoted over the two markets, while the latter has an advantage of direct access to the latest prices quoted in both markets, and posts his offer whenever a latency-induced arbitrage opportunity exists. Both types of traders have no learning abilities. The simulation results show that market fragmentation and the actions of a latency arbitrageur reduce market efficiency and liquidity. However, when a periodic call market is employed, the opportunities for latency arbitrage are removed, and the market efficiency is improved. Jacob Leal et al. [16] discuss the interplay between low-frequency traders (LFTs) who according to their relative profitability adopt a fundamentalist strategy or a chartist strategy, and high-frequency traders (HFTs) who use directional strategies. Their activations are based on the extent of price deviation. The simulation results indicate that the behavior of HFTs can bring about flash crashes. In addition, the higher that their rate of order cancellation is, the higher the frequency, and the shorter the duration of the crashes.

At the risk of being too general, the traders employed in these papers have no learning ability. In addition, the traders are either characterized as ZI traders or pre-specified as fundamentalists or chartists. However, such a design seriously restricts, if any, the emergent market dynamics as well as properties. Therefore, the conclusions made in these papers will possibly be limited. On the contrary, learning is an indispensable element of economic and financial behavior since in practice traders continuously learn from experience. Without taking into account this element, analyses would be biased. In [7], the authors point out the importance of learning to market properties. A market populated with ZI traders is unable to replicate the phenomena resulting from human traders. In [16], although the LFTs are endowed with learning and they can determine which type of strategy they would like to use, the HFTs employ pre-determined trading strategies without learning. Therefore, the purpose of this paper is to provide a framework enabling us to carry out a systemic investigation on the effects of HFT from a broader perspective. In our model, not only LFTs but also HFTs can learn to develop new trading strategies by means of a genetic programming (GP) algorithm. It is this hyperdimensional flexibility in strategy formation disciplined by survival pressure, not some ad hoc behavior rules as in the other heterogeneous agent models, that makes GP unique with regard to the resulting model complexity, which is way beyond what could be generated by the stochasticity of asset fundamentals, as agents collectively expand the strategy space while continuously explore and exploit it. The advantages of GP have been mentioned in [18] and [19]. The representation of GP not only can incorporate the traders' decision-making process and adaptive behavior, but also can open up the possibility of innovative behavior. Under this more general framework of an artificial stock market, we can have an in-depth discussion regarding the functions and effects of HFT.

The remainder of this paper is organized as follows. Section 2 describes the framework of the artificial stock market, traders' learning behavior, and the trading mechanisms. An analysis of the simulation outcomes and results is presented in Sect. 3. Section 4 concludes.

2 The Model

2.1 The Market Model

The market model considered in this paper is quite similar to that of [3] and [20]. It consists of traders whose preferences have the same constant absolute risk aversion (CARA) utility function, i.e. $U(W_{i,t}) = -\exp(-\lambda W_{i,t})$, where $W_{i,t}$ is trader i's wealth in period t, and λ is the degree of absolute risk aversion. Each trader has two assets for investment: a risky stock and riskless money. At the end of each period, the stock pays dividends which follow a stochastic process (D_t) unknown to traders.[2] The riskless money is perfectly elastically supplied with a constant interest rate per annum, r. Just as in the design proposed in [12], the interest rate can be measured over different time horizons, i.e. $r_d = r/K$ (where K stands for the number of

---

[2]While the dividend's stochastic process is unknown to traders, they base their derivation of the expected utility maximization on an Gaussian assumption. Under the CARA preference, the uncertainty of the dividends affects the process of expected utility maximization by being part of conditional variance.

transaction days over 1 year.[3]). Therefore, a gross return is equal to $R = 1 + r_d$. Trader i's wealth in the next period can be described as

$$W_{i,t+1} = RW_{i,t} + (P_{t+1} + D_{t+1} - RP_t)h_{i,t}, \quad (1)$$

where P_t is the price (i.e. the closing price) per share of the stock and $h_{i,t}$ is the number of stock shares held by trader i at time t. The second term in Eq. (1) accounts for the excess revenue from holding the stock over period $t + 1$ to earn the excess capital gain, $R_{t+1} = P_{t+1} + D_{t+1} - RP_t$.

Each trader myopically maximizes his one-period expected utility based on the wealth constraint shown in Eq. (1):

$$\max_h \{E_{i,t}(W_{t+1}) - \frac{\lambda}{2} V_{i,t}(W_{t+1})\}, \quad (2)$$

where $E_{i,t}(\cdot)$ and $V_{i,t}(\cdot)$ are his one-period ahead forecasts regarding the conditional expectation and variance at $t + 1$, respectively, on the basis of his information set, $I_{i,t}$, updated up to period t. Therefore, we can derive

$$E_{i,t}(W_{t+1}) = RW_{i,t} + E_{i,t}(P_{t+1} + D_{t+1} - RP_t)h_{i,t}$$
$$= RW_{i,t} + E_{i,t}(R_{t+1})h_{i,t}, \quad (3)$$
$$V_{i,t}(W_{t+1}) = h_{i,t}^2 V_{i,t}(P_{t+1} + D_{t+1} - RP_t) = h_{i,t}^2 V_{i,t}(R_{t+1}), \quad (4)$$

The optimal shares of the stock, $h_{i,t}^*$, for trader i at the beginning of each period are determined by solving Eq. (2),

$$h_{i,t}^* = \frac{E_{i,t}(R_{t+1})}{\lambda V_{i,t}(R_{t+1})} = \frac{E_{i,t}(P_{t+1} + D_{t+1}) - RP_t}{\lambda V_{i,t}(R_{t+1})}. \quad (5)$$

The difference between the optimal position and a trader's current stockholding is his demand for the stock:

$$d_{i,t} = h_{i,t}^* - h_{i,t}. \quad (6)$$

Therefore, a trader acts as a buyer (seller) if $d_{i,t} > 0$ ($d_{i,t} < 0$).

---

[3]For example, K = 1, 12, 52, 250 represents the number of trading periods measured by the units of a year, month, week and day, respectively.

2.2 Expectations

Based on Eq. (5), it is clear that each trader's optimal holding of the stock depends on his own conditional expectation and variance. The functional form for his expectation, $E_{i,t}(\cdot)$, is described as follows[4]:

$$E_{i,t}(P_{t+1}+D_{t+1}) = \begin{cases} (P_t + D_t)[1 + \theta_0(1 + \text{sech}(\beta \ln(1 + f_{i,t})))] & \text{if } f_{i,t} \geq 0.0, \\ (P_t + D_t)[1 + \theta_0(1 - \text{sech}(\beta \ln(-1 + f_{i,t})))] & \text{if } f_{i,t} < 0.0, \end{cases} \quad (7)$$

where θ_0 and β are coefficients of expectation transformation, and $f_{i,t}$ is the raw forecast value generated by the GP algorithm. P_t is the last closing price for LFTs and the last transaction price (P^T) for HFTs.

Each trader updates his active rule's estimated conditional variance at the end of each period. Let $\sigma_{i,t}^2$ denote $V_{i,t}(R_{t+1})$. Each trader's conditional variance is updated with constant weights θ_1, θ_2 and θ_3 by:

$$\sigma_{i,t}^2 = (1 - \theta_1 - \theta_2)\sigma_{i,t-1}^2 + \theta_1(P_t + D_t - u_{t-1})^2 + \theta_2[(P_t + D_t) - E_{i,t-1}(P_t + D_t)]^2, \quad (8)$$

where

$$u_t = (1 - \theta_1)u_{t-1} + \theta_1(P_t + D_t). \quad (9)$$

This variance update function is a variant of those used in [20] and [12]. In [12], they argue that part of the autocorrelation features and the power-law pattern found in financial markets may come from the geometric decay process of the sample mean and variance. As one of many alternatives to variance formulation away from rational expectations, our version is sufficient for representing the concept of bounded rationality as well as replicating several critical stylized facts.

To maximize one-period expected utility, each trader has to maximize his excess revenue by looking for a better forecast for the sum of the price and dividend of the next period, while at the same time minimizing his perceived conditional variance (uncertainty) of the excess return. In our model, each trader is endowed with a population of strategies (i.e. N_I models). Which strategy (or forecasting function) will be used in the next period crucially depends on its relative performance

---

[4]Our framework can incorporate other types of expectation functions; however, different specifications require different information available to traders, and the choice depends on the selected degree of model validation. Unlike [1] and [20], we do not assume that traders have the information regarding the fundamental value of the stock, i.e. the price in the equilibrium of homogeneous rational expectations. The functional form used in this paper extends the designs employed in [6, 20], and [26]. It allows for the possibility of a bet on the Martingale hypothesis if $f_{i,t} = 0$. For more details about applying the GP evolution to the function formation, we recommend the readers to refer to Appendix A in [27].

compared with the others up until the current period. The performance (strength) of each model is measured by its forecast accuracy, i.e. the negative value of its conditional variance,[5]

$$s_{i,j,t} = -\sigma^2_{i,j,t}, \tag{10}$$

where $s_{i,j,t}$ is the strength of trader i's jth model, $F_{i,j,t}$, up to period t.

The strategies of each trader evolve over time. The evolutionary process operates on the basis of the strengths of these strategies as follows. Each trader asynchronously iterates the evolutionary process to update his own strategy pool every N_{EC} periods (an evolutionary cycle).[6] At the beginning of the first period in each evolutionary cycle, each trader randomly selects N_T out of N_I models. Among these initial candidates, a trader selects the one with the highest strength to make his forecast, $f_{i,t}$ as shown in Eq. (7), in each period of this cycle. At the end of the last period in each cycle, each trader evaluates the strength of each model in his own strategy pool. The strategy with the lowest strength is replaced by a new model which is generated by one of the three operators: crossover, mutation, and immigration. The crossover operator implies the concept of combining the information existing in the strategy pool, while the mutation or the immigration operator represents a kind of innovation.

2.3 Trading Process

This paper adopts a simplified continuous double auction (CDA) process designed to mimic the intraday trading process. Unlike the existing literature applying ABM to the study of HFT in which traders are characterized as ZI or pre-specified fundamentalists or chartists, we consider two populations of uninformed and bounded-rational GP-learning traders, LFTs and HFTs, that are ignorant of the information regarding the fundamental value of the stock.[7] The numbers of LFTs and HFTs are N_{LF} and N_{HF}, respectively. The difference between LFTs and HFTs

---

[5] Similar designs were used in [1] and [20].
[6] Although N_{EC}, defining the time length of the periods between the occurrences of strategy evolution, is a parameter common to all traders, each trader has his first strategy evolution occurred randomly at one of the first N_{EC} period, so that traders are heterogeneous in asynchronous strategy learning. Because we focus on the effect of the speed difference between HFTs and LFTs, rather than the relative speed difference among HFTs nor how market dynamics drive the participants learn toward either kind of timing capability, on the market performance, a setting with endogenous evolution cycles though intriguing may blur, if any, the possible speed-induced effect by entangling the characteristics of the two trade-timing types.
[7] One alternative is including informed HFTs so as to examine their impacts. However, in order to focus on the effects resulting from high-frequency trading alone, both LFTs and HFTs are uniformly assumed to be uninformed traders.

is the speed of incorporating market information into their expectations as well as trading decisions. Although some instant market information, such as the best bid and ask, are observed for all traders, the LFTs can only form their expectations based on the information regarding the prices and dividends of the last k periods, i.e. $\{P_{t-1}, \ldots, P_{t-k}, D_{t-1}, \ldots, D_{t-k}\}$, and then make their trading decisions. By contrast, HFTs are assumed to be active speculators who can gain access to and quickly process the instant market information to form their expectations and trading decisions. To represent such a discrepancy, in addition to the information available to LFTs, HFTs are also kept updated on the last l transaction prices (i.e. $\{P^T, P^T_{-1}, \ldots, P^T_{-l+1}\}$) and the last l trading volumes (i.e. $\{V^T, V^T_{-1}, \ldots V^T_{-l+1}\}$), the highest l bids and the lowest l asks (i.e. $\{B_b, B_{b,-1}, \ldots B_{b,-l+1}; B_a, B_{a,-1}, \ldots, B_{a,-l+1}\}$),[8] on the order book. Therefore, their expectations and trading decisions may differ from one another as long as a new bid or ask is posted, even if they use the same forecasting model. For both LFTs and HFTs, the factors of the market information are incorporated into decision making by being part of the elements of the information set that are fed into the leaf nodes of GP trees as input. With the information input (leaf) nodes as the variables, a decision rule wrapped in tree-like representation of the generative GP syntax then could be decoded into a conventional mathematical equation.

At the beginning of each period, the orders in which both types of traders are queued to take turns to submit quotes are randomly shuffled. A period starts with m LFTs posting their orders, and all HFTs come after these LFTs to post their offers. Each HFT trader recalculates his expectation based on the latest instant information before submitting his order. After all HFTs have completed their submissions, the next m LFTs enter the market.[9] A trading period is finished when the last m LFTs and the ensuing HFTs have posted their orders; then, a new period begins.[10]

Each trader regardless of types, based on the best ask (B_a), the best bid (B_b) and his demand for the stock $(d_{i,t})$, determines his trading action, i.e. accepting (or posting) a bid or an ask, according to the following rules:

- If $d_{i,t} = 0$, he has no desire to make a quote.
- If $d_{i,t} > 0$, he decides to buy,

---

[8] P^T (V^T) is the last transaction price (trading volume), P^T_{-l+1} (V^T_{-l+1}) is the last lth transaction price (trading volume). $B_{b,-1}$ ($B_{a,-1}$) refers to the current second best bid (ask), and $B_{b,-l+1}$ ($B_{a,-l+1}$) is the current lth best bid (ask).

[9] Although the way we design the "jump-ahead" behavior of HFTs is not information-driven, it is neutral without subject interpretation of HFT behavior taking part in the assumptions. In light of the fact that the motives and the exact underlying triggering mechanisms of the vast variety of HFTs are still far from transparent, it would be prudent not to overly specify how HFTs would react to what information and thus have the results reflecting the correlation induced by ad hoc behavioral assumptions instead of the causality purely originated from the speed difference.

[10] The relative speeds of information processing for different types of traders are exogenous. One may vary the speed setting to investigate the effect of technology innovation in terms of the improvement in traders' reactions to market information.

- If B_b exists,

    If $B_b > E_{i,t}(\cdot)$, he will post a buy order at a price uniformly distributed in $(E_{i,t}(\cdot) - S_i, E_{i,t}(\cdot))$, where $S_i = \gamma \sigma_{i,t}^2$.
    If $B_b \leq E_{i,t}(\cdot)$, he will post a buy order at a price uniformly distributed in $(B_b, \min\{B_b + S_i, E_{i,t}(\cdot)\})$.

- If B_b does not exist but B_a exists,

    If $B_a > E_{i,t}(\cdot)$, he will post a buy order at a price uniformly distributed in $(E_{i,t}(\cdot) - S_i, E_{i,t}(\cdot))$.
    If $B_a \leq E_{i,t}(\cdot)$, he will post a market order, and buy at B_a.

- If neither B_b nor B_a exist,

    If $P_{LT} > E_{i,t}(\cdot)$, he will post a buy order at a price uniformly distributed in $(E_{i,t}(\cdot) - S_i, E_{i,t}(\cdot))$, where P_{LT} is the previous transaction price right before he submits his order.
    If $P_{LT} \leq E_{i,t}(\cdot)$, he will post a buy order at a price uniformly distributed in $(P_{LT}, P_{LT} + S_i)$.

- If $d_{i,t} < 0$, he decides to sell,

    - If B_a exists,

        If $B_a < E_{i,t}(\cdot)$, he will post a sell order at a price uniformly distributed in $(E_{i,t}(\cdot), E_{i,t}(\cdot) + S_i)$.
        If $B_a \geq E_{i,t}(\cdot)$, he will post a sell order at a price uniformly distributed in $(\max\{B_a - S_i, E_{i,t}(\cdot)\}, B_a)$.

    - If B_a does not exists but B_b exists,

        If $B_b < E_{i,t}(\cdot)$, he will post an sell order at a price uniformly distributed in $(E_{i,t}(\cdot), E_{i,t}(\cdot) + S_i)$.
        If $B_b \geq E_{i,t}(\cdot)$, he will post a market order, and sell at B_b.

    - If neither B_b nor B_a exist,

        If $P_{LT} < E_{i,t}(\cdot)$, he will post a sell order at a price uniformly distributed in $(E_{i,t}(\cdot), E_{i,t}(\cdot) + S_i)$.
        If $P_{LT} \geq E_{i,t}(\cdot)$, he will post a sell order at a price uniformly distributed in $(P_{LT} - S_i, P_{LT})$.

A transaction occurs when a trader accepts the existing best bid or ask, or when a pair of bid and ask cross. All unfulfilled orders that are left at the end of each period are canceled before the next period begins. The closing price of a period is defined as the last transaction price.

3 Simulation Results

To have a better understanding of the impacts of HFT, besides the benchmark market (BM) without HFTs, we experiment with a series of markets differing in the relative portions of LFTs and HFTs and in the speed (activation frequency) of HFTs. The activation frequency of HFTs is measured in the way that HFTs enter the market after m LFTs have posted their orders. A higher activation frequency of HFTs implies a lower value of m. Three different numbers of HFTs (i.e. $N_{HF} = 5, 10, 20$) and three different activation frequencies (i.e. $m = 40, 20, 10$) are considered. The scenarios of these markets are summarized in Table 1.

The model is calibrated to generate several prominent stylized facts observed in real financial markets, e.g. the unconditional returns characteristic of excess kurtosis, fat tails and insignificant autocorrelations as well as volatility clustering. Because our focus is on the general impact of HFTs, the calibration does not extend to the behavior of HFTs, i.e. the explicit trading tactics, and only aims at the daily rather than intraday stylized facts. This modeling strategy lessens the difficulty in isolating the pure effects of HFT activities, which would be very difficult to manage if we were to consider all possible behavioral traits of HFTs speculated in real financial markets.[11] The difficulty would be compounded as the arguable features of HFTs may vary with the evolutions of technology, communication techniques, and trading methods.

The parameters are presented in Table 2. The net supply of the stock is set to be zero. In addition, traders are allowed to sell short. Figure 1 presents the basic time series properties of a typical BM run. The top two panels of Fig. 1 are the price and return dynamics of the stock, respectively. The last two panels show the autocorrelation features of the raw returns and absolute returns, respectively.

Table 1 The market scenario

	m		
N_{HF}	40	20	10
5	A1	A2	A3
10	B1	B2	B3
20	C1	C2	C3

Table 2 Stylized facts of the calibrated model

	K	α	H_r	$H_{|r|}$
Minimum	13.98	2.49	0.49	0.94
Median	19.62	4.25	0.54	0.95
Average	22.83	4.31	0.54	0.95
Maximum	47.03	6.30	0.62	0.95

---

[11]The real characteristics of HFTs deserve further examination, in the hope of uncovering the causality between some specific trading behavior (not necessarily exclusive to HFTs) and market phenomena.

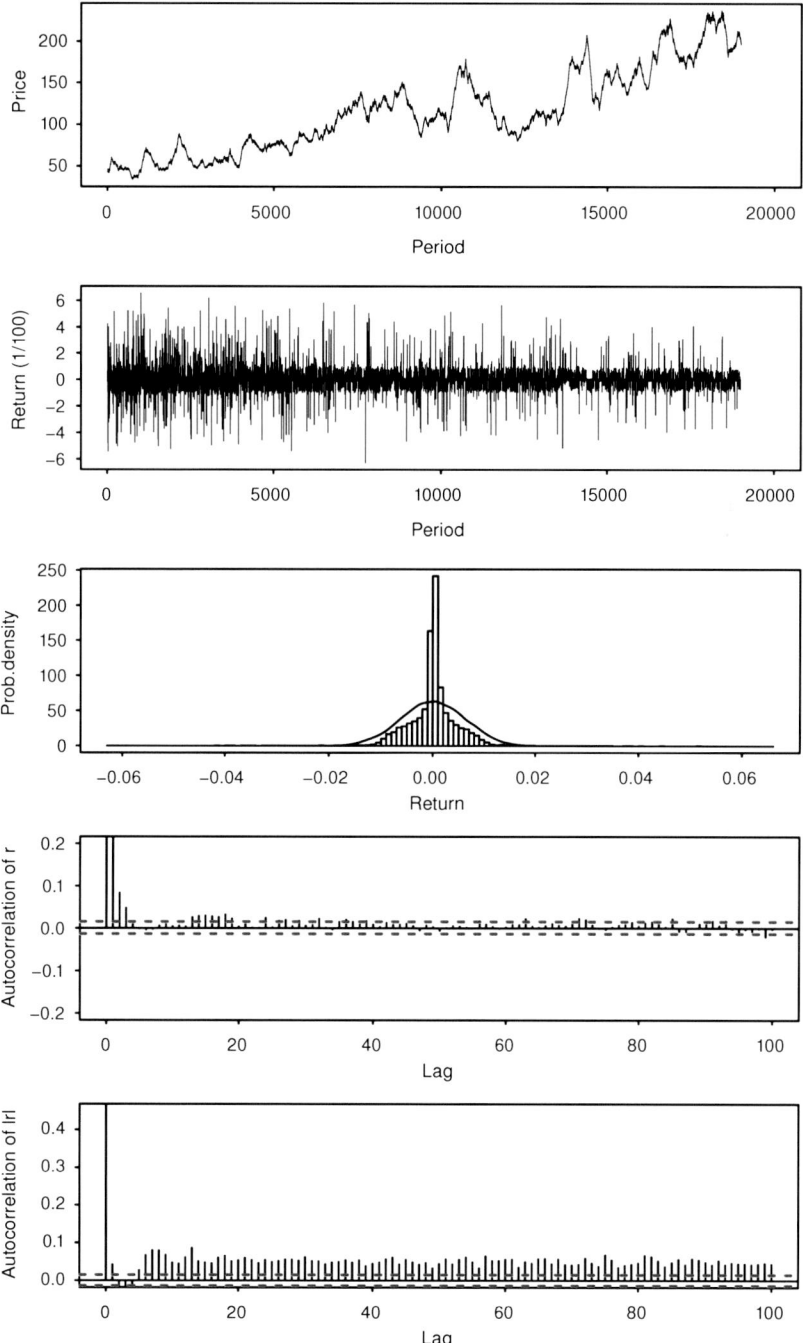

Fig. 1 Time series properties of the calibrated model

Table 3 Parameters for simulations

The stock market	
h_0	0
M_0	$100
(r, r_d)	(0.05, 0.0002)
D_t	$N(\overline{D}, \sigma_D^2) = N(0.02, 0.004)$
N_P	20000
Traders	
(N_{LF}, N_{HF})	(200, 5), (200, 10), (200, 20)
(N_T, N_I)	(5, 20)
(N_{EC}, N_{SC})	(3, 3)
λ	0.5
θ_0	0.5
β	0.05
(θ_1, θ_2)	(0.3, 0.3)
γ	0.5
Parameters of genetic programming	
Function set	{if-then-else;and,or,not; $\geq, \leq, =, +, -, \times, \%,$} sin,cos,abs,sqrt}
Terminal set for the LFTs	$\{P_{t-1}, \ldots, P_{t-5}, D_{t-1}, \ldots, D_{t-5}, R\}$
Terminal set for the HFTs	$\{P_{t-1}, \ldots, P_{t-5}, D_{t-1}, \ldots, D_{t-5},$ $P^T, \ldots, P^T_{-4}, V^T, \ldots, V^T_{-4},$ $B_b, \ldots, B_{b,-4}, B_a, \ldots, B_{a,-4}, R\}$
Probability of immigration (P_I)	0.1
Probability of crossover (P_C)	0.7
Probability of mutation (P_M)	0.2

R denotes the ephemeral random constant

Besides the first three lags, over most lag periods the autocorrelations of the raw returns are insignificant at the 5% significance level, while the autocorrelations of the absolute returns are significant for most lag periods. Table 3 reports some statistics for the returns for 20 BM runs. The second column is the excess kurtosis (K) of the raw returns. It is clear that the values of the excess kurtosis are larger than zero, which is an indication of fat tails. The tail phenomena are also manifested in the tail index α proposed in [14], which is a more appropriate statistic than the kurtosis. The index value is calculated based on the largest 5% of observations. The smaller the value is, the fatter is the tail. The empirical values usually range from two to five. The third column shows that the average value of the tail index is 4.31, which is consistent with the empirical results. The last two columns are the Hurst exponents of the raw returns and absolute returns, respectively. The Hurst exponent is used to measure the memory of a time series. Its value ranges between 0 and 1, and is 0.5 for a random series, above which a long memory process leading to persistence exists in a time series. Empirically, the Hurst exponent is about 0.5 for raw returns, and above that for absolute returns. It is evident that the values of the Hurst exponents of the simulated raw returns are close to 0.5, while those of the absolute returns are about 0.95. Therefore, these return series statistics indicate that our model is capable of replicating some salient empirical features.

In conducting the simulation based on this calibrated model, we perform twenty runs (each lasting 20,000 periods) with different random seeds for each market scenario. The statistical analyses of HFT are then based on the average of the 20 run averages for the whole period. First, we analyze the effects of HFT on the market volatility, price discovery, and trading volume. Following the design of [25], the market volatility (P_V) and price distortion (P_D) of a simulation run are measured by

$$P_V = \frac{100}{N_P - 1} \sum_{t=1}^{N_P} \left| \frac{P_t - P_{t-1}}{P_{t-1}} \right|, \quad P_D = \frac{100}{N_P} \sum_{t=1}^{N_P} \left| \frac{P_t - P_f}{P_f} \right|, \quad (11)$$

where P_f is the fundamental price of the stock under the assumption of homogeneous rational expectations.

$$P_f = \frac{1}{R-1}(\overline{D} - \lambda \sigma_D^2 h) \quad (12)$$

Therefore, the price distortion measures the extent to which the price deviates from the fundamental price. In our model, $P_f = 100.0$. Because the number of traders across different market scenarios is not the same, we consider the trading volume per trader for a fair comparison between the markets.

Figure 2 presents the relationship between the market volatility and the activation frequency of HFT participation when different numbers of HFTs are considered. The solid (black) line describes the average volatilities of 5 HFTs, while the dashed (blue) line and the chain (red) line represent those of 10 and 20 HFTs, respectively. When the markets have only 5 HFTs, the volatility pattern does not exhibit any significant difference between the markets either with or without the HFTs, regardless of the frequency of HFT participation. Basically, this phenomenon does not change much when the number of HFTs is 10. The volatility is slightly higher if the HFTs intervene for every 20 or 10 LFTs. However, such a property disappears when the number of HFTs is 20. It is evident that overall the market volatilities are greater for those markets with HFTs. In addition, more intense HFT activity in terms of a higher activation frequency of participation by HFTs results in greater volatility. The results reveal that, when the number of HFTs exceeds a threshold, the activity of HFTs gives rise to the more volatile market dynamics.

The influence of HFT on price distortion is presented in Fig. 3. Compared with the benchmark market, under the case of the number of HFTs being 5, the presence of HFTs results in lower price distortions. Besides, the price distortion decreases

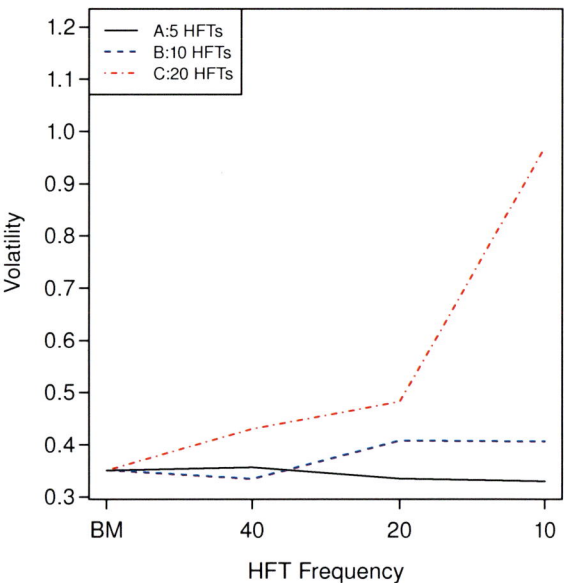

Fig. 2 Market volatility

with an increase in the activation frequency of HFT participation. When the number of HFTs increases from 5 to 10, such a pattern remains unchanged. Therefore, our results indicate that the existence of HFTs may help to improve price discovery. However, the increase in the number of HFTs still generates higher levels of price distortion, implying a subtle discrepancy between the effect of HFT population size (the scale) and that of HFT activation frequency (the intensity). When the number of HFTs is further raised from 10 to 20, the results change dramatically. The price distortions of the HFT markets become much larger than those in the BM except in the case of low HFT activation intensity where the HFTs intervene every 40 LFTs. Moreover, the price distortion increases with an increase in the activation frequency of HFT participation. This phenomenon indicates that, as long as the markets are dominated by the HFT activity, the HFT will hinder the price discovery process.

Figure 4 displays the patterns of volume. The trading volume largely increases when the HFTs exist, regardless of their population size. When only 5 HFTs exist, the trading volume does not significantly further increase as the HFT activity intensifies. This phenomenon is the same in the case of 10 HFTs with the activation frequencies for HFT participation at 10 and 20. However, when the number of HFTs further increases from 10 to 20, the trading volume exhibits a relatively clear upward trend as the activation frequency of HFT participation increases. Therefore, our results indicate that the activity of HFTs helps to enhance market liquidity. In addition, the impact of the number of HFTs seems to be greater than that of the activation frequency of HFT participation.

Fig. 3 Price distortion

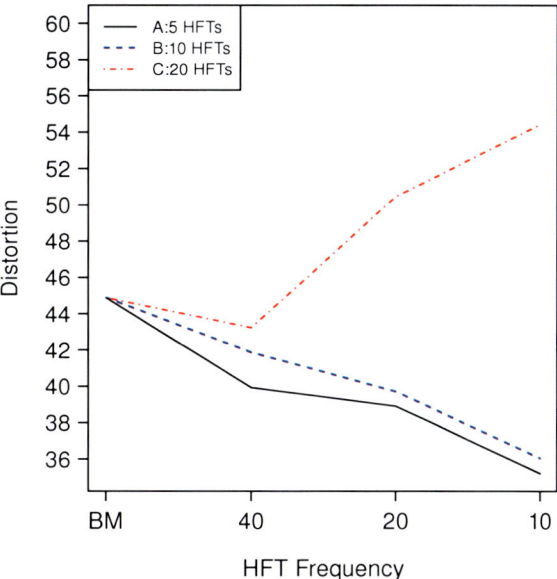

Fig. 4 Trading volume

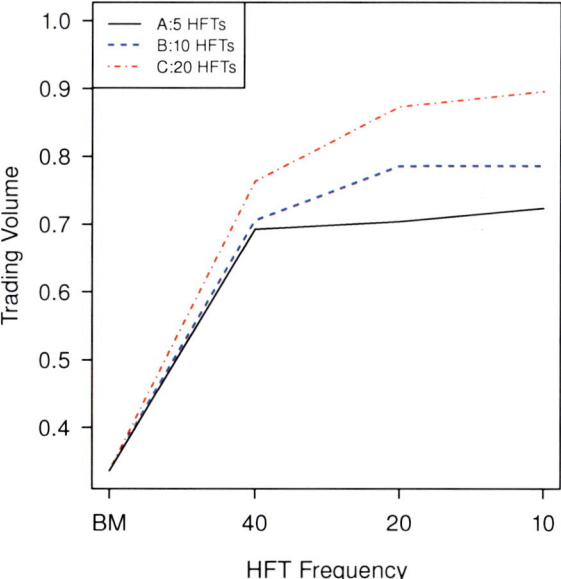

Fig. 5 Allocative efficiency loss

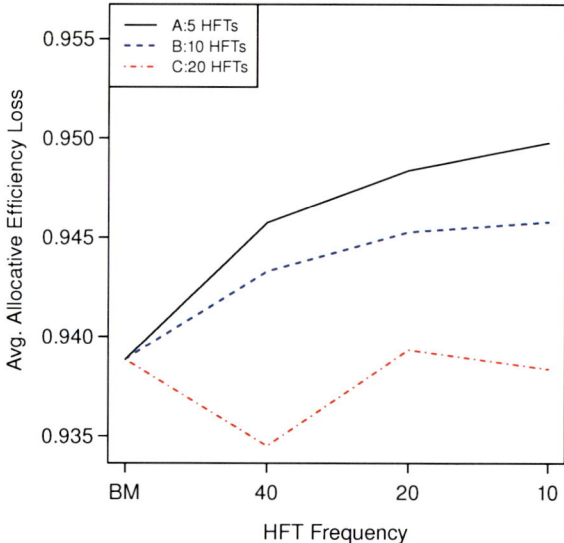

To understand how HFT may affect market efficiency, we adopt the measure of the allocative efficiency loss proposed in [2]. The allocative efficiency loss for trader i in period t is defined as:

$$L_{i,t} = 1 - \frac{1}{1 + |h_{i,t}(P_t) - h_{i,t}|P_t}, \qquad (13)$$

where $h_{i,t}(P_t)$ is trader i's desired quantity of the stock at price P_t, and $h_{i,t}$ is the number of stock shares held at the end of period t. Therefore, this measure is calculated based on the difference between a trader's realized stockholding and his optimal position at the end of each period. The results of the allocative efficiency loss are presented in Fig. 5. It is clear that, when the markets have only 5 HFTs, the level of allocative efficiency loss increases accompanied by an increase in the activation frequency of HFT participation. The growth rate of this trend is lower when the number of HFTs is 10. Such a pattern disappears when more HFTs, such as 20, are added to the markets. In addition, the level of allocative efficiency loss decreases as the number of HFTs increases. This implies that, when the number of HFTs is below some threshold, the presence of HFTs results in a higher level of allocative efficiency loss. However, as long as the number of HFTs is large enough, the allocative efficiency losses in the markets with HFTs display no clear trend in terms of the deviations from the loss in the BM case. As the chain (red) line shows, except in the case where the activation frequency of HFT participation is 40, the magnitudes of the losses do not differ much from that for the BM.

4 Conclusion

This paper proposes an agent-based artificial stock market to examine the effects of HFT on the market. In this model, both HFTs and LFTs are adaptive, and are able to form very diverse trading strategies. The differences between LFTs and HFTs are the speeds of order submission and information processing capacity (i.e. the ability to incorporate up-to-the-minute order book information in expectations). Our analysis covers the market volatility, price distortion, trading volume, and allocative efficiency loss as a result of the interplay between the scale (the size of the population) and the intensity (the activation frequency of the participation) of the HFT activity. The simulation results indicate that the presence of HFT may give rise to both positive and negative effects on the markets. HFT does help in improving market liquidity. However, its impacts on market volatility, price distortion, and allocative efficiency loss vary with the number of HFTs. Therefore, evaluations of the influence of HFT ought to take into consideration both the intensity of participation by HFTs and the number of HFTs.

Acknowledgements Research support from MOST Grant no. 103-2410-H-155-004-MY2 is gratefully acknowledged.

References

1. Arthur, W. B., Holland, J., LeBaron, B., Palmer, R., & Tayler, P. (1997). Asset pricing under endogenous expectations in an artificial stock market. In W. B. Arthur, S. Durlauf, & D. Lane (Eds.), *The economy as an evolving complex system II* (pp. 15-44). Reading, MA: Addison-Wesley.
2. Bottazzi, G., Dosi, G., & Rebesco, I. (2005). Institutional architectures and behavioral ecologies in the dynamics of financial markets. *Journal of Mathematical Economics, 41*(1–2), 197–228.
3. Brock, W. A., & Hommes. C. H. (1998). Heterogeneous beliefs and routes to chaos in a simple asset pricing model. *Journal of Economic Dynamics and Control, 22*(8–9), 1235–1274.
4. Carrion, A. (2013). Very fast money: High-frequency trading on the NASDAQ. *Journal of Financial Markets, 16*(4), 680–711.
5. Cartea, Á., & Penalva, J. (2012). Where is the value in high frequency trading? *Quarterly Journal of Finance, 2*(3), 1250014-1–1250014-46.
6. Chen, S.-H., & Yeh, C.-H. (2001). Evolving traders and the business school with genetic programming: A new architecture of the agent-based artificial stock market. *Journal of Economic Dynamics and Control, 25*(3–4), 363–393.
7. Cliff, D., & Bruten, J. (1997). *Zero is Not Enough: On the Lower Limit of Agent Intelligence for Continuous Double Auction Markets.* HP Technical Report, HPL-97-141.
8. Gsell, M. (2008). *Assessing the Impact of Algorithmic Trading on Markets: A Simulation Approach.* Working paper.
9. Hagströmer, B., & Lars Nordén, L. (2013). The diversity of high-frequency traders. *Journal of Financial Markets, 16*(4), 741–770.
10. Hanson, T. A. (2011). *The Effects of High frequency Traders in a Simulated Market.* Working paper.

11. Hasbrouck, J., & Saar, G. (2013). Low-latency trading. *Journal of Financial Markets, 15*(4), 646–679.
12. He, X.-Z., & Li, Y. (2007). Power-law behaviour, heterogeneity, and trend chasing. *Journal of Economic Dynamics and Control, 31*(10), 3396–3426.
13. Hendersgott, T., Jones, C. M., & Menkveld, A. J. (2011). Does algorithmic trading improve liquidity? *The Journal of Finance, 66*(1), 1–33.
14. Hill, B. M. (1975). A simple general approach to inference about the tail of a distribution. *Annals of Statistics, 3*(5), 1163–1174.
15. Hommes, C. H. (2006) Heterogeneous agent models in economics and finance. In L. Tesfatsion & K. L. Judd (Eds.), *Handbook of computational economics* (Vol. 2, Chap. 23, pp. 1109–1186). Amsterdam: Elsevier.
16. Jacob Leal, S., Napoletano, M., Roventini, A., & Fagiolo, G. (2014). Rock around the clock: An agent-based model of low- and high-frequency trading. In: *Paper Presented at the 20th International Conference on Computing in Economics (CEF'2014)*.
17. Jarrow, R. A., & Protter, P. (2012). A dysfunctional role of high frequency trading in electronic markets. *International Journal of Theoretical and Applied Finance, 15*(3), 1250022-1–1250022-15.
18. Kirman, A. P. (2006). Heterogeneity in economics. *Journal of Economic Interaction and Coordination, 1*(1), 89–117.
19. LeBaron, B. (2006). Agent-based computational finance. In L. Tesfatsion & K. L. Judd (Eds.), *Handbook of computational economics* (Vol. 2, Chap. 24, pp. 1187–1233). Amsterdam: Elsevier.
20. LeBaron, B., Arthur, W. B., & Palmer, R. (1999). Time series properties of an artificial stock market. *Journal of Economic Dynamics and Control, 23*(9–10), 1487–1516.
21. Prewitt, M. (2012). High-frequency trading: should regulators do more? *Michigan Telecommunications and Technology Law Review, 19*(1), 1–31.
22. Sornette, D., & von der Becke, S. (2011). *Crashes and High Frequency Trading*. Swiss Finance Institute Research Paper No. 11–63.
23. Wah, E., & Wellman, M. P. (2013). Latency arbitrage, market fragmentation, and efficiency: A two-market model. In *Proceedings of the Fourteenth ACM Conference on Electronic Commerce* (pp. 855–872). New York: ACM.
24. Walsh, T., Xiong, B., & Chung, C. (2012). *The Impact of Algorithmic Trading in a Simulated Asset Market*. Working paper.
25. Westerhoff, F. (2003). Speculative markets and the effectiveness of price limits. *Journal of Economic Dynamics and Control, 28*(3), 493–508.
26. Yeh, C.-H. (2008). The effects of intelligence on price discovery and market efficiency. *Journal of Economic Behavior and Organization, 68*(3-4), 613–625.
27. Yeh, C.-H., & Yang, C.-Y. (2010). Examining the effectiveness of price limits in an artificial stock market. *Journal of Economic Dynamics and Control, 34*(10), 2089–2108.
28. Zhang, X. F. (2010). *High-Frequency Trading, Stock Volatility, and Price Discovery*. Working paper.

Modelling Price Discovery in an Agent Based Model for Agriculture in Luxembourg

Sameer Rege, Tomás Navarrete Gutiérrez, Antonino Marvuglia, Enrico Benetto, and Didier Stilmant

Abstract We build an ABM for simulation of incentives for maize to produce bio-fuels in Luxembourg with an aim to conduct life cycle assessment of the additional maize and the consequent displacement of other crops in Luxembourg. This paper focuses on the discovery of market price for crops. On the supply side we have farmers who are willing to sell their produce based on their actual incurred costs and an expected markup over costs. On the demand side, we have buyers or middlemen who are responsible for quoting prices and buying the output based on their expectation of the market price and quantity. We have N buyers who participate in the market over R rounds. Each buyer has a correct expectation of the total number of buyers in each market. Thus in each round, the buyer bids for a quantity $q_b^r = \frac{Q_b^e}{N \times R}$, where Q_b^e is the expected total output of a crop. The buyer at each round buys $\min(q_b^r, S_t^r)$, the minimum of the planned purchase at each round r and the total supply S_t^r by farmers in the round at a price p_b^r. The market clears over multiple rounds. At each round, the buyers are sorted by descending order of price quotes and the highest bidder gets buying priority. Similarly the farmers are sorted according to the ascending order of quotes. At the end of each round, the clearance prices are visible to all agents and the agents have an option of modifying their bids in the forthcoming rounds. The buyers and sellers may face a shortfall which is the difference between the target sale or purchase in each round and the actual realised sale. The shortfall is then covered by smoothing it over future rounds (1–4). The more aggressive behaviour is to cover the entire shortfall in the next round, while a more calm behaviour leads to smoothing over multiple (4) rounds. We find that there

---

S. Rege (✉)
ModlEcon S.à.r.l-S, Esch-sur-Alzette, Luxembourg, Luxembourg

T. Navarrete Gutiérrez · A. Marvuglia · E. Benetto
Luxembourg Institute of Science and Technology (LIST), Belvaux, Luxembourg
e-mail: tomas.navarrete@list.lu; antonino.marvuglia@list.lu; enrico.benetto@list.lu

D. Stilmant
Centre Wallon de Recherches Agronomiques (CRA-W), Libramont, Belgium
e-mail: d.stilmant@cra.wallonie.be

© Springer Nature Switzerland AG 2018
S.-H. Chen et al. (eds.), *Complex Systems Modeling and Simulation in Economics and Finance*, Springer Proceedings in Complexity, https://doi.org/10.1007/978-3-319-99624-0_5

is a statistically distinct distribution of prices and shortfall over smoothing rounds and has an impact on the price discovery.

Keywords Agent based models (ABM) · Agriculture · Biofuels · Price discovery · Life cycle assessment (LCA) · Luxembourg

1 Introduction

Life cycle assessment (LCA) is a standardised methodology used to quantify the environmental impacts of products across their whole life cycle [6]. LCA is nowadays an assessment methodology recognised worldwide, although with some persisting limitations and certain barriers, as witnessed by the survey of [3]. It has been used at the fine grain level (at the product level) or at a more global level (policy).

The MUSA (MUlti-agent Simulation for consequential LCA of Agro-systems) project http://musa.tudor.lu, aims to simulate the future possible evolution of the Luxembourgish farming system, accounting for more factors than just the economy oriented drivers in farmers' decision making processes. We are interested in the behavioural aspects of the agricultural systems, including the green conscience of the farmers. The challenges facing the farming system are manifold. Dairy and meat production have been the financial base of the national agricultural landscape with production of cereals and other crops as a support to the husbandry. This sector is fraught with multiple rules and regulations including restrictions on milk production via quotas as dictated by the Common Agricultural Policy of the EU. There is also a complex set of subsidies in place to enable the farmers to be more competitive. The quotas are set to disappear adding to increased pressure on the bottom line for dairy farmers. Just as a chain is as strong as its weakest link any model is as robust as its weakest assumption. Model building is a complex exercise but modelling behaviour is far more complex. Statistical and optimisation models fail to account for vagaries of human behaviour and have limited granularity. Preceding the MUSA project, in [9] we have built a partial equilibrium model for Luxembourg to conduct a consequential LCA of maize production for energy purposes (dealing with an estimated additional production of 80,000 tons of maize) using non-linear programming (NLP) and positive mathematical programming (PMP) approaches. PMP methodology [5] has been the mainstay of modelling methodology for agriculture models relating to cropping patterns based on economic fundamentals. This approach converts a traditional linear programming (LP) into an NLP problem by formulating the objective function as a non-linear cost function to be minimised. The objective function parameters are calibrated to replicate the base case crop outputs. This approach is useful as a macro level such as countries or regions where one observes the entire gamut of crops planted at a regional level. To increase the granularity to investigate the impacts of policy on size of farms is still possible provided each class of farms based on size exhibits plantation of all crops in the system. When the granularity increases to the farm level, the crop rotation takes priority for the farmer and then one observes only a subset of the entire list of crops in the system. It is at this level that the PMP approach fails as the objective function is calibrated to the crops observed at a specific point in time. This limitation is overcome by the agent based model (ABM) approach. To investigate the possible set of outcomes due to human responses to financial and natural challenges, ABM are a formal mechanism to validate the range of outcomes due to behavioural differences.

In the work we report here, we investigate the different price discovery mechanisms based on agent behaviour to assess convergence of prices in thin markets which are typically agriculture markets.

In Sect. 2 we present a brief tour of different agent-based models dealing with agricultural systems as well as LCA research done using agent-based models. Then, in Sect. 3 we describe our model. We present the initial results of the simulations in Sect. 4 and present our conclusions in Sect. 6.

2 Literature Survey

Happe et al. [4] use an ABM called AgriPoliS, for simulation of agricultural policies. The focus is on simulating the behaviour of farms by combining traditional optimisation models in farm economics with agent based approach for the region of Hohenlohe in southwest Germany. The paper classifies farms into size classes and assumes all farms in a specific class size to have the same area. The variation in the farms is on account of individual asset differences (both financial and physical). Farms are classified only as grasslands or arable lands with no further classification by crops or rotation schemes. Berger [1] is another application on the lines of [4] with an integration of farm based linear programming models under a cellular automata framework, applied to Chile. The model explicitly covers the spatial dimension and its links to hydrology with a policy issue of investigating the potential benefits of joining the Mercosur agreement. Le et al. [8] is an application of agent based modelling framework to assess the socio-ecological impacts of land-use policy in the Hong Ha watershed in Vietnam. The model uses in addition to farms as agents, the landscape agents that encapsulate the land characteristics. Additional sub models deal with farmland choice, forest choice, agricultural yield dynamics, forest yield dynamics, agent categoriser that classifies the households into a group and natural transition to enable transformation between vegetation types. Berta et al. [2] is a similar study to simulate structural and land use changes in the Argentinian pampas. The model is driven by agent behaviour that aims to match the aspiration level of the farmer to the wealth level. The farmer will change behaviour until the two are close. In addition to 6 equal sized plots in the farm, the farmer has a choice of planting 2 varieties of wheat, maize and soybean. The model studies the potential penetration of a particular crop and find that soybean is being cultivated on a much larger scale.

Despite most of the previously cited related works include a high number of economical aspects in their models, they lack the inclusion of aspects related to price discovery.

3 Data and Model Structure

In the model, the farmers are represented via entities called "agents" who take decisions based on their individual profiles and on a set of decision (behavioural) rules defined by the researchers on the basis of the observation of real world (e.g. the results of a survey) and on the interactions with other entities.

Luxembourg provides statistics about the economy of the agricultural sector through STATEC [11] and Service d'Economie Rural [10]. The statistics deal with the area under crops over time, farm sizes and types of farms by output classification, use of fertilisers, number of animals by type (bovines, pigs, poultry, horses) and use (meat, milk). The latest year for which one obtains a consistent data set across all the model variables was 2009.[1] In 2009 there were 2242 farms with an area of 130,762 ha under cultivation that included vineyards and fruit trees and pastures amongst other cereal and leaf crops.

Table 1 shows the distribution of farm area by size of farms in Luxembourg for the year 2009. We also know the total area of crops planted under each farm type. Figure 1 shows the initial proportions of each crop present in 2009. From [7] we know the different rotation schemes for crops. Rotation schemes are used by farmers to maintain the health of the soil and rotate cereals and leaves on the same field in a specified manner. We randomly assign a rotation scheme to each farm and then randomly choose a crop from cereals or leaves. As to crops that are neither cereals nor leaves, they are a permanent cultivation such as vineyards, fruits, meadows or pastures. Once the random allocation of crops has been made to the farms, we scale the areas so as to match the total area for the crop under a specific farm type.

3.1 Model Calibration

In the absence of real data on field ownership by farmers and the crops planted on those fields, one has to find proxy methods that can be used to allocate the farms to farmers such that the allocation of land to each crop for each farm type (from A to I) matches the data in 2009. In reality farmers plant crops on fields based

Table 1 Distribution of farms by size (ha) in Luxembourg in 2009

		Farm type								
		A	B	C	D	E	F	G	H	I
	Total	<2	2–4.9	5–9.9	10–19.9	20–29.9	30–49.9	50–60.9	70–99.9	100+
Number	2242	230	165	217	186	116	246	263	398	421
Area (ha)	130,762	131	598	1533	2667	2890	9956	15,743	33,583	63,661
Average size	58.32	0.57	3.62	7.06	14.34	24.91	40.47	59.86	84.38	151.21

---

[1]The details of the data are available in [9].

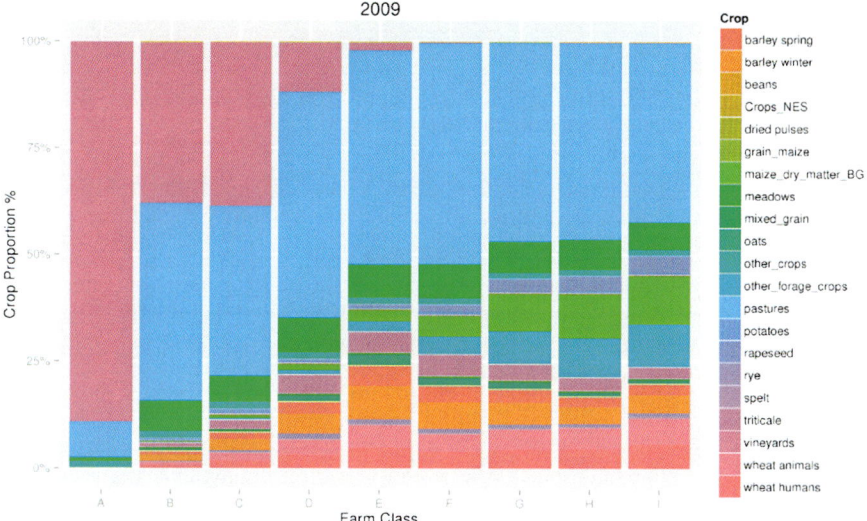

Fig. 1 Initial proportions of crops planted in 2009

on seasons and in many cases plant more than one crop on the same field in a calendar year. This leads to area under crops being greater than the actual amount of land area available for cultivation. Fortunately for the case of Luxembourg, the winter (September/October to July/August) and summer or spring (March/April to August/September) crops are not amenable for planting multiple times on the same plot of land as the winter crops are still growing when the summer crops need to be planted. Another important aspect related to the calibration is data collection by the statistical services. In the case of Luxembourg, the annual data on the crops planted come from a survey administered once annually. Since the number of farmers and farms is small, it covers very well the population. Farmers involved in rearing animals also grow crops for feed but harvest them before maturity for conversion into silage. In such instances one can encounter multiple crops on the same plot of land where in the winter crop is harvested before the summer season begins. The occurrence of such instances is quite low, but there is no statistical data of this phenomenon in the case of Luxembourg.

The objective of the calibration is to assign farm area to each of the 2242 farmers based on the farm type they belong and assign crops to the farms such that they match the randomly drawn rotation schemes for each farm and also match the aggregate area under cultivation for each crop. The ideal situation would be to have information on each farmer's land holdings by field type and choice of a rotation scheme so as to proceed with the actual modelling of behaviour in a more realistic and inclusive manner, but as previously mentioned, this is not the case. In absence of this information we proceed with the calibration as follows:

1. Randomly allocate farm size to each farmer between the minimum and maximum based on the farm type (A to I) as shown in Table 1
2. Each farmer has a rotation scheme associated with a field. We assume that the farmer consistently uses the same rotation scheme across all the fields (owned or rented) under cultivation. There are six rotation schemes in use, LCC, LCCC, LCCLC, LC, LLCC, LLLLC, where L stands for leaf crops and C for cereals. So a rotation scheme LLCC means that the field has a leaf crop, which is substituted by another leaf crop, which is then substituted by a cereal and finally by another cereal preferably different from the earlier cereal crop. The field then returns to a leaf crop to begin the cycle.

    A uniform random number between 1 and 6 (inclusive) is chosen to assign a rotation scheme to each farmer. Once a scheme has been allotted to a farmer, we assume that it remains in place till the end of the simulation over the necessary time periods.
3. (a) We initially use the naïve assumption that the number of farms planting a crop will be in proportion to the share of that crop's area in the total area for that crop's type. So to compute the number of farms planting triticale winter for farm type I, we take the area under triticale winter which is a cereal and divide it by the total area under all cereals for a specific farm type (in this case I) and use that share as a proportion of the total number of farms (here 421 as witnessed by Table 1) for that farm type (here I) to arrive at a preliminary number of farms planting that crop.

    With this procedure one could in principle land up with a very small number of farms that are planting a specific crop. In order to overcome this we use the following modification to ensure a substantial number of farms planting a crop.

---

if $selectNumberOfFarms \leq 20$ **then**
    $selectNumberOfFarms =$
    $\max(farmID.size()/2, selectNumberOfFarms)$;
else
    $selectNumberOfFarms =$
    $\min(farmID.size(), (int)(selectNumberOfFarms \times 1.5))$;
end

Algorithm 1: Selecting number of farms planting crops

where $farmID.size()$ is the total number of farms in each farm type and *selectNumberOfFarms* is the initial number of farms to plant a given crop.

  (b) After choosing the number of farms, we randomly choose unique farms from the list of farms for the specific farm type (A to I). This then completes the list of farms planting the chosen crop and same procedure is repeated for all crops.

  (c) To allocate the area for each crop under each farm, multiply the total crop area under each farm type by the fraction of the total farm area of selected farms.

Table 2 Area (ha) under crop by farm type (A to I)

cropName	Type	A	B	C	D	E	F	G	H	I
wheat_winter	C	0	4	25	81	124	355	684	1590	3712
wheat_summer	C	0	4	27	85	129	370	715	1660	3877
Spelt	C	0	0	2	5	8	22	42	97	226
rye_winter	C	0	1	7	31	27	90	141	221	585
barley_winter	C	0	8	39	118	200	573	792	1375	2759
barley_spring	C	0	4	23	70	119	343	474	823	1651
Oats	C	0	3	5	36	65	169	253	314	539
mixed_grain_winter	C	0	1	2	3	3	9	15	32	59
mixed_grain_spring	C	0	0	1	2	2	9	14	31	58
grain_maize	L	0	2	5	9	9	31	49	106	198
triticale_winter	C	0	5	29	108	124	450	580	1047	1712
other_forage_crops	L	0	1	7	29	59	388	1196	3148	6515
maize_dry_matter _BG	L	0	1	8	32	66	434	1339	3526	7296
dried_pulses	L	0	1	4	7	6	23	37	79	148
Beans	L	0	0	1	2	2	6	9	20	37
Potatoes	L	0	3	15	12	5	66	37	138	328
Rapeseed	L	0	0	3	11	23	153	473	1246	2577
clover_grass_mix	L	2	9	24	40	39	138	221	481	896
Meadows	O	1	38	90	206	196	729	1113	2445	4207
Pastures	O	10	249	584	1329	1268	4714	7194	15,804	27,191
Vineyards	O	106	203	564	293	47	10	1	16	60
crops_NES	O	0	2	4	7	7	24	38	83	155

crops_NES are crops not specified

(d) Finally we update the area of each farm as the sum of the area under each crop under plantation. This completes the crop allocation aspect of the model.

We have to resort to the Algorithm 1 due to the unavailability of actual data and the ground realities of cropping activities in the agriculture sector. The ideal situation would have been complete information on the field size by farm, the crop planted on each field and the rotation scheme for each of the fields. Also the data released in public statistics is an outcome of an annual survey conducted by the government based on a sample of farms with information reported on a specific day in the year. In case we did not use the heuristics as outlined in Algorithm 1 we would have had to contend with a larger than normal area allocated to a particular crop in randomly chosen farms. To be more precise, from Table 2, the area under cropping for winter barley for farm type C is 39 ha. There are 217 farms totaling 1533 ha for this type C. When we calibrate, and have a very small number of farms, the average area allocated to winter barley per farm is larger than should be, even though the calibration in terms of area and crop allocation is correct, the distribution is biased in favour of larger farm sizes. This implicitly increases the risk associated with a particular crop for the farmer and has potential implications for the crop rotation scheme. The larger the area devoted to each crop, the greater the swings in cropping patterns and hence larger volatility of output. This volatility is only on account of calibration. In order to mitigate this problem, we artificially increase the number of farms planting the crops, from the naïve value. The other option would have been to allocate every single crop to every farm but that again would bias the estimates to smaller sizes. in absence of information, the heuristic mentioned above was the best possible and the approach is flexible enough to cover various degrees of farm sizes.

4. Table 3 shows the average data on yield (t/ha), price (€/t), output (t), various costs in €/ha for seeds, crop protection, miscellaneous others, rentals, machine, labour, area and building and data on the kg/ha of fertiliser of N, P and K.

To initialise the data across all farms we take these numbers as the mean for all farms and use a standard deviation of 15% to allocate the different heads across all farms. The cost of fertilisers is expected to be the same for all farmers, while the change is on the amount applied.

The summation of all costs leads to the cost of production for each crop planted for each farmer. The markup over cost is a random number between 5 and 15% and each farmer is assigned a random markup over cost. Sorting the farmers' supply price in an ascending order generates the supply curve for each crop.

3.2 Price Discovery in Rounds

We have N buyers who participate in the market over R rounds. Each buyer has a correct expectation of the total number of buyers in each market. In each round, the buyer bids for a quantity $q_b^r = \frac{Q_b^e}{N \times R}$, where Q_b^e is the expected total output of a crop. The buyer at each round buys $\min(q_b^r, S_t^r)$, the minimum of the planned purchase at each round r and the total supply S_t^r by farmers in the round at a price p_b^r. The market clears over multiple rounds, following the six steps described next.

1. Initialise buyer and seller (farmer) rounds
2. Buyer b enters the quantity $(q_b^{r_b})$ and price $(p_b^{r_b})$ for quantity q_b in buyer round r_b. For example, say for buyer $b = 5$, in buyer round $r_b = 3$ out of a maximum buyer rounds 5, $q_b^{r_b} = 20$, where 20 tons is q_b in round $r_b = 3$
3. The market maker sorts the buyer bids in descending order with the highest bidder getting the right to purchase first.
4. Seller (farmer) s, enters the quantity $(q_s^{r_s})$ and price $(p_s^{r_s})$ for quantity q_s in round r_s. For example, say for farmer (seller) $s = 1234$, in seller round $r_s = 2$ out of a maximum seller rounds 5, $q_s^{r_s} = 3$, where 3 tons is q_s in round $r_s = 2$
5. The market maker sorts the seller bids in ascending order with the lowest priced seller getting the chance to sell first.
6. Once the buyers and sellers are sorted, the first buyer gets a chance to purchase the quantity desired. Two things can happen for the buyer

    (a) The entire quantity bid is available and the market clears at price $p_b^{r_b}$

Table 3 Crop details

cropName	T	S	E	Season	Yield	Price	Output	Seed	Protection	Other	Rentals	Machine	Labour	Area	Building	N	P	K
wheat_winter	C	10	8	Winter	6.6557	145.74	43,761	48.44	41.52	55.36	55.36	55.36	55.36	13.84	13.84	147.9	11.5	11.3
wheat_summer	C	2	8	Summer	6.6187	105.76	45,451	51.44	30.86	41.15	41.15	41.15	61.73	10.29	10.29	147.9	11.5	11.3
Spelt	C	10	8	Winter	4.6425	208.94	1866	69.20	69.20	69.20	69.20	69.20	69.20	13.84	13.84	147.9	11.5	11.3
rye_winter	C	10	8	Winter	6.2888	80.30	6937	35.11	35.11	35.11	35.11	35.11	35.11	7.02	7.02	103.1	9.5	8.5
barley_winter	C	10	7	Winter	6.1477	87.02	36,050	34.86	34.86	34.86	34.86	34.86	34.86	6.97	6.97	134.5	11.5	9.0
barley_spring	C	3	8	Summer	5.2335	90.76	18,354	30.97	30.97	30.97	30.97	30.97	30.97	6.19	6.19	134.5	11.5	9.0
Oats	C	3	8	Summer	5.2001	87.68	7197	67.01	33.50	33.50	100.51	33.50	33.50	6.70	6.70	103.1	9.5	8.5
mixed_grain_winter	C	10	8	Winter	5.2562	87.68	652	70.72	35.36	35.36	106.09	35.36	35.36	7.07	7.07	103.1	9.5	8.5
mixed_grain_spring	C	3	8	Summer	5.2562	87.68	615	67.73	33.87	33.87	101.60	33.87	33.87	6.77	6.77	103.1	9.5	8.5
grain_maize	L	4	11	Summer	5.9976	134.12	2453	96.91	48.46	48.46	145.37	48.46	48.46	9.69	9.69	134.5	19.1	25.4
triticale_winter	C	10	8	Winter	6.2676	86.16	25,415	82.87	41.44	41.44	124.31	41.44	41.44	8.29	8.29	103.1	9.5	8.5
other_forage_crops	L	4	10	Summer	13.6743	98.57	155,108	133.70	66.85	66.85	200.55	66.85	66.85	13.37	13.37	88.2	4.5	4.0
maize_dry_matter_BG	L	4	10	Summer	13.6743	98.57	173,691	188.02	94.01	94.01	282.02	94.01	94.01	18.80	18.80	134.5	19.1	25.4
dried_pulses	L	3	8	Summer	3.9541	25.29	1206	9.96	4.98	4.98	14.94	4.98	4.98	1.00	1.00	22.4	11.5	12.7

(continued)

Table 3 (continued)

cropName	T	S	E	Season	Yield	Price	Output	Seed	Protection	Other	Rentals	Machine	Labour	Area	Building	N	P	K
Beans	L	1	1	All	3.5195	125.00	271	43.81	21.91	21.91	65.72	21.91	21.91	4.38	4.38	22.4	11.5	12.7
Potatoes	L	4	10	Summer	33.1854	179.14	20,044	625.46	312.73	312.73	938.18	312.73	312.73	62.55	62.55	12.5	26.7	48.1
Rapeseed	L	9	7	Winter	3.9170	259.84	17,572	141.97	70.99	70.99	212.96	70.99	70.99	14.20	14.20	67.2	7.6	4.2
clover_grass_mix	L	9	7	Winter	53.1337	29.26	98,297	137.45	68.72	68.72	206.17	68.72	68.72	13.74	13.74	22.4	11.5	12.7
Meadows	O	1	1	All	8.2248	163.53	74,229	118.91	59.46	59.46	178.37	59.46	59.46	11.89	11.89	88.2	4.5	4.0
Pastures	O	1	1	All	8.2251	222.87	479,877	162.07	81.03	81.03	243.10	81.03	81.03	16.21	16.21	88.2	4.5	4.0
Vineyards	O	1	1	All	10851.3740	1.97	14,106,786	2042.74	1021.37	1021.37	5106.85	3064.11	3064.11	204.27	204.27	22.4	11.5	12.7
crops_NES	O	1	1	All	6.2023	330.04	1985	226.24	113.12	113.12	339.36	113.12	113.12	22.62	22.62	22.4	11.5	12.7

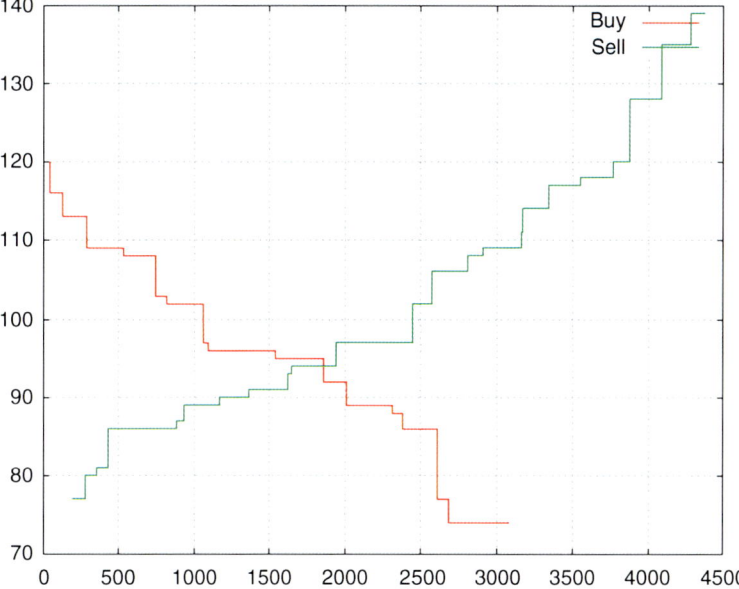

Fig. 2 Market clearance. In x axis, the quantity in tons for a crop, in y axis the price in € per ton

(b) A partial quantity is available for sale and the market clears at price $p_b^{r_b}$ but only with lesser quantity $q_a^{r_b}$. There is a shortfall of $q_b^{r_b} - q_a^{r_b} > 0$, which is then added to the remaining rounds.

For the seller something similar

(a) The entire quantity offered by the farmer (seller) is sold at price $p_b^{r_b}$
(b) Only a partial quantity is sold $q_a^{r_s}$ and there is a shortfall of $q_s^{r_s} - q_a^{r_s} > 0$, which is then added to the remaining rounds.

The market clearance takes place as illustrated in Fig. 2. The farmers are sorted in the ascending order of their price bids and a supply curve is generated. For the demand curve each buyer at each round is willing to buy a quantity $q_b^{r_b}$ at price $p_b^{r_b}$. The demand curve is a horizontal line at price $p_b^{r_b}$ which is $q_b^{r_b}$ wide.

The behaviour aspects enter the process after each round. Each agent (buyer or seller individual) evaluates his/her inventory of the unsold or non procured stock. Based on their behaviour, each agent is willing to modify their quantity and price quotes.

We follow one strict procedure as long as all buyers are not done with the purchase in a specific round. Modifications to the pricing and quantity behaviour by previously successful buyers in the same round is disallowed. For example, in buyer round 2, if buyer 4 was the highest bidder to procure 10 tons of maize at 300€/t, buyer 4 is not allowed to purchase additional quantity of maize even though the revised bid may be the highest, until all buyers have finished their purchase for that round. All the remaining buyers can be unsuccessful.

Similar procedure is adopted for farmers with a round being complete before moving to the next.

The price convergence dynamics enters the market via the behaviour when buyers and sellers panic or are cool and modify their price and quantity quotes.

The behaviour part is divided into impatient and patient buyers and sellers. Impatience can be applied to price or quantity or both.

1. For an impatient buyer b, if shortfall is s_b^{rb}, then the modified price is $p_n = p_o \times \left[1 + \frac{s_b^{rb}}{q_b^{rb}}\right]$. Increase the bid price in the next round by the extent of shortfall.
2. For a buyer b, with quantity impatience, if shortfall is s_b^{rb}, then the amount to be purchased in the next round $r^b + 1$ is sum of the existing shortfall s_b^{rb} and the original bid in round $r^b + 1$: q_b^{rb+1}. Hence $q_b^{rb+1} = q_b^{rb+1} + s_b^{rb}$.
3. For a medium level of quantity impatience, the shortfall could be spread over the remaining rounds in an equal fashion leading to a smooth buying profile. If there are k rounds left, then the shortfall for the each of the remaining round is $\frac{s_b^{rb}}{k}$, which is added to the bid quantity of the future rounds.
4. For a patient buyer, the bid price hike could be in steps of the remaining rounds with $p_n = p_o \times \left[1 + \frac{s_b^{rb}}{k \times q_b^{rb}}\right]$.

Similar behaviour applies to the sellers but in a reverse manner with lower quotes on prices in case they are unable to sell their output at the quoted price. The impatience on the quantity however would remain similar as both buyers and sellers want an increasing amount of quantity traded over successive rounds.

3.3 Scheduling

The buyers are sorted according to the descending order of prices quoted, with the highest bidder having priority. At each round, the buyer with the highest price completes the purchase before another buyer is permitted to begin buying. The buyers are free to modify their purchase price and quantities during future rounds. Similarly the sellers or farmers are sorted according to the ascending order of price with the farmer quoting the lowest getting the priority to sell. The farmers are also allowed to modify their markups and quantities for sale.

4 Experiments and Results

The agriculture system in Luxembourg does not lend itself to open access to information on price discovery of farm products. One can confidently assume that the markets are thin, in terms of volume turnover, and the amount of trade in any commodity is sparse. Most of the crops grown are used as feed for animals and may not be traded in the open market but carries a shadow price. This is determined by the opportunity cost of planting or not planting a particular crop at any season. To compound the lack of data, the main stay of the sector is animal products like milk and meat, with milk being the dominant commodity. Milk is procured from the farmers by the co-operative of farmers themselves and thus price setting is opaque to the outside world. In such a scenario, we proposed a generic tool to discover the price in any market wherein buyers and sellers engage in trading over multiple rounds. We assume that the last traded price is available to all the market participants.[2]

From a perspective broader than the work we present here, the ultimate goal is to study the evolution of cropping patters when maize is given some sort of incentive to be used as an input in biogas plants. This warrants a dynamic model wherein the past behavioural responses determine the future evolution of the agricultural landscape. One could in principle use two approaches.

1. Use an exogenous model for price evolution of all crops based on time series methodology
2. Determine prices endogenously based on price discovery and enable the introduction of a wide genre of behaviours ranging from using own and cross price elasticity of supply for crop by farmers to risk appetite modelled via impatience exhibited in maintaining the minimum possible stock for each crop.

We use the latter approach in the current study as it enables endogenous modelling of behaviour that is consistent over time and also helps in determining the impact of agricultural practices such as crop rotation on the price of crops.

4.1 Modelling Behaviour

Individuals are different and exhibit a variety of responses to the same stimuli. This is true of markets too. Some have a low capacity to risk while others are high risk takers. When the information dimension is expanded to include wealth, one could then further classify rich and poor traders who are risk averse and who exhibit a high appetite for risk. From our model perspective we have kept the risk behaviour uni-dimensional in that we do not differentiate based on wealth of agents. Agents only decide the extent to which they would like to smooth the shortfall that they incur in trading at each round. The shortfall is determined by difference between their expected purchase or sale quantity and the actual amount of quantity purchased or

---

[2]This assumption is violated in many markets wherein each buyer and seller negotiate a price in private and is unknown to all, but to the buyer and seller. Each buyer and seller goes through a protracted buying and selling process with multiple individuals before forming an opinion of the supply and demand in the market.

sold. A shortfall for the buyer implies that the quoted price was low while that for a seller implies that the quoted price was on the higher side. Based on the proclivity of the buyer or seller towards risk, a buyer or seller might want to modify the price quote for the next round besides modifying the target quantity for sale or purchase. The approach to smooth the shortfall over the remaining rounds implicitly defines the impatience of the agents. We arbitrarily set the maximum rounds to smooth to shortfall to 4, but this could be easily extended to an arbitrary number S_r. We run simulations for buyers and sellers changing the number of rounds to smooth the shortfall from 1 to 4. If the response to shortfall made no difference then we would observe similar distributions of shortfalls over the different smoothing responses.

The moot point is that the price quoted by buyers is based on some previous price that is available in the base data but there is no additional source of information that will enable the buyers to deviate substantially from that price. The same applies to the sellers. This is a crucial aspect in the time evolution of the model as the lack of formal methods to identify a price trend or speculation does not lead to substantial swings in pricing. 2009 was a bad year in which prices were almost at an all time low and they picked up in 2010. Given the lack of information on supply and demand in other markets available to both buyers and sellers, there is no a priori case to arbitrarily set a price different from the existing one. From a pure academic perspective, one could set extreme prices (high and low) that would only lead to a particular buyer either meeting the target earlier at a higher cost or not being able to meet the target on account of lower prices. From the farmers' perspective as long as the prices cover the cost of cultivation, the only impact is on the profits. Agriculture markets are predominantly buyer markets with prices set by buyers or middlemen. The farmers have little or no market power and are left to the vagaries of not only nature but also market elements. In such a situation where there is no alternative of access to other markets or limited ability to store output across seasons, there is no choice but to sell at the prevailing price.[3]

4.2 Experiments

We have 2242 farmers, 10 buyers (randomly chosen number), 4 rounds for farmers to choose smoothing (1–4), 4 rounds for buyers to choose smoothing (1–4) for 41 simulations for 22 crops. In sum, there are an arbitrary maximum of 19 trading rounds for the buyers, times 2242 sellers, times 10 buyers, times 16 smoothing options (4 for sellers, 4 for buyers), times 22 crops = $19 \times 2242 \times 10 \times 16 \times 22 = 149{,}944{,}960$. All this repeated 41 times, yielding $6{,}147{,}743{,}360$ market clearance rounds.

---

[3]The downward pricing pressure on the French farmers after the tit-for-tat sanctions imposed by Russia on agriculture produce from NATO countries is a classic case of the above.

In the following experiments we have restricted the simulations to only 1 time period as the objective was to study the impact of various behaviours on price discovery. The model presented in Sect. 3 was coded in java and the full simulation was run in an average PC with 4 GigaBytes of ram and a Centrino II 2010 microprocessor. The full simulation took around 48 h. No specific agent programming libraries were used, only code developed explicitly for the study and libraries for reading and writing files.

4.3 Results

We focus on the price discovery for maize_dry_matter_BG (BG for biogas) in the figures. The simulations for price discovery are run for all crops. Figure 3 shows the box plots of market clearing price (PE) for each buyer (0–9) for different smoothing options (1–4). Figure 4 shows the box plots of shortfall for all buyers across different smoothing options. The mean of the shortfall for smoothing rounds 1–4 are $\mu_s = [153.66, 152.90, 169.88, 189.45]$ with a standard deviation of $\sigma_s = [77.62, 71.61, 77.33, 73.79]$, respectively. The mean and standard deviation for prices is as follows $\mu_{pe} = [81.61, 81.46, 80.30, 80.38]$ and $\sigma_{pe} = [6.85, 6.84, 6.50, 6.18]$. We find that the prices are similar due to the fact that the buyers set the price and these prices do not lead to a loss for the farmers. In addition the choice available to the farmers is to vary the quantity supplied and implicitly accept a markup over costs that does not result in losses.

Figures 5 and 6 show the cumulative distribution functions of market clearing price (PE) and shortfall, respectively, for each buyer (0–9) for different smoothing options (1–4).

Tables 4 and 5 show the computed values for the Anderson-Darling (AD) and Kolmogorov-Smirnov (KS) test to compare the similarities between the distributions of the prices and shortfalls across different smoothing options (1–4).

What we find is that barring smoothing choice over 1 or 2 rounds, the cumulative distribution of shortfall and prices do not statistically exhibit similarity as shown by the AD and KS test. The null hypothesis that all smoothing over rounds makes no difference to the overall market outcome does not hold. We find that both quantities and prices show different distributions.

4.3.1 Price Convergence

In theory we can plot 16 graphs of variations in the selling prices of sellers for 4 smoothing rounds of buyers and sellers with variations across rounds. Figure 7 shows the variation in quoted selling price by sellers aggregated across all rounds for different seller smoothing behaviours. One of this variation for seller smoothing round and buyer smoothing round both equaling 1 is shown in Fig. 8. We largely observe a smaller variation across price quotes as the smoothing rounds increases

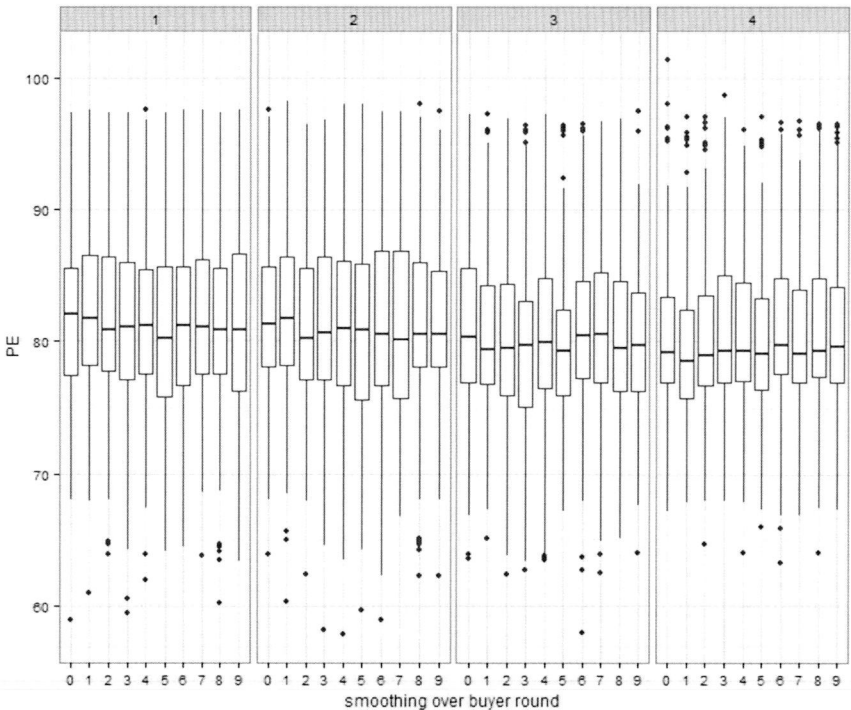

Fig. 3 Boxplot of prices for different smoothing rounds of buyers

from $b = 1$ to $b = 4$. The choice of smoothing has a bearing on the quantity procured by each buyer in the subsequent rounds and smoothing over a single round exhibits aggressive behaviour to cover the shortfall at the earliest possible time, with an aggressive price bid. This effect is countered by the price quote mechanism used in the model that uses an arithmetic mean of the average of previous market prices and the price quote of the buyer.

5 Discussion

Globally farmers face risk on multiple fronts. There is little they can do to mitigate environmental risks and risks attributed to weather patterns. The possibility of market risk can be hedged to a certain extent by planting multiple crops and having access to the market.

The market maker in these markets is the middleman and larger the number of middlemen in the chain from the farm gate to food plate, the greater is the gap between the realised price for the farmer and the market price paid by the consumer.

Modelling Price Discovery in an Agent Based Model for Agriculture in Luxembourg 107

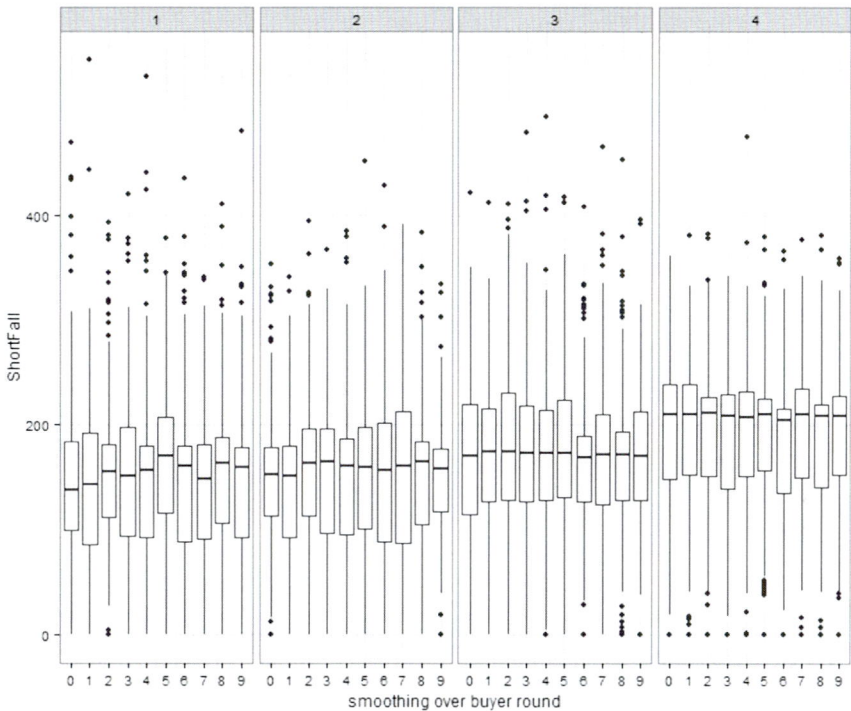

Fig. 4 Boxplot of shortfall for different smoothing rounds of buyers

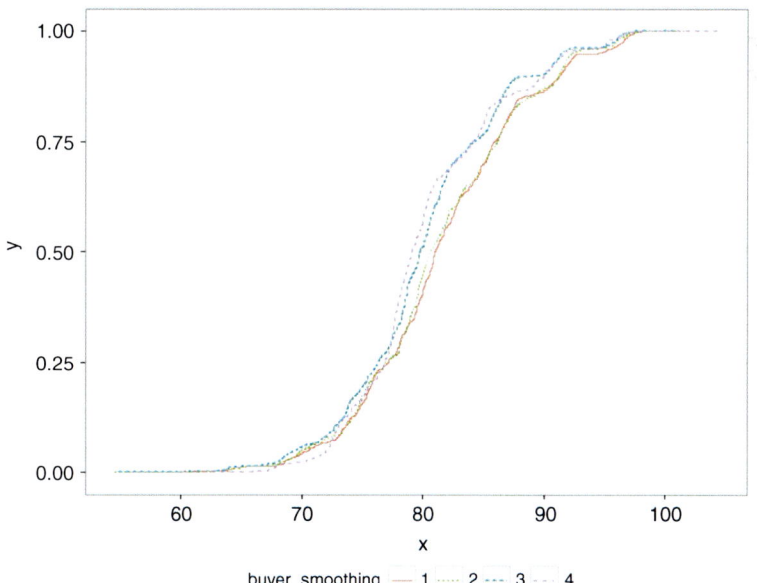

Fig. 5 Cumulative distribution function of prices for different smoothing rounds of buyers

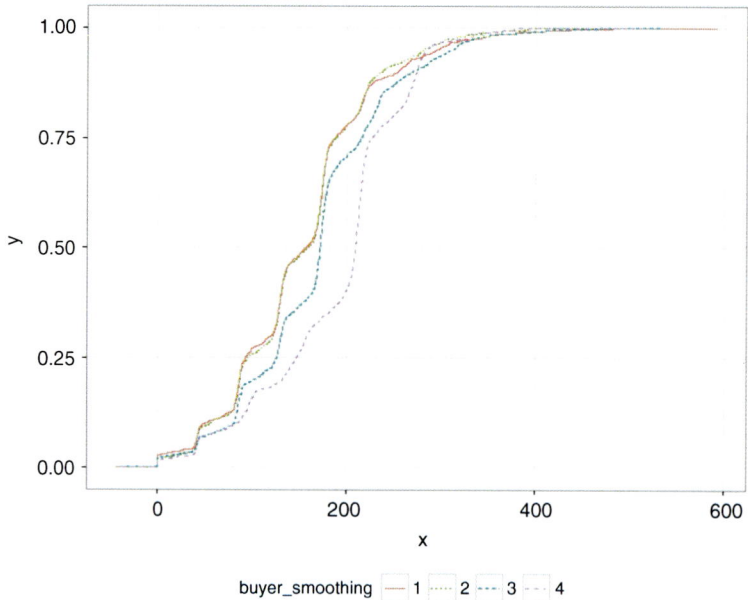

Fig. 6 Cumulative distribution function of shortfall for different smoothing rounds of buyers

Table 4 AD & KS test for difference in price convergence across smoothing

CDF	AD test	AD p value	KS test	KS p value
s1:s2	1.0013	0.35617	0.042262	0.0995
s1:s3	20.69	1.3652e−11	0.125	7.958e−12
s1:s4	33.688	9.8185e−19	0.18274	2.2e−16
s2:s3	16.252	3.7518e−09	0.10595	1.29e−08
s2:s4	28.329	8.6558e−16	0.14405	1.443e−15
s3:s4	6.1219	0.00081821	0.078571	6.263e−05

Table 5 AD & KS test for difference in shortfall across smoothing

CDF	AD test	AD p value	KS test	KS p value
s1:s2	0.6523	0.59941	0.02381	0.7277
s1:s3	22.796	9.4979e−13	0.12381	1.309e−11
s1:s4	153.39	1.6255e−84	0.38988	2.2e−16
s2:s3	22.892	8.4196e−13	0.11786	1.467e−10
s2:s4	152.8	3.4214e−84	0.38452	2.2e−16
s3:s4	72.966	2.5516e−40	0.31964	2.2e−16

Buyers and sellers implement strategies to get the maximum return on their investment. The buyers want to procure the output at the least possible price as opposed to the farmers who want the maximum. In absence of information these markets are supposed to be operating in perfect competition as far as farmers are concerned.

Modelling Price Discovery in an Agent Based Model for Agriculture in Luxembourg   109

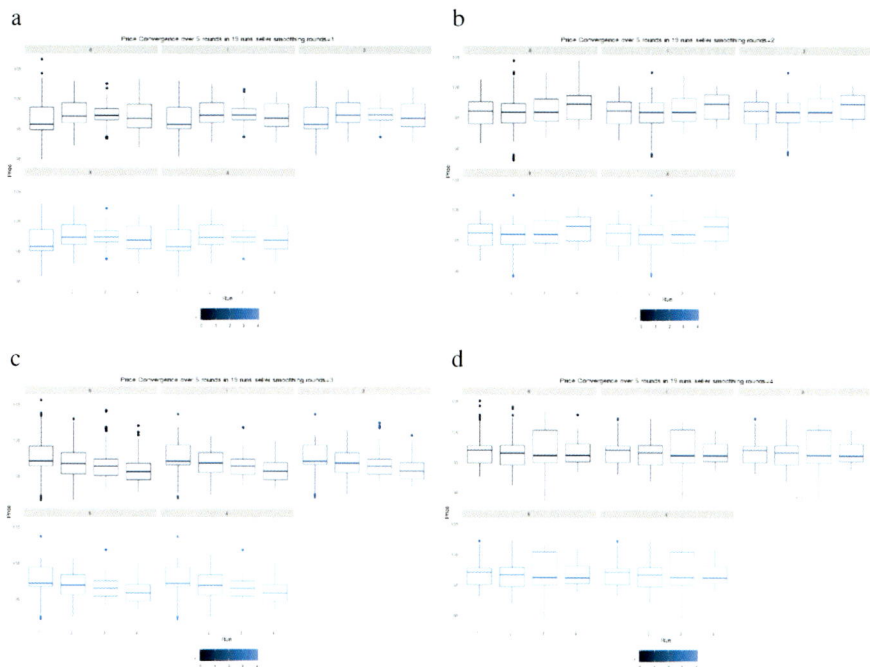

Fig. 7 Boxplot of prices for different smoothing rounds of sellers aggregated across all rounds. (**a**) Seller smoothing round = 1, (**b**) Seller smoothing round = 2, (**c**) Seller smoothing round = 3, (**d**) Seller smoothing round = 4

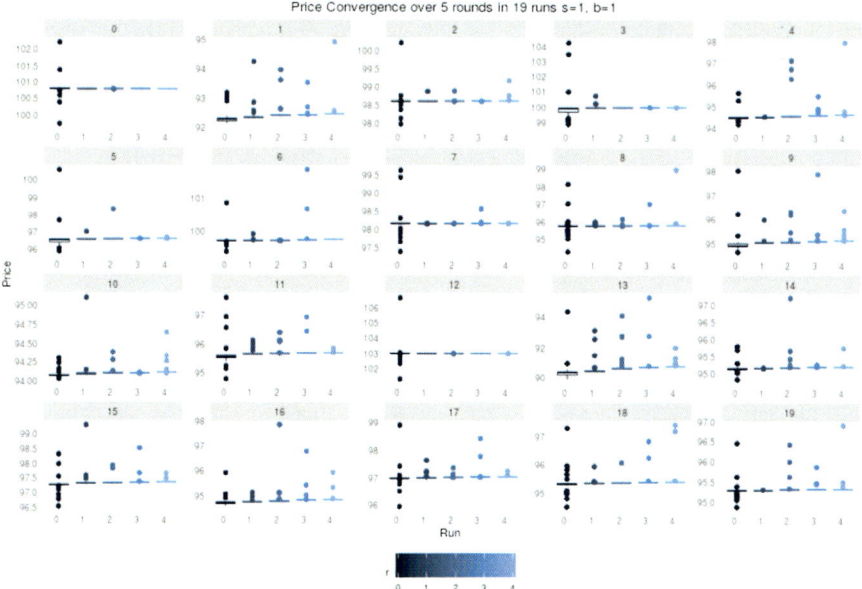

Fig. 8 Boxplot of prices for different smoothing rounds of buyers

Our model is an agent based model which though classified as a bottom-up model,[4] the driver of the model is still top-down.[5] In standard economic models such as computable general equilibrium or partial equilibrium models, there is the law of a single price. However in reality, markets operate through a double auction process and there is a variation in the price and one normally assumes that the market clearing price is the mean of the prices observed during the day.

What we have embarked upon is to discover if there is a strategy in smoothing inventory of unsold stock in perishable commodities as exhibited in the agriculture sector. Quoting lower price would mean lower returns but possible lower or zero inventory, while higher price would mean the opposite. In our model there is no way to control the initial opening quotes of the buyers and it is possible that the opening quote can be lower than the production cost of the farmer, thus causing a loss to the farmer. This is a phenomenon not unheard of in agriculture markets.

We find that price does converge to a "mean" value as exhibited by a lower variation in the box plots of prices over buyer and seller rounds. The case where both the buyers and sellers are smoothing over multiple rounds enables the price discovery process to exhibit lower volatility captured by the lower extreme values in the box plots.

We are assuming that the buyers are unable to exhibit market power or collude in order to dictate the price and that the initial quote is above the cost of production of the farmer for a specific crop.

Empirical validation is the best possible test for any agent based model. However in Luxembourg, given the legislation on data confidentiality, it is impossible to obtain information on the price and profitability of the agriculture operations. We have carried out a survey of farmers with limited success and it was difficult to obtain information across all data fields to generate meaningful statistical distribution to estimate parameters for the agent based model. Despite this setback, the model has been used to generate numbers that are close to reality. Given the vagaries in the agriculture sector it is difficult if not impossible to use agent based models as a forecasting tool for price evolution and use these forecasts to evaluate the potential cropping pattern changes induced by policy interventions.

6 Conclusions

We have developed a generic model for any agriculture system with pure cropping and endogenous price discovery of crops. In absence of market information, lack of time series for exogenously forecasting prices, this approach has the benefit of endogenous price discovery. The behaviour to smooth buying and selling across multiple agents introduced the element of patience as far as purchasing or selling

---

[4]Decisions are made by individual agents.
[5]The price quotes of the buyers are exogenously given before each simulation.

decisions are concerned. In the current simulation we have used a single layer of buyers interacting directly with farmers. We believe that given a sufficient number of layers of buyers and sellers, where within each layer the buyer has market power, one can show evidence of the price gap between food plate and farm gate, leading to low realised prices for farmers.

From an architecture perspective for agent based models, the model is in a standard reactive framework as far as price discovery is concerned. A cognitive architecture that would take into account memory and different beliefs and intentions of agents, for example, could be considered in future work. For the current state, given the limited time period for simulation, the changes in the agriculture system in Europe, and lack of more precise data, make it difficult if not impossible to implement.

From an economic price discovery perspective, this is an empirical demonstration of a bottom-up model using agent based models. The limitations of all bottom-up models are relevant for this approach too. The information asymmetry built into bottom-up models as opposed to top-down models prevents the agents from setting prices that are not generated from the model simulations.

Finally smoothing behaviour of covering shortfall for buyers and sellers used as a proxy for impatience does exhibit statistically significant difference as far as distribution of prices is concerned over different smoothing rounds.

Acknowledgements This work was done under the project MUSA (C12/SR/4011535) funded by the Fonds National de la Recherche (FNR), Luxembourg. We thank Romain Reding and Rocco Lioy from CONVIS (4 Zone Artisanale Et Commerciale, 9085 Ettelbruck, Grand-duchy of Luxembourg) for their valuable insight and for the participation to the discussions for the definition of the project MUSA, and of the data collection survey. We thank professors Shu-Heng Chen, Ye-Rong Du, Ragu Pathy, Selda Kao and an anonymous referee for their valuable comments. This paper was presented at the 21st International Conference on Computing in Economics and Finance June 20–22, 2015, Taipei, Taiwan. Sameer Rege gratefully acknowledges the FNR funding for the conference participation.

References

1. Berger, T. (2001). Agent-based spatial models applied to agriculture: A simulation tool for technology diffusion, resource use changes and policy analysis. *Agricultural Economics, 25*, 245–260.
2. Berta, F. E., Podestá, G. P., Rovere, S. L., Menéndez, A. N., North, M., Tatarad, E., et al. (2011). An agent based model to simulate structural and land use changes in agricultural systems of the argentine pampas. *Ecological Modelling, 222*, 3486–3499.
3. Cooper, J. S., & Fava, J. A. (2006). Life-cycle assessment practitioner survey: Summary of results. *Journal of Industrial Ecology, 10*(4), 12–14.
4. Happe, K., Kellermann, K., & Balmann, A. (2006). Agent-based analysis of agricultural policies: An illustration of the agricultural policy simulator AgriPoliS, its adaptation, and behavior. *Ecology and Society, 11*(1), 329–342. Available from: http://www.ecologyandsociety.org/vol11/iss1/art49/.

5. Howitt, R. E. (1995). Positive mathematical programming. *American Journal of Agricultural Economics, 77*, 329–342.
6. ISO. (2010). Environmental management — Life cycle assessment — Principles and framework. Geneva, Switzerland: International Organization for Standardization. 14040:2006.
7. KTBL. (2006). Faustzahlen für die Landwirtschaft (in German). Darmstadt, Germany: Kuratorium für Technik und Bauwesen in der Landwirtschaft.
8. Le, Q. B., Park, S. J., Vlek, P. L. G., & Cremers, A. B. (2010). Land-use dynamic simulator (LUDAS): A multi-agent system model for simulating spatio-temporal dynamics of coupled human–landscape system. I. Structure and theoretical specification. *Ecological Informatics, 5*, 203–221.
9. Rege, S., Arenz, M., Marvuglia, A., Vázquez-Rowe, I., Benetto, E., Igos, E., et al. (2015). Quantification of agricultural land use changes in consequential Life Cycle Assessment using mathematical programming models following a partial equilibrium approach. *Journal of Environmental Informatics*. https://doi.org/10.3808/jei.201500304
10. SER. Available from: http://www.ser.public.lu/
11. STATEC. Available from: http://www.statistiques.public.lu/en/actors/statec/index.htm

Heterogeneity, Price Discovery and Inequality in an Agent-Based Scarf Economy

Shu-Heng Chen, Bin-Tzong Chie, Ying-Fang Kao, Wolfgang Magerl, and Ragupathy Venkatachalam

Abstract In this chapter, we develop an agent-based Scarf economy with heterogeneous agents, who have private prices and adaptively learn from their own experiences and those of others through a meta-learning model. We study the factors affecting the efficacy of price discovery and coordination to the Walrasian Equilibrium. We also find that payoff inequality emerges endogenously over time among the agents and this is traced back to intensity of choice (a behavioural parameter) and the associated strategy choices. Agents with high intensities of choice suffer lower payoffs if they do not explore and learn from other agents.

Keywords Non-tâtonnement processes · Coordination · Learning · Agent-based modeling · Walrasian general equilibrium · Heterogeneous agents

1 Introduction

How do market economies achieve coordination, even if only imperfectly, among millions of actors performing a multitude of actions without centralization? Research in General Equilibrium theory over many decades has provided many interesting insights. The notion of *coherence* is in this literature interpreted in terms

---

S.-H. Chen (✉) · Y.-F. Kao
AI-ECON Research Center, Department of Economics, National Chengchi University, Taipei, Taiwan

B.-T. Chie
Tamkang University, Tamsui, Taipei, Taiwan

W. Magerl
Vienna University of Technology, Vienna, Austria

AI-ECON Research Center, Department of Economics, National Chengchi University, Taipei, Taiwan

R. Venkatachalam
Institute of Management Studies, Goldsmiths, University of London, London, UK

© Springer Nature Switzerland AG 2018
S.-H. Chen et al. (eds.), *Complex Systems Modeling and Simulation in Economics and Finance*, Springer Proceedings in Complexity,
https://doi.org/10.1007/978-3-319-99624-0_6

of an *equilibrium* phenomenon. Furthermore, the discovery of equilibrium prices that balance the demands and supplies of various actors in the aggregate is assumed to be mediated through a fictional, centralized authority. This authority, known as the *Walrasian auctioneer*, supposedly achieves this discovery through a process of trial and error (tâtonnement). In this framework, trading happens, if it at all happens, only in equilibrium.

What happens when agents are out of this equilibrium configuration? Whether and *how* these agents manage to coordinate back to equilibrium over time through decentralized exchanges is an important question. The tâtonnement narrative is highly divorced from how agents go about achieving coordination by searching and learning to discover efficient prices in reality. Therefore, it becomes necessary to go beyond the tâtonnement process and find plausible descriptions of behaviour that respect the cognitive and informational limitations that human agents face. Research on non-tâtonnement processes [8, 10, 25] that permit disequilibrium trading have been developed over the years. Agent-based models provide a conducive setup to understand decentralized, disequilibrium dynamics of market economies.

Many aggregate economic outcomes such as growth and fluctuations are often explained as being driven by the expectations of agents. There is often a diversity of expectations among agents and in the face of this heterogeneity, the aggregate outcomes that result from the interaction among agents are complex to understand. But how these expectations come into being and evolve over time has not yet been completely understood. Similarly, their psychological underpinnings and relation to agent-level characteristics have not been sufficiently clarified. The mechanics of expectation formation are intimately linked to how agents learn from their experiences and the environment. Hence, it may not be sufficient to just point to the existence of a diversity of expectations, but it may be necessary to go deeper into understanding their origins and dynamics.

Even in a simple decentralized economy with barter, there is no a priori reason to believe that heterogeneous agents can successfully learn to align their expectations regarding future prices. There is no guarantee that they will eventually discover the equilibrium prices, where all mutually beneficial exchanges will be exhausted. Even if we suppose that they do, how the gains from these exchanges will be distributed among these agents and their relation to the structure of expectations is not obvious.

In this chapter, we investigate these issues through an agent-based, Scarf economy, where the agents are characterized by Leontief payoff functions. Agents attempt to maximize their payoffs, and engage in bilateral trade with others based on their initial endowment and a subjective perception of "fair" prices. Trade failures, either in the form of unsatisfied demand or involuntary inventory, signal misperceptions in the economy. No one person has complete knowledge of the entire system and the economy is best viewed as a complex adaptive process that dynamically evolves. In order to successfully coordinate in such a system, agents may need to learn both from their own experience (individual learning) and from each other (social learning). In our model, agents *consciously* decide when to engage in social and individual. This choice is formulated as meta-learning and we apply reinforcement learning to model this meta-learning. Heterogeneity among agents is in terms of differences in their learning process, more specifically, the intensity of choice parameter of the meta-learning model. Within this set-up, we ask the following questions:

- Can heterogeneous agents successfully discover the (unique) equilibrium prices of the model?
- How do current and accumulated payoffs vary among agents who are identical in all respects except for their intensity of choice?
- If there is a substantial variation in payoffs, what are the factors that drive this inequality?

The rest of this chapter is organized as follows: Sect. 2 develops the agent-based Scarf economy, and Sect. 3 describes the individual, social and meta-learning mechanisms that the agents in the model employ. The simulation design is explained in Sects. 4 and 5 presents the results. Section 6 provides a discussion of the results.

2 The Scarf Economy and the Non-Tâtonnement Process

Herbert Scarf [20] demonstrated an economy in which prices may not converge to the Walrasian general equilibrium under the tâtonnement process, even when the equilibrium is unique. This paper has since led to a lot of research on the issue of the stability of the general equilibrium. Recently, [9] developed an agent-based model and showed that when agents hold private prices and imitate other agents who have been successful, prices converge to the unique general equilibrium of the Scarf-like economy. The Scarf economy has a simple structure that is amenable to investigation and has been widely studied. Previous studies in this area have not tackled the issue of heterogeneity among agents in a Scarf-like set-up. In the remainder of the section, we will develop a heterogeneous, agent-based model of the Scarf economy.

2.1 An Agent-Based Model of the Scarf Economy

Following [20], we consider a pure exchange economy composed of N agents. This economy has three goods, denoted by $j = 1, 2, 3$, and correspondingly, N agents are grouped into three different 'types', τ_j, $j = 1, 2, 3$, where $\tau_1 \equiv \{1, \ldots, N_1\}$; $\tau_2 \equiv \{N_1 + 1, \ldots, N_1 + N_2\}$; $\tau_3 \equiv \{N_1 + N_2, \ldots, N\}$. Agents who belong to τ_j (type-j) are initially endowed with w_j units of good j, and zero units of the other two goods. Let $\mathbf{W_i}$ be the endowment vector of agent i:

$$\mathbf{W_i} = \begin{cases} (w_1, 0, 0), & i \in \tau_1, \\ (0, w_2, 0), & i \in \tau_2, \\ (0, 0, w_3), & i \in \tau_3. \end{cases} \quad (1)$$

All agents are assumed to have a Leontief-type payoff function.

$$U_i(x_1, x_2, x_3) = \begin{cases} \min\{\frac{x_2}{w_2}, \frac{x_3}{w_3}\}, & i \in \tau_1, \\ \min\{\frac{x_1}{w_1}, \frac{x_3}{w_3}\}, & i \in \tau_2, \\ \min\{\frac{x_1}{w_1}, \frac{x_2}{w_2}\}, & i \in \tau_3. \end{cases} \qquad (2)$$

Given the complementarity feature of this payoff function, we populate equal numbers of agents in each type. This ensures that the economy is balanced and that no free goods appear.

We assume that agents have their own subjective expectations of the prices of different goods,

$$\mathbf{P_i^e}(t) = (P_{i,1}^e(t), P_{i,2}^e(t), P_{i,3}^e(t)), \quad i = 1, \ldots, N. \qquad (3)$$

where $P_{i,j}^e(t)$ is agent i's price expectation of good j at time t.[1]

Given a vector of the subjective prices (private prices), the optimal demand vector, $\mathbf{X}^* = \Psi^*(\mathbf{W})$ can be derived by maximizing payoffs with respect to the budget constraint.

$$\mathbf{X_i^*} = (x_{i,1}^*, x_{i,2}^*, x_{i,3}^*) = \begin{cases} (0, \psi_i^* w_2, \psi_i^* w_3), & i = 1, \ldots, N_1, \\ (\psi_i^* w_1, 0, \psi_i^* w_3), & i = N_1 + 1, \ldots, N_1 + N_2, \\ (\psi_i^* w_1, \psi_i^* w_2, 0), & i = N_1 + N_2 + 1, \ldots, N. \end{cases} \qquad (4)$$

where the multiplier

$$\psi_i^* = \begin{cases} (P_{i,1}^e w_1)/(\sum_{j=2,3} P_{i,j}^e w_j), & i = 1, \ldots, N_1 \\ (P_{i,2}^e w_2)/(\sum_{j=1,3} P_{i,j}^e w_j), & i = N_1 + 1, \ldots, N_1 + N_2, \\ (P_{i,3}^e w_3)/(\sum_{j=1,2} P_{i,j}^e w_j), & i = N_1 + N_2 + 1, \ldots, N. \end{cases} \qquad (5)$$

Note that the prices in the budget constraint are 'private' prices. Rather than restricting the prices within a certain neighbourhood (for instance, a unit sphere in [20]), we follow [2] and set one of the prices (P_3) as the numéraire. The Walrasian Equilibrium (WE) for this system is $P_1^* = P_2^* = P_3^* = 1$, when the endowments for each *type* are equal and symmetric.[2] However, in this model agents may have own price expectations that may be very different from this competitive equilibrium price, and may base their consumption decision on their private price expectations $P_{i,j}^e$. To facilitate exchange, we randomly match agents in the model with each other and they are allowed to trade amongst each other if they find it to be beneficial.

---

[1] More specifically, t refers to the whole market day t, i.e., the interval $[t, t-1)$.
[2] Given this symmetry, the type-wise grouping of agents does not qualify as a source of real heterogeneity in terms of characteristics. This, as we will see in the later sections, will be characterized in terms of the agents' learning mechanisms.

For this, we need to specify a precise bilateral trading protocol and the procedures concerning how agents dynamically revise their subjective prices.

2.2 Trading Protocol

We randomly match a pair of agents, say, i and i'. Let i be the *proposer*, and i' be the *responder*. Agent i will initiate the trade and set the price, and agent i' can accept or decline the offer. We check for the double coincidence of wants, i.e., whether they belong to the same type. If they do not, they will be rematched. If not, we will then check whether the agents have a positive amount of endowment in order for them to engage in trade. Let m_i be the commodity that i is endowed with ($m_i = 1, 2, 3$). Agent i, based on his subjective price expectations, proposes an exchange to agent i'.

$$x^*_{i,m'_i} = \frac{P^e_{i,m_i} x_{i',m_i}}{P^e_{i,m_{i'}}}, \tag{6}$$

Here, the proposer (i) makes an offer to satisfy the need of agent i' up to x_{i',m_i} in exchange for his own need $x^*_{i,m_{i'}}$.

Agent i' will evaluate the 'fairness' of the proposal using his private, subjective expectations and will consider the proposal to be interesting provided that

$$P^e_{i',m_i} x_{i',m_i} \geq P^e_{i',m_{i'}} x^*_{i,m_{i'}}; \tag{7}$$

otherwise, he will decline the offer. Since the offer is in the form of *take-it-or-leave-it* (no bargaining), this will mark the end of trade.

Agent i' will accept the proposal if the above inequality (7) is satisfied, and if $x_{i',m_i} \leq x^*_{i',m_i}$. This *saturation condition* ensures that he has enough goods to trade with i and meet his demand. However, only if the saturation condition is not satisfied, will the proposal still be accepted, but the trading volume will be adjusted downward to $x_{i,m_{i'}} < x^*_{i,m_{i'}}$. Agents update their (individual) excess demand and as long as they have goods to sell, they can continue to trade with other agents. The agents are rematched many times to ensure that the opportunities to trade are exhausted. Once the bilateral exchange is completed, the economy enters the consumption stage and the payoff of each agent $U_i(\mathbf{X_i(t)}), i = 1, 2, \ldots, N$, is determined. Note that $\mathbf{X_i(t)}$, the realized amount after the trading process may not be the same as the planned level $\mathbf{X^*_i(t)}$. This may be due to misperceived private prices and a sequence of 'bad luck', such as running out of search time (number of trials) before making a deal, etc. Based on the difference between $\mathbf{X_i(t)}$ and $\mathbf{X^*_i(t)}$, each agent i in our model will adaptively revise his or her private prices through a process of learning.

3 Learning

Agents have to *learn* to coordinate, especially in dynamic, disequilibrium environments and non-convergence to the most efficient configuration of the economy can be interpreted as a coordination failure. In this section, we introduce two modes of learning that are frequently used in agent-based computational economics, namely, individual learning and social learning. In the individual learning mode, agents learn from their own past experiences. In social learning, agents learn from the experiences of other agents with whom they interact. These two modes of learning have been characterized in different forms in the literature and have been extensively analysed (see: [3, 14, 21, 22]). The impact of individual and social learning on evolutionary dynamics has been analysed in [18, 26] and more recently in [5]. We describe the learning procedures in more detail below.

3.1 Individual Learning

Under individual learning, an agent typically learns from analysing his own experience concerning the past outcomes and strategies. The mechanism that we employ can be thought of as a modified agent-level version of the Walrasian price adjustment equation. Let the optimal and actual consumption bundles be denoted by the vectors $\mathbf{X_i^*(t)}$ and $\mathbf{X_i(t)}$, respectively. Agent i can check his excess supply and excess demand by comparing these two vectors component-wise. Agents then reflect on how well their strategy (private price expectations) has performed in the previous trading round and adjust their strategies based on their own experience of excess demand and supply. We employ a gradient descent approach to characterize individual learning and in a generic form it can be written as:

$$\mathbf{P_i^e(t+1)} = \mathbf{P_i^e(t)} + \underbrace{\Delta \mathbf{P_i(X_i(t), X_i^*(t))}}_{\text{gradient descent}}, \quad i = 1, 2, \ldots, N \qquad (8)$$

We shall detail its specific operation as follows.

Let $\mathbf{m}_i^y$ and $\mathbf{m}_i^c$ denote the production set and consumption set of agent i. In the Scarf economy, $\mathbf{m}_i^y \cap \mathbf{m}_i^c = \emptyset$ and in this specific 3-good Scarf economy, $\mathbf{m}_i^y = \{m_i\}$. At the end of each market period, agent i will review his expectations for all commodities, $P_{i,j}^e(t), \forall j$. For the good that the agent 'produces',[3] $j \in \mathbf{m}_i^y$), the price expectations $P_{i,j}^e(t)$ will be adjusted downward if m_i is not completely sold out (i.e., when there is excess supply). Nonetheless, even if m_i has been completely sold out, it does not mean that the original price expectation will be sustained. In fact, under these circumstances, there is still a *probability* that $P_{i,j}^e(t)$ may be adjusted *upward*. This is to allow agent i to explore or experiment with whether

---

[3] More precisely, the good that he is endowed with.

his produced commodity might deserve a better price. However, so as not to make our agents over-sensitive to zero-inventory, we assume that such a tendency for them to be changing their prices declines with the passage of time. That is to say, when agents constantly learn from their experiences, they gradually gain confidence in their price expectations associated with zero-inventory. Specifically, the time-decay function applied in our model is exponential, which means that this kind of exploitation quickly disappears with time. For those goods in the vector $\mathbf{X}_i$ that are a part of the consumption set of the agent, i.e., $j \in \mathbf{m}_i^c$, the mechanism will operate in exactly the opposite manner by increasing the price expectations if there is excess demand. The individual learning protocol is summarized below.

3.1.1 Protocol: Individual Learning

1. At the end of the trading day, agent i examines the extent to which his planned demand has been satisfied. Let

$$\Delta x_{i,j}(t) = \begin{cases} x^*_{i,j}(t) - x_{i,j}(t), & \text{if } j \in \mathbf{m}_i^c, \\ 0 - x_{i,j}(t), & \text{if } j \in \mathbf{m}_i^y \end{cases} \quad (9)$$

2. The subjective prices $P^e_{i,j}$ of all three goods will be adjusted depending on $|\Delta x_{i,j}(t)|$.
3. If $|\Delta x_{i,j}(t)| > 0$ (i.e., $|\Delta x_{i,j}(t)| \neq 0$),

$$P^e_{i,j}(t+1) = \begin{cases} (1 + \alpha(|\Delta x_{i,j}(t)|))P^e_{i,j}(t), & \text{if } j \in \mathbf{m}_i^c. \\ (1 - \alpha(|\Delta x_{i,j}(t)|))P^e_{i,j}(t), & \text{if } j \in \mathbf{m}_i^y. \end{cases} \quad (10)$$

where $\alpha(.)$ is a *hyperbolic tangent function*, given by:

$$\alpha(|\Delta x_{i,j}(t)|) = \tanh(\varphi |\Delta x_{i,j}(t)|) = \frac{e^{(\varphi|\Delta x_{i,j}(t)|)} - e^{(-\varphi|\Delta x_{i,j}(t)|)}}{e^{(\varphi|\Delta x_{i,j}(t)|)} + e^{(-\varphi|\Delta x_{i,j}(t)|)}} \quad (11)$$

4. If $|\Delta x_{i,j}(t)| = 0$,

$$P^e_{i,j}(t+1) = \begin{cases} (1 - \beta(t))P^e_{i,j}(t), & \text{if } j \in \mathbf{m}_i^c. \\ (1 + \beta(t))P^e_{i,j}(t), & \text{if } j \in \mathbf{m}_i^y. \end{cases} \quad (12)$$

where β is a random variable, and is a function of time.

$$\beta = \theta_1 \exp\frac{-t}{\theta_2}, \quad \theta_1 \sim U[0, 0.1], \quad (13)$$

where θ_2 is a time scaling constant.

3.2 Social Learning

Social learning broadly involves observing the actions of others (peers, strangers or even members of different species) and acting upon this observation. This process of acquiring relevant information from other agents can be decisive for effective adaptation and evolutionary survival. Imitation is one of the most commonly invoked, simplest forms of social learning. Players imitate for a variety of reasons and the advantages can be in the form of lower information-gathering costs and information-processing costs, and it may also act as a coordination device in games [1]. Although the idea of imitation is fairly intuitive, there are many different forms in which agents and organisms can exhibit this behaviour ([11], section G, 115–120.).

We adopt a fairly basic version of imitation behaviour, where agents exchange their experiences regarding payoffs with other agents, with whom they are randomly matched. This can be thought of as a conversation that takes place with friends in a café or a pub at the end of a trading day, where they share their experiences regarding their amount of and pleasure associated with consumption as well as their price expectations. An agent with a lower payoff can, based on observing others, replace his own price expectations with those of an agent with a higher payoff. This is assumed to be done in the hope that the other agent's strategy can perform better. If they both have the same payoffs, then the agent chooses between them randomly. If the agent ends up meeting someone who has performed worse than him, he does not imitate and retains his original price expectations.

3.2.1 Protocol: Social Learning

1. At the end of each day, each agent consumes the bundle of goods that he has obtained after trading, and derives pleasure from his consumption $U_i(t)$ ($i = 1, \ldots, N$).
2. Agents are matched randomly, either with other agents of the same type or with agents who are of different types. This is achieved by randomly picking up a pair of agents (i, i') *without replacement* and they are given a chance to interact.
3. Their payoffs are ranked, and the price expectations are modified as follows:

$$\mathbf{P}_i^e(t+1) = \begin{cases} \mathbf{P}_{i'}^e(t), & \text{if } U_i(t) < U_{i'}(t), \\ \mathbf{P}_i^e(t), & \text{if } U_i(t) > U_{i'}(t), \\ \text{Random}(\mathbf{P}_i^e(t), \mathbf{P}_{i'}^e(t)), & \text{if } U_i(t) = U_{i'}(t). \end{cases} \quad (14)$$

The protocol makes it easier for the agents to meet locally and enables them to exchange information. An agent who has performed well can influence someone who hasn't performed as well by modifying his perception of the economy (i.e., price expectations).

3.3 Meta Learning

In our model, an agent can consciously choose between social learning or individual learning during each period. In such a setting, it is necessary to specify the basis on which—i.e., how and when—such a choice is made. Agents may choose between these strategies either randomly, or based on the relative expected payoffs by employing these strategies. In our model, adaptive agents choose individual or social learning modes based on the past performance of these modes. In other words, the focus is on *learning how to learn*. This meta-learning framework is not restricted to two modes alone, but we begin with the simplest setting. We formulate this choice between different learning modes as a two-armed bandit problem.

3.3.1 Two-Armed Bandit Problem

In our market environment, an agent repeatedly chooses between two learning modes, individual learning and social learning. Denote the action space (feasible set of choice) by Γ, $\Gamma = \{a_{il}, a_{sl}\}$, where a_{il} and a_{sl} refer to the actions of individual learning and social learning, respectively. Each action chosen at time t by agent i yields a payoff $\pi(a_{k,t})(k = il, sl)$. This payoff is uncertain, but the agent can observe this payoff ex-post and this information is used to guide future choices. This setting is analogous to the familiar *two-armed bandit* problem. In the literature, reinforcement learning has been taken as a standard behavioural model for this type of choice problem [4] and we follow the same approach.

3.3.2 Reinforcement Learning

Reinforcement learning has been widely investigated both in artificial intelligence [23, 27] and economics [7, 19]. The intuition behind this learning scheme is that better performing choices are reinforced over time and those that lead to unfavourable or negative outcomes are not, or, alternatively, the better the experience with a particular choice, the higher is the disposition or the propensity towards choosing it in the future. This combination of *association* and *selection*, or, equivalently, *search* and *memory*, or the so-called *Law of Effect*, is an important aspect of reinforcement learning [23, chapter 2].

In our model, each agent reinforces only two choices, i.e., individual learning and social learning ($\Gamma = \{a_{il}, a_{sl}\}$). In terms of reinforcement learning, the probability of a mode being chosen depends on the (normalized) propensity accumulated over time. Specifically, the mapping between the propensity and the choice probability is represented by the following Gibbs-Boltzmann distribution:

$$\text{Prob}_{i,k}(t+1) = \frac{e^{\lambda_i \cdot q_{i,k}(t)}}{e^{\lambda_i \cdot q_{i,a_{il}}(t)} + e^{\lambda_i \cdot q_{i,a_{sl}}(t)}}, \quad k \in \{a_{il}, a_{sl}\}, \tag{15}$$

where $\text{Prob}_{i,k}(t)$ is the choice probability for learning mode k ($k = a_{il}, a_{sl}$); we index this choice probability by i (the agent) and t (time) considering that different agents may have different experiences with respect to the same learning mode, and that, even for the same agent, experience may vary over time. The notation $q_{i,k}(t)$ ($k = a_{il}, a_{sl}$) denotes the propensity of the learning mode k for agent i at time t. Again, it is indexed by t because the propensity is revised from time to time based on the accumulated payoff. The propensity updating scheme applied here is the one-parameter version of [19].

$$q_{i,k}(t+1) = \begin{cases} (1-\phi)q_{i,k}(t) + U_i(t), & \text{if } k \text{ is chosen.} \\ (1-\phi)q_{i,k}(t), & \text{otherwise.} \end{cases} \quad (16)$$

where $U_i(t) \equiv U_i(\mathbf{X_i(t)})$, $k \in \{a_{il}, a_{sl}\}$, and ϕ is the so-called recency parameter, which can be interpreted as a memory-decaying factor.[4] The notation λ is known as the intensity of choice. With higher λs, the agent's choice is less random and is heavily biased toward the better-performing behavioural mode; in other words, the degree of exploration that the agent engages in is reduced. In the limit as $\lambda \to \infty$, the agent's choice is degenerated to the greedy algorithm which is only interested in the "optimal choice" that is conditional on the most recent updated experience; in this case, the agent no longer explores.

3.4 Reference Points

We further augment the standard reinforcement learning model with a reference-point mechanism to decide when Eq. (15) will be triggered. Reference dependence in decision making has been made popular by *prospect theory* [13], where gains and losses are defined relative to a reference point. This draws attention to the notion of *position* concerning the stimuli and the role it plays in cognitive coding that describes the agent's perception of the stimuli.

We augment our meta-learning model with reference points, where agents are assumed to question the appropriateness of their 'incumbent' learning mode *only* when their payoffs fall short of the reference point (along the lines of [6], p. 152–153). Let $U_i(t) (\equiv U_i(\mathbf{X_i(t)}))$ be the payoff of an agent i at time t. Let his reference point at time t be $R_i(t)$. The agent will consider a mode switch *only* when his realized payoff $U_i(t)$ is lower than his reference point $R_i(t)$.

---

[4]Following [4], a normalization scheme is also applied to normalize the propensities $q_{i,k}(t+1)$ as follows:

$$q_{i,k}(t+1) \leftarrow \frac{q_{i,k}(t+1)}{q_{i,a_{il}}(t+1) + q_{i,a_{sl}}(t+1)}. \quad (17)$$

The reference points indexed by t, $R_i(t)$, imply that they need not be static or exogenously given; instead, they can endogenously evolve over time with the experiences of the agents. Based on the current period payoffs, the reference point can be revised up or down. This revision can be symmetric, for example, a simple average of the current period payoffs and the reference point. A richer case as shown in Eq. (18) indicates that this revision can be asymmetric; in Eq. (18), the downward revisions are more pronounced than the upward revisions.

$$R_i(t+1) = \begin{cases} R_i(t) + \alpha^+ \ (U_i(t) - R_i(t)), & if \ \ U_i(t) \geq R_i(t) \geq 0, \\ R_i(t) - \alpha^- \ (R_i(t) - U_i(t)), & if \ \ R_i(t) > U_i(t) \geq 0. \end{cases} \quad (18)$$

In the above equation, α^- and α^+ are revision parameters and $\alpha^-, \alpha^+ \in [0, 1]$. The case with $\alpha^- > \alpha^+$ would indicate that the agents are more sensitive to negative payoff deviations from their reference points.[5] Note that this is similar to the idea of loss aversion in prospect theory [24], where the slopes of the values function on either side of the reference point are different. For the rest of the simulations in this paper, we have utilized the asymmetric revision case.

4 Simulation Design

4.1 A Summary of the Model

4.1.1 Scale Parameters

Table 1 presents a summary of the agent-based Scarf model which we describe in Sects. 2 and 3. It also provides the values of the control parameters to be used in the rest of this paper. The Walrasian dynamics of the Scarf economy run in this model has the following fixed scale: a total of 270 agents ($N = 270$), three goods ($M = 3$), and hence 90 agents for each type of agent $N_1 = N_2 = N_3 = 90$. The initial endowment for each agent is 10 units ($w_i = 10$), which also serves as the Leontief coefficient for the utility function. Our simulation routine proceeds along the following sequence for each market day (period): production process (endowment replenishment), demand generation, trading, consumption and payoff realization, learning and expectations updating, propensity and reference point updating, and learning mode updating. Each market day is composed of 10,000 random matches ($S = 10,000$), to ensure that the number of matches is large enough to exhaust the possibilities of trade. At the beginning of each market day, inventories perish ($\delta = 1$) and the agents receive fresh endowments (10 units). Each single run of the simulation lasts for 2500 days ($T = 2500$) and each simulation series is

---

[5]The results do not vary qualitatively for perturbations of these parameters.

Table 1 Table of control parameters

Parameter	Description	Value/Range
N	Number of agents	270
M	Number of types	3
N_1, N_2, N_3	Numbers of agents per type (4), (5)	90, 90, 90
w_i ($i = 1, 2, 3$)	Endowment (1)	10 units
	Leontief coefficients (2)	10
δ	Discount rate (Perishing rate)	1
S	Number of matches (one market day)	10,000
T	Number of market days	2500
$P_i^e(0)$ ($i = 1, 2, 3$)	Initial price expectations (3)	$\sim$ Uniform[0.5, 1.5]
φ	Parameter of price adjustment (11)	0.002
θ_1	Parameter of price adjustment (13)	$\sim$ Uniform[0,0.1]
θ_2	Parameter of price adjustment (13)	1
K	Number of arms (Sect. 3.3.1)	2
λ	Intensity of choice (15)	$\sim$ Uniform(0,8), $\sim$Uniform(0,15), $\sim$Normal(4,1), $\sim$Normal(8,1)
$POP_{a_{il}}(0)$	Initial population of a_{il}	1/2
$POP_{a_{sl}}(0)$	Initial population of a_{sl}	1/2
ϕ	Recency effect (16)	0
α^+	Degree of upward revision (18)	0.1
α^-	Degree of downward revision (18)	0.2

The numbers inside the parentheses in the second column refer to the number of the equations in which the respective parameter lies

repeated 50 times.[6] These scale parameters will be used throughout all simulations in Sect. 5.

4.1.2 Behavioural Parameters

The second part of the control parameters is related to the behavioural settings of the model, which begins with initial price expectations and price expectation adjustment, followed by the parameters related to the meta-learning models. As noted earlier, we set good 3 as a numéraire good, whose price is fixed as one. Most of these parameters are held constant throughout all simulations, as indicated in Table 1, and specific assumptions concerning distributions are specified in places where necessary.

We can systematically vary the values of the different parameters. The focus of this paper is on the intensity of choice (λ) (Table 1). The initial price vector, $\mathbf{P_i^e(0)}$,

---

[6]We have examined the system by simulating the same treatments for much longer periods and we find that the results are robust to this.

for each agent is randomly generated where the prices lie within the range [0.5,1.5], i.e., having the WE price as the centre. Except in the cases involving experimental variations of initial conditions, agents' learning modes, namely, individual learning (innovation) and social learning (imitation) are initially uniformly distributed ($POP_{a_{il}}(0) = POP_{a_{sl}}(0) = 1/2$).

4.2 Implementation

All simulations and analysis were performed using NetLogo 5.2.0 and Matlab R2015a. To comply with the current community norm, we have made the computer program available at the OPEN ABM.[7] We classify the figure into five blocks. The first block (the left-most blocks) is a list of control bars for uses to supply the values of the control parameters to run the economy. The parameters are detailed in Table 1, and include N, M, S, T, φ, θ_1, θ_2, K, λ, POP_{RE}, and $POP_{a_{il}}(0)$. In addition to these variables, other control bars are for the selection of the running model, and include individual learning (only), social learning (only), the exogenously given market fraction, and the meta-learning model. For the exogenously given market fraction, $POP_{a_{il}}(0)$ needs to be given in addition.

5 Results

5.1 Heterogeneity and Scaling Up

Unlike the representative agent models in mainstream economic theory, agent-based models are capable of incorporating heterogeneity in terms of the varying characteristics of the agents. The importance of heterogeneity in agent-based models is further underscored because of its potential to break down linear aggregation due to interaction effects. In this context, we can ask whether the quantitative phenomena observed in a heterogeneous economy are simply a linear combination of the respective homogeneous economies. If the aggregation of the effects is linear, the behaviour of the heterogeneous model can be simply deduced from the homogeneous economies. If not, studying a heterogeneous version of the economy becomes a worthwhile exercise to gain additional insights.

In our model, the only non-trivial source of heterogeneity comes from the intensity of choice (λ) that the agents possess. For simplicity, we simulate a heterogeneous economy with only two (equally populated) groups of agents with λ_i values of 1 and 7, respectively. To compare these, first, we need to determine the variable which will form the basis of comparison between the homogeneous

---
[7]https://www.openabm.org/model/4897/

and heterogeneous economies. Second, we need a method for determining the values of the above chosen variable across different homogeneous economies, which correspond to the heterogeneous economy in question. For the first, we consider the price deviations from the WE, measured in terms of the Mean Absolute Percentage Error (MAPE) of good 1, calculated based on the last 500 rounds of the simulation. The MAPE of the homogeneous economies, each with distinct values of λ is shown in the left panel of Fig. 1.

For the second, there are at least two ways to compare a heterogeneous economy (HE) (where 50% of the agents have $\lambda = 1$ and the other half have $\lambda = 7$) with homogeneous economies. Let M^{HE} represent the MAPE value in the heterogeneous economy. Let M_i^{HO} be the MAPE in a homogeneous economy, where all the agents have the same intensity of choice, i and Ω is the weighting constant. We can then consider an average of the MAPE values between the homogeneous economies with $\lambda = 1$ and 7, respectively. i.e.,

$$\bar{M}_{1,7}^{HO} = \Omega M_1^{HO} + (1 - \Omega) M_7^{HO} \qquad (19)$$

For the case of a simple average that weights the MAPE values of the two economies equally, we have:

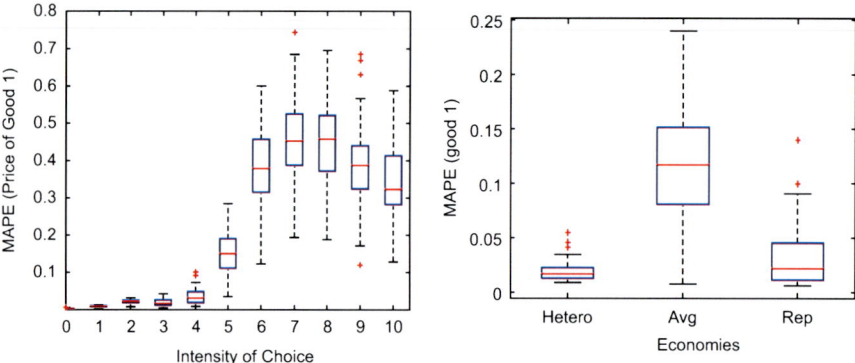

Fig. 1 Homogeneity vs. heterogeneity and aggregation: the above figures compare the values of the Mean Absolute Percentage Error (MAPE) of good 1 across different economies. The MAPE is calculated based on the last 500 periods of each run and it is defined as MAPE(P_j) = $\frac{1}{500} \sum_{t=2001}^{2500} |P_j(t) - P_j^*|$ ($j = 1, 2$), where $P_j^* = 1$. The left panel indicates the price deviation for homogeneous economies according to λ. The right panel shows the comparisons between the heterogeneous and relevant homogeneous economies. The three economies that are compared are: (1) an economy consisting of agents with heterogeneous intensities of choice (**Hetero**), (2) an economy constructed by averaging the MAPE values of two homogeneous economies with $\lambda = 1$ and 7(**Avg**), and (3) a homogeneous economy with agents holding an 'average' value of intensity of choice (i.e., $\lambda = 4$, which is the average of 1 and 7.).(**Rep**). The boxplot shows the variability of the MAPE across 50 different runs for each economy

$$\bar{M}^{HO}_{1,7} = \frac{M^{HO}_1 + M^{HO}_7}{2} \tag{20}$$

Another option would be to choose a single, homogeneous economy (**Rep**) that is *representative*, given the λ values in the heterogeneous counterpart. Instead of averaging the values of the two economies as shown above, we can choose MAPE values corresponding to the homogeneous economy with $\lambda = 4$, which is the midpoint of 1 and 7. We now have three different 'economies' to compare: (1) An economy with heterogeneous intensities of choice—$\lambda_i = 1, 7$(**Hetero**), (2) an economy representing the *average* MAPE values of two homogeneous economies with $\lambda = 1$ and 7(**Avg**), and (3) a *representative* homogeneous economy with agents holding an 'average' value of intensity of choice (i.e., $\lambda = 4$, the average of 1 and 7) (**Rep**). The boxplot in the right panel of Fig. 1 compares the MAPE values across these three versions. The whiskerplot shows the variability of MAPE across 50 different runs for each economy.

We find that the values of MAPE corresponding to the heterogeneous version are remarkably different from the other two versions in terms of the median values and its dispersion across runs. In particular, MAPE values of the **Avg** economy are much higher compared to the **Hetero** economy, indicating the absence of linear scaling up. From the right panel of Fig. 1, we observe that the MAPE across homogeneous economies is nonlinear with respect to λ. This partly explains the inadequacy of the analysis that solely relies on the information from homogeneous economies to understand the macro level properties. This nonlinearity, combined with potential interaction effects, indicates the breakdown of a linear scale-up and merits an independent investigation of the heterogeneous Scarf economy.

5.2 Price Convergence

In this section, we explore the ability to coordinate and steer the economy to the WE, starting from out-of-equilibrium configurations. To do so, we simulate the agent-based Scarf economy outlined in the previous section, where agents adaptively learn from their experiences using a meta-learning model. Given that the heterogeneity amongst agents is characterized in terms of their varying intensities of choice, we explore the influence of this parameter in detail. We simulate an economy with parameters indicated in Table 1. The agents differ in terms of the intensity of choice parameter (the exploration parameter, λ) and the intensities are uniformly distributed across agents[8] as $\lambda_i \in U(0, 8)$. The bounds of the said uniform distribution are chosen based on the initial insights gained from simulations with homogeneous agents (in terms of λ), where the behaviour of the economy is qualitatively similar

---

[8]The granularity of λ values is chosen in increments of 0.1, and hence there are 81 distinct values of λ_i corresponding to U(0,8).

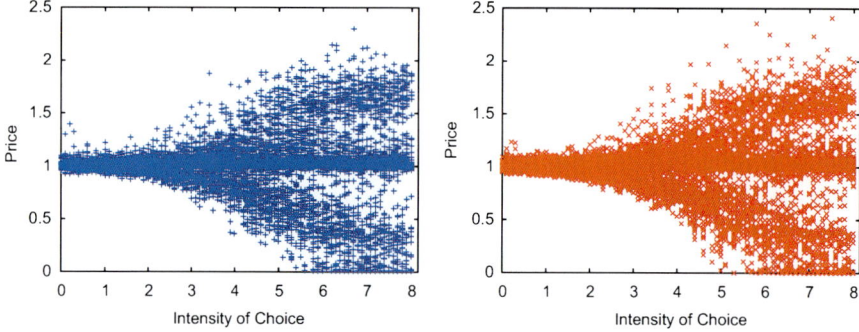

Fig. 2 Price convergence and heterogeneity: the above figure indicates the association between the mean prices for good 1 among different agents and their associated intensities of choice. The intensity of choice ranges between 0 and 8 and is distributed uniformly across agents in the economy, i.e., $\lambda_i \in U(0, 8)$. The prices of goods 1 and 2 reported here are the mean market prices for each agent which are calculated based on the last 500 periods of each run, i.e., $P_{ij} = \frac{1}{500} \sum_{t=2001}^{2500} P_{ij}(t)$, $(j = 1, 2)$. The mean prices of some agents in the economy diverge away from the neighbourhood of the WE prices ($P_{ij}^* = 1$, for $j = 1, 2$) for values of intensity of choice greater than 3

for higher values of the parameter. We simulate the economy for 2500 periods for each run and conduct 50 such repetitions. The data on prices from these repetitions are pooled together and analysed in order to obtain robust results.

Figure 2 shows the relationship between the prices of goods 1 and 2 among different agents and their corresponding intensities of choice. The prices of goods 1 and 2 reported here are the mean market prices for each agent, which are calculated by considering the last 500 periods of each run, i.e., $P_{ij} = \frac{1}{500} \sum_{t=2001}^{2500} P_{ij}(t)$, $(j = 1, 2)$. The agent-level data on mean prices across 50 repetitions are pooled together and plotted. We find that these mean prices resulting from the interaction and adaptive learning reveal some distinctive patterns. In particular, Fig. 2 shows a range of values of intensity of choice, roughly between 0 and 3, for which the agents remain in the neighbourhood of the WE. Beyond this range, the agents with higher values of λ do not necessarily have their mean prices confined to the neighbourhood of the WE. Even though some agents stay in the neighbourhood of the WE for $\lambda_i \in (3, 8)$, we observe a greater dispersion of mean prices on both sides of the WE. This is illustrated better in terms of the 3-D plot (Fig. 3) where the frequency of agents corresponding to different prices is shown on the z-axis. Although there is a dispersion on both of its sides, the WE still remains the predominant neighbourhood around which the mean prices gravitate. We also observe that there is a sizeable number of agents (more than 100) who hover close to zero as their mean price for the values of intensity of choice approaches 8.

Notice that the distribution of the intensity of choice among agents is uniform and that the distribution is split between two regions, one with less dispersion in terms of mean prices with the WE as the centre and the other with higher dispersion ($\lambda \in (4, 8)$). We test whether the WE remains as the predominant neighbourhood

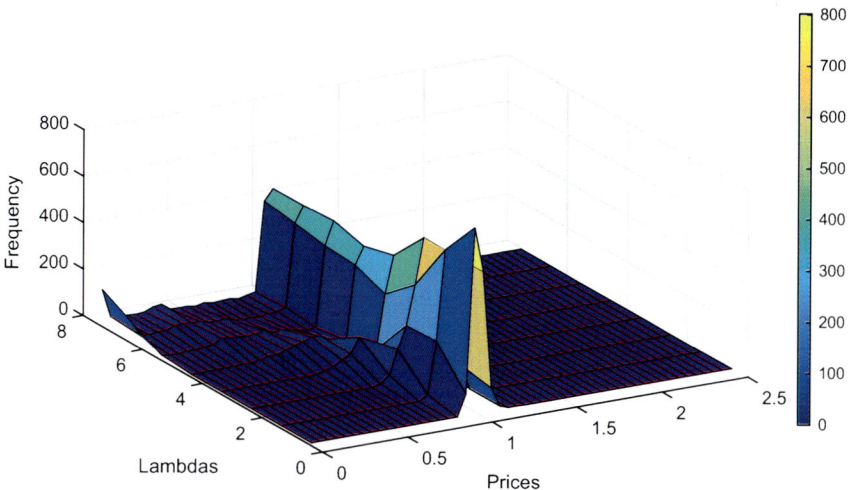

Fig. 3 Intensity of choice and price convergence for $\lambda_i \in U(0, 8)$: the above figure shows both the prices of good 1, the frequency of agents with those prices and the associated intensities of choice of each agent, where $\lambda_i \in U(0, 8)$. The prices of good 1 reported here are the mean market prices for each agent, calculated based on the last 500 periods of each run, i.e., $P_{i1} = \frac{1}{500} \sum_{t=2001}^{2500} P_{i1}(t)$. We carry out 50 repetitions of the simulation and pool together the data from all repetitions on the mean prices for all agents ($270 \times 50 = 13{,}500$) for the above plot. The agents are grouped into eight equally spaced bins based on their intensity of choice and the mean prices are grouped into 50 equally spaced bins

with which heterogeneous agents coordinate by altering the proportion of agents between high and low dispersion regions in Fig. 2. To do so, we alter the bounds of the uniform distribution according to which agents are assigned different intensities of choice in the economy. We increase the upper bound of this distribution to 15, $\lambda_i \in U(0, 15)$ and examine how the mean prices vary according to λ. Figure 4 illustrates this relationship. We find that the results are qualitatively similar and that the WE continues to be the attraction point for the agents with an increasing dispersion with increasing λ. As in the previous case, a group of agents seems to cluster close to zero, which seems to be the dominant attraction point other than the WE for the agents in the economy.

In addition to the normal distribution, we investigate the behaviour of the prices in economies with alternative distributional assumptions concerning the intensity of choice. Table 2 summarizes the prices of goods 1 and 2 at the aggregate level and at the level of agent types for 2 variations of normal and uniform distributions. For the case where $\lambda_i \in U(0, 8)$, the aggregate prices do stay within the 6% neighbourhood around the WE. They remain in a slightly larger neighbourhood (8%) for $\lambda_i \in U(0, 15)$, albeit with relatively higher variations among different runs. In the case of a normal distribution where $\lambda_i \in N(4, 1)$, the mean intensity of choice among agents lies where the increasing dispersion of prices takes off. The balance of stable and unstable tendencies explains the extremely high proximity of

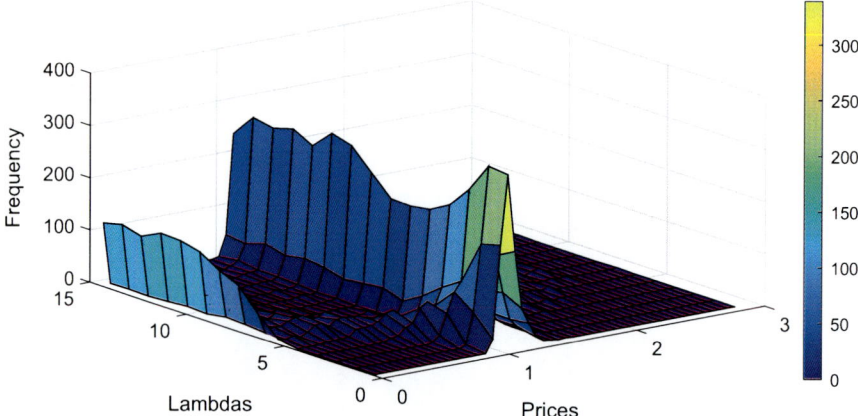

Fig. 4 Intensity of choice and price convergence for $\lambda_i \in U(0, 15)$: the above figure shows both the prices of good 1, the frequency of agents with those prices and the associated intensities of choice of each agent, where $\lambda_i \in U(0, 15)$. The prices of good 1 reported here are the mean market prices for each agent, calculated based on the last 500 periods of each run, i.e., $P_{i1} = \frac{1}{500} \sum_{t=2001}^{2500} P_{i1}(t)$. We carry out 50 repetitions of the simulation and pool together the data from all repetitions on the mean prices for all agents ($270 \times 50 = 13{,}500$) for the above plot

Table 2 Prices of Goods 1 and 2 are shown at the aggregate level and according to the types of agents

Prices/Distrib.	Aggregate		Type 1		Type 2		Type 3	
	P_1	P_2	P_1	P_2	P_1	P_2	P_1	P_2
U [0,8]	0.943	0.995	0.889	0.922	0.873	0.931	1.068	1.132
	(0.004)	(0.014)	(0.152)	(0.133)	(0.173)	(0.148)	(0.008)	(0.026)
U [0,15]	1.080	1.044	0.924	0.888	0.948	0.925	1.370	1.320
	(0.014)	(0.030)	(0.324)	(0.280)	(0.280)	(0.299)	(0.086)	(0.072)
N (4,1)	1.000	1.001	0.932	0.949	0.939	0.928	1.130	1.127
	(0.004)	(0.005)	(0.169)	(0.152)	(0.166)	(0.174)	(0.016)	(0.008)
N(8,1)	1.407	1.510	1.197	1.296	1.109	1.270	1.914	1.964
	(0.089)	(0.008)	(0.404)	(0.573)	(0.571)	(0.528)	(0.434)	(0.019)

The prices reported are the mean market prices calculated based on the last 500 periods of each run, which are averaged over all agents. For the aggregate case, $P_{ij} = \frac{1}{N} \sum_{i=1}^{N} \frac{1}{500} \sum_{t=2001}^{2500} P_{ij}(t)$, where $j = 1, 2$ and $N = 270$. For the type wise prices, the mean prices are averaged over 90 agents. We carry out 50 repetitions of the simulation and the standard deviations across 50 runs are indicated in the parenthesis

aggregate prices to the WE in this case. However, when we place the mean of λ_i in the normal distribution at 8, which is in the unstable region (i.e., $\lambda_i > 4$), the aggregate prices can be as much as 51% more than the WE.

Remember that agents with excess supply have to adjust the price expectations of production goods downward and adjust the price expectations of consumption goods upward (Eq. (10)). The upward bias in the divergence of prices stems from the fact that the agents endowed with the numerairé good who are dissatisfied can only adjust the price expectations of consumption goods upward and, unlike the agents endowed with other two goods, they cannot adjust their prices of their endowed goods downward. Those other agents with unsatisfied demand naturally bid up their prices for the goods in question. Therefore, all commodities receive more upward-adjusted potentials than downward-adjusted potential (2/3 vs 1/3 of the market participants), except for commodity 3, which serves as a numéraire. Hence, there is a net pulling force for the prices of commodities 1 and 2, leading them to spiral up.

Although we observe a high variation in mean prices among agents for larger values of λ_i, we still observe that a huge proportion of agents continue to remain in the neighbourhood of the WE for these high values of λ_i (see Figs. 3 and 4). This motivates us to search for other discriminating factors that might shed light on the dispersion of prices. In order to see whether the learning strategies chosen by the agents, viz., innovation or imitation, hold any advantage in enhancing price coordination, we classify the agents based on their strategy choices during the course of the simulation. We define *Normalized Imitation Frequency* (NIF), which indicates the ratio of the number of periods in which an agent has been an imitator over the entire course (2500 periods) of the simulation. If the agent has almost always been an innovator throughout the simulation, his NIF will be close to zero. On the contrary, if the agent has been an imitator throughout, the NIF will be equal to 1 and the intermediate cases lie between 0 and 1. The mean prices of good 1 held by the agents are plotted against the NIF in Fig. 5. The left and right panels in the figure correspond to the distributions $\lambda_i \in U(0, 8)$ and $\lambda_i \in U(0, 15)$, respectively. The entries in red correspond to agents with $\lambda_i > 4$ and those in blue denote agents with $\lambda_i < 4$.

We observe that the there are two distinct clusters for agents in red ($\lambda_i > 4$). Agents with high intensities of choice who happen to be 'almost always innovators' (with an NIF close to 1) can be seen to remain in the neighbourhood of WE, while the high dispersion in prices seems to be coming from innovating agents with a high λ_i. In our meta-learning model characterized by the reinforcement learning mechanism, agents reduce their tendency to explore for higher values of λ_i and get locked into one of the two learning strategies. These 'immobile agents' populate either corner of the NIF scale. Figure 5 indicates that the price strategy to which agents get locked in while cutting down their exploration seems to have a decisive impact on whether or not agents can coordinate themselves in the neighbourhood of the WE.

To summarize, in this section, we find that the ability of the agents to steer themselves toward the WE on aggregate depends on their tendency to explore that is captured by the intensity of choice. The distributional assumptions concerning the intensity of choice among agents influence whether or not the aggregate prices are closer to the WE. More specifically, the relative proportion of agents with λ_i corresponding to less and more volatile regions, roughly on either side of $\lambda_i = 4$ is crucial and the larger the proportion of the latter, the higher is the deviation from the WE in the aggregate. Despite the high variability in prices among agents with

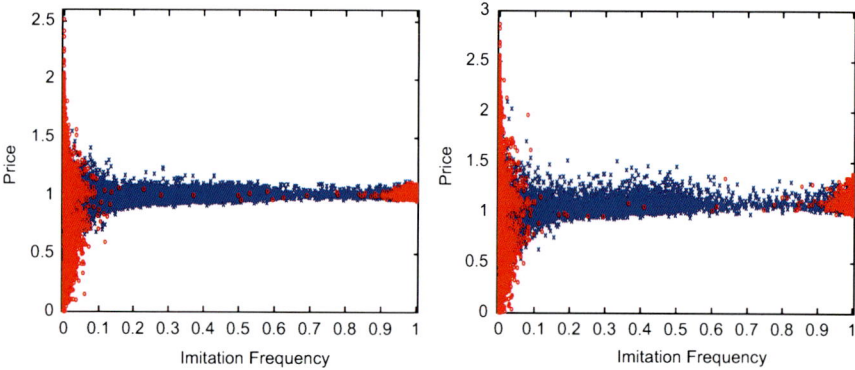

Fig. 5 Mean price of Good 1 and imitation frequency (normalized): the figure indicates the price distribution among all agents and their associated imitation frequencies. The intensity of choice (λ) is distributed uniformly across agents in the economy and the left and the right panel denote economies with distributions $\lambda_i \in U(0, 8)$ and $\lambda_i \in U(0, 15)$. Agents with $\lambda_i \leq 4$ are denoted by blue and those in red denote agents with $\lambda_i > 4$. The prices of good 1 reported here are the mean market prices for each agent, calculated based on the last 500 periods of each run, i.e., $P_{i1} = \frac{1}{500} \sum_{t=2001}^{2500} P_{i1}(t)$. We carry out 50 repetitions of the simulation and pool together the data from all repetitions on mean prices for all agents ($270 \times 50 = 13,500$) for the above plot. The imitation frequency (normalized) denotes the ratio of the number of periods in which an agent has been an imitator over the entire course (2500 periods) of each repetition of the simulation

a high λ, a sizeable number still remain in the neighbourhood of the WE. Agents who do not explore and remain predominantly as innovators for all periods exhibit a higher dispersion in their mean prices.

5.3 Current and Accumulated Payoffs

In this section, we investigate whether there are any significant differences amongst agents in terms of their current and the accumulated payoffs.[9]

Figure 6 shows the relationship between the accumulated payoffs (or 'lifetime wealth') of agents and their corresponding intensity of choice (λ_i). While there is a noticeable variability in accumulated payoffs for agents with higher intensities of choice ($\lambda_i > 4$), however, λ alone does not seem to sufficiently explain the cause. For instance, by increasing λ_i beyond 4, some (but not all) agents seem to be increasingly worse off in terms of their accumulated payoffs that keep falling. However, some other agents with the same values of λ_i enjoy much higher payoffs,

---

[9]In our model, there is no material wealth or savings since all goods are consumed entirely within each period. However, it is possible to interpret the accumulated payoffs of each agent over the simulation period as a proxy for their 'quality of life' or 'lifetime wealth'.

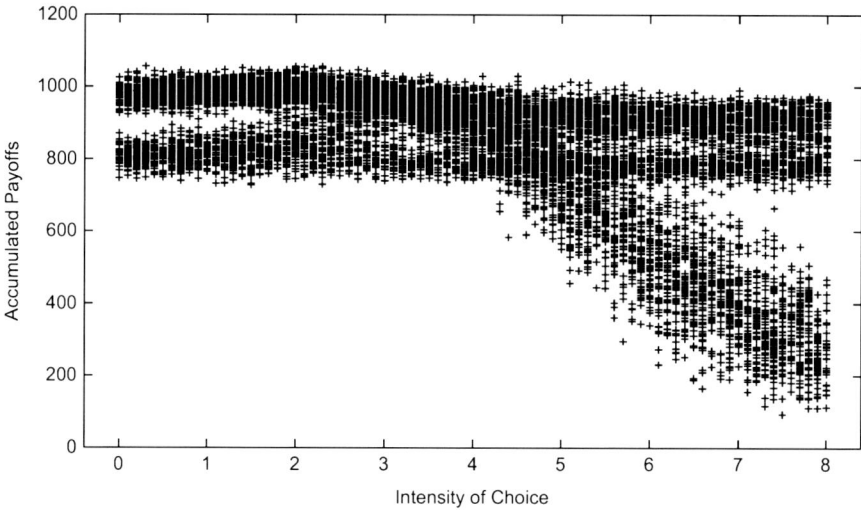

Fig. 6 Accumulated payoffs: the figure shows the accumulated payoffs for all agents (i), where $\lambda_i \in U(0, 8)$. The variability in accumulated payoffs is not entirely explained by variations in their intensity of choice alone. Notice that there are remarkable deviations between agents in terms of their accumulated payoffs for $\lambda_i > 4$

which are comparable to those with $\lambda_i <$. In other words, there is a wealth inequality that emerges among agents over time. We need to identify the characteristics of agents that are responsible for these payoff differences.

Since our agents are similar in terms of all relevant characteristics other than λ, potential indirect channels of influence stemming from λ need to be examined. Strategy choices by agents—to innovate or imitate—over the period constitute one such channel since they are governed by the λ parameter in the meta-learning (reinforcement) model. In the previous section, we learnt that imitating agents managed to coordinate to the WE on average, despite having a high λ. Along the same lines, we examine whether imitation and innovation also impact the payoffs among agents. Figure 7 shows how the current payoffs of agents vary according to their NIF for $\lambda_i \in U(0, 8)$ and $\lambda_i \in U(0, 15)$ in the left and right panels, respectively. Agents with $\lambda_i \leq 4$ are shown in blue and those with $\lambda_i > 4$ are shown in red. This indicates that agents who are predominantly innovators (for a high λ) seem to have high variability among themselves in terms of their current payoffs and agents who are predominantly imitators seem to enjoy relatively higher payoffs with considerably less variability.

In Fig. 7 we infer a clearer idea of the underlying phenomena. The left and right panels denote the relationship between intensity of choice and normalized frequency of imitation for $\lambda_i \in U(0, 8)$ and $\lambda_i \in U(0, 15)$, respectively. As the intensity of choice increases, there is a polarization in terms of NIF: for lower values of λ (roughly, $0<\lambda < 3$), NIF values span a wide range and they are not clustered close to the end of the interval (0,1). These intermediate values indicate that agents

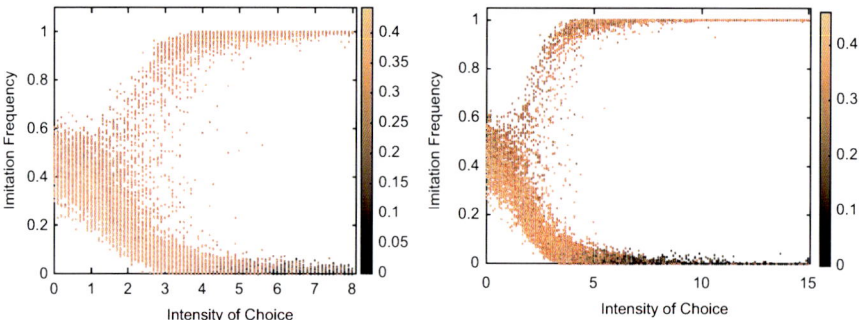

Fig. 7 Intensity of choice, imitation frequency and current payoffs: the figures in the left and right panels denote the relationship between λ and the normalized mean value imitation frequency (NIF) for all agents based on 50 repetitions, corresponding to $\lambda_i \in U(0, 8)$ and $\lambda_i \in U(0, 15)$, respectively. NIF values of 0 and 1 indicate that the agent is an innovator and imitator for 100% of the time, respectively. The heat map denotes the associated mean current payoffs, which are calculated based on the last 500 periods

switch between their strategies (innovation and imitation) and are 'mobile', thus balancing between exploration and exploitation. By contrast, for λ values of 4 and beyond, we observe that there is a drastic fall in the number of agents and the agents are either predominantly innovators or imitators. This phenomenon can be explained by the *power law of practice* that is well recognized in psychology.[10] The heat map in Fig. 7 indicates that the distribution of current payoffs varies among the agents. Mobile agents (with intermediate values of NIF) and agents who are predominantly imitators enjoy higher current payoffs compared to agents who are predominantly innovators (close to the X-axis). This pattern persists despite changes in assumptions regarding the distributions indicated above.

5.3.1 Accumulated Payoffs

We now turn to the differences among agents in terms of their accumulated payoffs that we observed in Fig. 6. We group agents into different groups based on their accumulated payoffs and analyse specific characteristics associated with agents in each of these groups that could explain the differences in their accumulated payoffs.

Figure 8 shows the agents clustered into four different groups (Very High, High, Medium, Low) based on their accumulated payoffs for $\lambda_i \in U(0, 8)$ and $U(0, 15)$ in the left and right panels, respectively. We use the K-means technique (Lloyd's algorithm) to group the agents, which allows us to group agents into K mutually

---

[10]In the psychology literature, the *power law of practice* states that the subjects' early learning experiences have a dominating effect on their limiting behaviour. It is characterized by initially steep but then flatter learning curves.

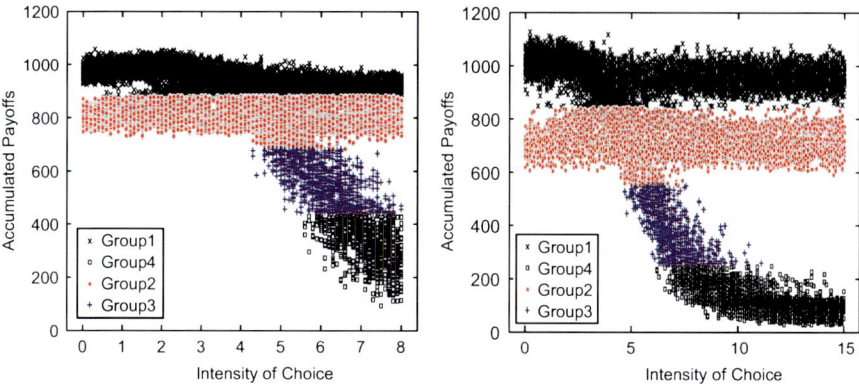

Fig. 8 Accumulated payoffs—clustered: the figures on the left and right show the accumulated payoffs for all agents (i), where the intensities are distributed uniformly as $\lambda_i \in U(0,8)$ and $\lambda_i \in U(0,15)$, respectively. Based on their accumulated payoffs, the agents are clustered in four distinct groups using the K-means clustering technique

Table 3 Accumulated payoffs of heterogeneous agents: table compares four different clusters of agents, who are pooled from 50 different runs of the simulation

	U(0,8)				U(0,15)			
	N	Π	AIF	$\bar{\lambda}$	N	Π	AIF	$\bar{\lambda}$
Low	891	320.29	24.797	7.175	2006	123.71	9.485	11.266
Medium	902	558.42	53.396	6.038	725	383.83	40.662	6.906
High	4329	805.11	1215.298	4.127	4911	714.08	1243.968	7.322
Very High	7378	957.85	965.453	3.311	5858	968.42	1607.920	6.293

These are the number of agents in each group (N) and the group-wise averages of the accumulated payoffs (Π), intensity of choice ($\bar{\lambda}$), and imitation frequency (AIF)

exclusive clusters based on the distance from the K different centroids. We use $K = 4$ based on visual heuristics and it provides the best classification for the intended purpose. Higher values of K do not show improvements in terms of their silhouette values that measure the degree of cohesion with other points in the same cluster compared to other clusters.

We showed earlier that eventual strategy choices associated with a high λ could explain variations in current payoffs. We examine whether agents (with high intensity of choice) who are predominantly imitators end up with significantly higher accumulated payoffs compared to innovating agents. Table 3 shows the four different groups consisting of agents pooled from 50 different runs of the simulation. The table also indicates the number of agents in each group and the group-wise averages of accumulated payoffs, intensity of choice, and imitation frequency. Agents with relatively better average accumulated payoffs for the two high groups differ significantly in terms of their (AIF) compared to those for the medium and low groups. Average Imitation Frequency measures the average number of periods during which agents have been imitators in the 2500 periods of the simulation.

From Fig. 1, we see that agents who have $\lambda > 4$ are those that constitute the medium and low payoff clusters. From Table 3, by comparing low and high clusters in U(0,8), we observe that agents with higher AIF (predominantly imitators) have relatively higher accumulated payoffs. However, the significance of AIF comes into play largely for agents with $\lambda > 4$. For lower values of λ the strategy choices are not tilted predominantly towards innovation or imitation. In this range, both innovating and imitating agents have a chance to obtain medium or high accumulated payoffs.

The strength of AIF as a discriminating factor becomes evident when comparing different clusters in U(0,15), where the number of agents with a high λ are in relative abundance. This relative abundance helps for a robust understanding of payoff inequality among agents with a high λ. Table 3 denotes the monotonic and increasing relationship between accumulated payoffs and AIF between clusters. The average accumulated payoff for agents from the 'Very High' cluster is 968.4, compared to 123.7 for those in the 'Low' cluster. Their corresponding AIF are 9.5 and 1607.9. Agents with abysmal accumulated payoffs for comparable λ values are explained by their strategy choice—reluctance to imitate.

6 Discussion

We have examined the possibility of coordination among agents in a stylized version of a decentralized market economy. We have analysed whether these agents can reach the equilibrium (on average) in a Scarf-like economy starting from disequilibrium states. In the presence of heterogeneity, Sect. 5.1 demonstrated that it may not be adequate to look only at corresponding economies with homogeneous agents and extrapolate patterns concerning price deviations. A straightforward linear aggregation breaks down in our model due to potential interaction effects and more importantly, due to the presence of a non-linear relationship between intensity of choice—the sole source of heterogeneity in our model—and the MAPE. The consequences of diversity among agents in shaping macroeconomic outcomes can be readily investigated using agent-based models compared to equation-based models.

Agents can and do succeed in coordinating themselves without a centralized mechanism, purely through local interaction and learning. They learn to revise private beliefs and successfully align their price expectations to those of the WE. However, this is not guaranteed in all cases and the coordination to the WE critically depends on a single behavioural parameter, viz., intensity of choice (λ). The level of intensity indirectly captures the degree or the frequency with which agents engage in exploration by trying different strategies. Lower levels of λ are associated with a higher degree of exploration and vice versa. When the intensity of choice is zero, it is equivalent to the case where agents choose between innovation and imitation with equal probability. As the λ values rise, agents increasingly stick to the 'greedy' choice and explore less and less. As a consequence, we have fewer agents who are switching between their strategies at higher λ values. They remain immobile and the strategy choices are thus polarized with AIF being either 0 or 1.

Since agents differ only in terms of λ and coordination success is linked to λ, the *extent* and nature of diversity among agents in an economy is crucial. It is hard to speak about the possibility of success in coordination independently of the distributional assumptions regarding λ. As pointed out in Sect. 5.2, if a relatively large proportion of agents have high λ values, the coordination to equilibrium is less likely to be successful, when compared to an economy with most agents holding lower intensities of choice. In sum, we find that (a) the balance between exploration and exploitation is crucial for each agent to reach the neighbourhood of equilibrium prices, and (b) the level of heterogeneity (distributions of λ) among agents determines the proportion of agents who explore little. This, in turn, determines the extent of the average price deviation from the WE for the economy as a whole (see Table 2).

The influence of a high λ (or 'strong tastes') of agents, although important, is not the whole story. Similarly, sole reliance on meso-level explanations that look at the fixed proportion of agents holding different strategies (market fraction) is also inadequate in explaining the success or failure of coordination. From Fig. 5, we find that individuals can reach the neighbourhood of the WE prices for a large intermediate range ($0 < NIF < 1$) where the agents are mobile. We also find that having a strong affinity to a particular strategy that performs well during the early periods is not detrimental per se. Instead, the *nature* of the strategy to which an agent is committed matters more. Notice from the figure that agents who are pure imitators ($NIF \approx 1$) also reach the neighbourhood of the WE, just as mobile agents do. Thus, even with high intensity of choice, agents who explore (social learning) are more effective in correcting their misperceptions unlike those who learn only from their own experiences.

We observe that inequality emerges among agents in their accumulated payoffs. There have been different explanations concerning the emergence of income and wealth inequality among individuals and nations, such as social arrangements that deny access to resources and opportunities to some groups vis-à-vis others, differences in initial conditions like endowments or productivity [12], and financial market globalization [15], to name a few. Other scholars see wealth and income inequality as a by-product of the capitalistic system and due to differential rewards to various factors of production [17].[11] In addition to these explanations, we point to another possible source of inequality over time: differences in learning behaviour. Our analysis shows that payoff inequality can potentially stem from diversity in expectations, which is traced back to a micro-level parameter that governs the learning behaviour of agents. Even though agents have the same amount of endowments, some become relatively worse off over time due to their choice of learning strategy. Agents who do not explore enough and learn from others end up receiving drastically lower payoffs and thus remain poor.

---

[11] See also: [16].

7 Conclusion

We have examined the possibility of price discovery in a decentralized, heterogeneous agent-based Scarf economy and whether or not agents can coordinate themselves to the WE starting from disequilibrium. We find that the coordination success is intimately tied to learning mechanisms employed by agents and, more precisely, how they find a balance between exploration and exploitation. A reduction in or an absence of exploration has an adverse effect on accumulated payoffs and coordination possibilities, endogenously giving rise to inequality. Social learning has often been considered to be inferior and there has been a relatively bigger focus on individual learning. It turns out that there is plenty that we can learn from others after all. Neither of these learning modes is in itself sufficient to ensure efficient outcomes and hence it may be necessary to balance them both. Although our model is highly stylized, with the results lacking the desired generality, it brings us a step closer towards comprehending the role of heterogeneity, and learning dynamically shapes various aggregate outcomes.

Acknowledgements Shu-Heng Chen and Ragupathy Venkatachalam are grateful for the research support provided in the form of the Ministry of Science and Technology (MOST) grants, MOST 103-2410-H-004-009-MY3 and MOST 104-2811-H-004-003, respectively.

References

1. Alós-Ferrer, C., & Schlag, K. H. (2009). Imitation and learning. In P. Anand, P. Pattanaik and C. Puppe (Eds.), The handbook of rational and social choice. New York: Oxford University Press.
2. Anderson, C., Plott, C., Shimomura, K., & Granat, S. (2004). Global instability in experimental general equilibrium: The Scarf example. *Journal of Economic Theory, 115*(2), 209–249.
3. Arifovic, J., & Ledyard, J. (2004). Scaling up learning models in public good games. *Journal of Public Economic Theory, 6*, 203–238.
4. Arthur, B. (1993) On designing economic agents that behave like human agents. *Journal of Evolutionary Economics 3*(1), 1–22.
5. Bossan, B., Jann, O., & Hammerstein, P. (2015). The evolution of social learning and its economic consequences. *Journal of Economic Behavior & Organization, 112*, 266–288.
6. Erev, I., & Rapoport, A. (1998). Coordination, "magic," and reinforcement learning in a market entry game. *Games and Economic Behavior, 23*, 146–175.
7. Erev, I., & Roth, A. (1998). Predicting how people play games: Reinforcement learning in experimental games with unique, mixed strategy equilibria. *American Economic Review, 88*(4), 848–881.
8. Fisher, F. M. (1983). *Disequilibrium foundation of equilibrium economics*. Cambridge, UK: Cambridge University Press.
9. Gintis, H. (2007). The dynamics of general equilibrium. *Economic Journal, 117*(523), 1280–1309.
10. Hahn, F. H., & Negishi, T. (1962). A theorem on non-tâtonnement stability. *Econometrica, 30*(3), 463–469.
11. Hoppitt, W., & Laland, K. N. (2008). Social processes influencing learning in animals: A review of the evidence. *Advances in the Study of Behavior, 38*, 105–165.

12. Hopkins, E., & Kornienko, T. (2010). Which inequality? The inequality of endowments versus the inequality of rewards. *American Economic Journal: Microeconomics, 2*(3), 106–37.
13. Kahneman, D., & Tversky, A. (1979). Prospect theory: An analysis of decision under risk. *Econometrica, 47*, 263–291.
14. Laland, K. (2004). Social learning strategies. *Animal Learning & Behavior, 32*(1), 4–14.
15. Matsuyama, K. (2004). Financial market globalization, symmetry-breaking, and endogenous inequality of nations. *Econometrica,* 72(3), 853–884.
16. Piketty, T. (2000), Theories of persistent inequality and intergenerational mobility. Chapter 8, *Handbook of Income Distribution* (vol. 1, pp. 429–476). Amsterdam: Elsevier.
17. Piketty, T. (2014), *Capital in the Twenty-First Century*. Cambridge: Harvard University Press.
18. Rendell, L., Boyd, R., Cownden, D., Enquist, M., Eriksson, K., Feldman, M. W., et al. (2010). Why copy others? Insights from the social learning strategies tournament. *Science, 328*(5975), 208–213.
19. Roth, A., & Erev, I. (1995). Learning in extensive-form games: Experimental data and simple dynamic models in the intermediate term. *Games and Economic Behaviour, 8*, 164–212.
20. Scarf, H. (1960). Some examples of global instability of the competitive economy. *International Economic Review, 1*(3), 157–172.
21. Schlag, K. (1998). Why imitate, and if so, how? A boundedly rational approach to multi-armed bandits. *Journal of Economic Theory, 78*(1), 130–156.
22. Schlag, K. (1999). Which one should I imitate? *Journal of Mathematical Economics, 31*(4), 493–522.
23. Sutton, R., & Barto, A. (1998). *Reinforcement learning: An introduction*, Cambridge, MA: MIT Press.
24. Tversky, A., & Kahneman, D. (1991). Loss aversion in riskless choice: A reference-dependent model. *Quarterly Journal of Economics, 106*, 1039–1061.
25. Uzawa, H. (1960). Walras' tâtonnement in the theory of exchange. *The Review of Economic Studies, 27*(3), 182–194.
26. Vriend, N. (2000). An illustration of the essential difference between individual and social learning, and its consequences for computational analyses. *Journal of Economic Dynamics & Control, 24*(1), 1–19.
27. Wiering, M., & van Otterlo, M. (2012). *Reinforcement learning: State of the art*. Heidelberg: Springer.

Rational Versus Adaptive Expectations in an Agent-Based Model of a Barter Economy

Shyam Gouri Suresh

Abstract To study the differences between rational and adaptive expectations, I construct an agent-based model of a simple barter economy with stochastic productivity levels. Each agent produces a certain variety of a good but can only consume a different variety that he or she receives through barter with another randomly paired agent. The model is constructed bottom-up (i.e., without a Walrasian auctioneer or price-based coordinating mechanism) through the simulation of purposeful interacting agents. The benchmark version of the model simulates homogeneous agents with rational expectations. Next, the benchmark model is modified by relaxing homogeneity and implementing two alternative versions of adaptive expectations in place of rational expectations. These modifications lead to greater path dependence and the occurrence of inefficient outcomes (in the form of suboptimal over- and underproduction) that differ significantly from the benchmark results. Further, the rational expectations approach is shown to be qualitatively and quantitatively distinct from adaptive expectations in important ways.

Keywords Rational expectations · Adaptive expectations · Barter economy · Path dependence · Agent-based modeling

1 Introduction

The dynamic stochastic general equilibrium approach ("DSGE") approach and its representative agent ("RA") version have been widely criticized on the basis of empirical, theoretical, and methodological objections in a longstanding, large, and rich body of literature. To name just three of numerous recent papers, [9, 21], and [17], that summarize and explain many of the criticisms leveled against DSGE models and advocate the use of agent-based modeling ("ABM") as a preferred

S. Gouri Suresh (✉)
Davidson College, Davidson, NC, USA
e-mail: shgourisuresh@davidson.edu

alternative. Various papers such as [14] and [18] have explicitly constructed models to compare the two approaches in a variety of different ways. This paper also compares the two approaches by constructing a basic illustrative benchmark DSGE model embedded within an ABM framework that also allows for certain typical DSGE assumptions to be selectively relaxed. Results confirm that changes in assumptions regarding homogeneity and rational expectations lead to quantitatively and qualitatively different outcomes even within a model designed to resemble a simplified real business cycle ("RBC") environment.

Of the many assumptions required in typical DSGE models, this paper focuses broadly on three:

1. Rational expectations hypothesis ("REH")—According to Frydman and Phelps in [11, p. 8]:

    > REH presumes a special form of expectational coordination, often called the 'rational expectations equilibrium' ("REE"): except for random errors that average to zero, an economist's model—'the relevant economic theory'—adequately characterizes each and every participant's forecasting strategy.

    However, notably, in the same chapter, [11, p. 2] the authors cite some of their prior work to emphasize that

    > REH, even if viewed as a bold abstraction or approximation, is grossly inadequate for representing how even minimally reasonable profit seeking participants forecast the future in real-world markets.

    The primary goal of this paper is to compare outcomes with and without the assumption of rational expectations in an illustrative ABM framework by constructing agents who could either have rational expectations or one of two different types of adaptive expectations.

2. Homogeneous stochastic productivity levels—Stiglitz and Gallegati [21, p. 37] argue that

    > First and foremost, by construction, [in the RA approach] the shock which gives rise to the macroeconomic fluctuation is uniform across agents. The presumption is that idiosyncratic shocks, affecting different individuals differently, would 'cancel out.' But in the real world idiosyncratic shocks can well give rise to aggregative consequences...

    More specifically, within the DSGE framework, it can typically be demonstrated that if complete and frictionless insurance markets exist, only aggregate shocks matter for macroeconomic consequences. However, in practice, the real world includes a large number of uninsured idiosyncratic shocks. In order to examine the consequences of this critical assumption, the model in this paper allows shocks to either be homogeneous or heterogeneous across agents.

3. Walrasian equilibrium—[4] and the references included therein provide a detailed and critical analysis of the concept of Walrasian equilibrium. In particular, [22] states that four conditions (Individual Optimality, Correct Expectations, Market Clearing, and the Strong Form of Walras' Law) must hold for Walrasian equilibrium and then proceeds to carefully construct a non-Walrasian agent-based model by plucking out the Walrasian Auctioneer or

the implicit price setting mechanism employed in the DSGE framework. The model developed in this paper also does away with the Walrasian Auctioneer as transactions between agents take place through barter. A decentralized trading system has been implemented in [12] and various other influential papers in the literature. This decentralized trading system is straightforward to implement in an ABM framework and explicitly allows for situations where markets do not clear or equilibrium is not attained in the sense that some bartering agents remain unsatisfied with the volume exchanged.

The model in this paper is inspired in part by a rudimentary RBC model where the primary source of fluctuations is a stochastic productivity shock. The reason for choosing simplicity and parsimony is to illustrate how consequential various DSGE assumptions can be even within elementary models. Despite its simplicity, the model in this paper allows for the implementation of a typical rational expectations equilibrium with homogeneous agents as well as the optional inclusion of heterogeneous random productivity levels, incomplete and decentralized markets with the potential for disequilibria, adaptive expectations, price stickiness, and asymmetric information. Results suggest that multiple equilibria plague the rational expectations solution as is common in numerous papers in the extensive literature on coordination failures—see, for instance, [5–7], and [2]. Under adaptive expectations, the model has richer path dependence as well as convergence to a particular Pareto-inferior rational expectations equilibrium. Moreover, some forms of adaptive expectations are found to be particularly sensitive to the scale of the model.

The remainder of the chapter is organized as follows. Section 2 sets up the model mathematically and describes some of its features. Section 3 provides results from the analyses of various versions of the model and compares the outcomes of the benchmark homogeneous version with the outcomes of various rational expectations and adaptive expectations versions. The paper concludes with further discussion in Sect. 4.

2 Model Development

In this model economy, although there is essentially only one good, it comes in two varieties, red and blue. Half the agents in the economy produce the blue variety and consume the red variety, whereas the other half of the agents in the economy produce the red variety and consume the blue variety. The number of agents in the economy equals N where N is restricted to be an even number. Each agent is endowed with three units of time each period which the agent can allocate in discrete quantities to leisure and labor. All agents receive disutility from labor and, as stated previously, agents who produce the red variety derive utility from consuming the blue variety and vice versa. The Cobb Douglas utility functions for red and blue producers are given by (1) and (2):

$$U_i = B_i^\alpha (3 - L_i)^{1-\alpha} \qquad (1)$$

$$U_j = R_j^\alpha (3 - L_j)^{1-\alpha} \qquad (2)$$

The per period utility for Agent i, assuming that Agent i is a red variety producer, is given by U_i, where B_i is the number of units of the blue variety consumed by Agent i and L_i is the number of units of time devoted by the Agent i towards labor. The utility function for Agent j, a blue variety producer, is similar except for the fact that Agent j derives utility from consuming the red variety. The parameter α reflects the relative weight assigned to consumption over leisure and it is assumed that agents accord a higher weight to consumption than leisure or $0.5 < \alpha < 1$. Notice that no agent would ever devote all 3 of their units of time towards labor and the specified preference structure reflects non-satiation.

Each agent conducts her own private production process that is unobservable to other agents based on the following constant returns to scale Leontief production functions, (3) and (4):

$$R_i = Min(A_i, L_i) \qquad (3)$$

$$B_j = Min(A_j, L_j) \qquad (4)$$

The quantity of the red variety produced by the Agent i, (again, assuming that Agent i is a red variety producer) is given by R_i, where A_i is the idiosyncratic private stochastic productivity level experienced by Agent i, and L_i once again is the number of units of time devoted by the Agent i towards labor. The production function for Agent j, a blue variety producer, is similar except for the fact that Agent j is subject to her own idiosyncratic stochastic productivity level, A_j. The stochastic productivity level can take on a value of either 1 (the low value) or 2 (the high value) and evolves independently for each agent according to the same Markov process described through the Markov transition matrix Π.

After the production process is completed privately, an agent who produces the red variety is randomly paired with another agent who produces the blue variety. The two agents barter as much of their produce as possible, at the fixed exchange rate of one blue variety good for one red variety good. No further barter opportunities are allowed for these two agents in the period and any excess goods they possess that were not bartered are discarded. This barter process replaces the Walrasian Auctioneer and allows for decentralized transactions. The barter process can be represented by (5) and (6):

$$B_i = Min(R_i, B_j) \qquad (5)$$

$$R_j = Min(R_i, B_j) \qquad (6)$$

In other words, B_i, the quantity of the blue variety of good that Agent i, the red variety producer obtains through barter, is equal to the lesser of R_i and B_j, the quantities produced by Agent i and Agent j, respectively. Similarly, the quantity of the red variety of good that Agent j, the blue variety producer obtains through barter, is also equal to the lesser of R_i and B_j, the quantities produced by Agent i and Agent j, respectively.

In an analytically isomorphic formulation, all agents could be thought of as producers of one of two colors of an intermediate good which are then perfect complements in the production process for the final good with the production function, $F = 2 \times Min(R_i, B_j)$, and where the agents then subsequently split the final good thus obtained equally between themselves. Alternatively, this model can also be thought of as a discrete variant of the coconut model in [7] adapted for the ABM framework.

Notice that there are two distinct sources of randomness in this model—the first is the randomness associated with the stochastic productivity level and the second is the randomness inherent in the pairwise matching for barter. The following outcomes are of interest in this model: (1) per capita output or average output, (2) per capita utility or average utility, (3) per capita overproduction or the average quantity of goods discarded due to mismatches in barter (for example, if one agent has produced 1 unit, while the other agent has produced 2 units then 1 unit produced by the latter agent is considered overproduction), and (4) per capita underproduction or the average difference between the quantity of goods actually produced and the maximum quantity of goods that could have been consumed by a pair given the stochastic productivity levels of its members (if either agent has a low productivity level, then there is no underproduction because the most that each agent could have consumed equals one; however, if both agents have high productivity levels, then there is underproduction of one unit if one agent produces two goods and the other produces one and there is underproduction of two units when both agents produce only one good each).

Table 1 provides the value of the preference parameter α, the values that comprise the Markov transition matrix Π, and the long-term probabilities implied by the values in Π.[1]

Table 1 Parameter values

Preference parameter		Value	
α		0.8	
Markov transition matrix Π			
		Future state	
		Low	High
Current state	Low	0.6	0.4
	High	0.1	0.9
Long-term probabilities due to Π			
Low		0.2	
High		0.8	

---

[1]Results are somewhat quantitatively sensitive to the chosen parameters but the broader implications do not change. This will be discussed further in the next section.

Table 2 lists all the theoretical outcomes that are possible for any pair of bartering agents given the discrete nature of their stochastic productivity levels and labor inputs.

The model constructed above is rather simple and unrealistic compared to typical agent-based models. This, however, is a feature rather than a shortcoming since the goal of this paper is not to replicate the real world but rather to demonstrate the consequences of common DSGE assumptions within the ABM framework even in a simple model that contains only the most basic RBC elements as specified by Plosser in [20, p. 54]:

> ...many identical agents (households) that live forever. The utility of each agent is some function of the consumption and leisure they expect to enjoy over their (infinite) lifetimes. Each agent is also treated as having access to a constant returns to scale production technology for the single commodity in this economy... In addition, the production technology is assumed to be subject to temporary productivity shifts or technological changes which provide the underlying source of variation in the economic environment to which agents must respond.[2]

Despite its obvious simplicity, the model contains some useful features:

1. Heterogeneous stochastic productivity levels—The model can be solved under the assumption that each period all agents face the same stochastic productivity level (i.e., idiosyncratic shocks are assumed to be eliminated through the existence of perfect insurance) or under the assumption that each agents faces an uninsurable idiosyncratic stochastic productivity level each period.
2. Private information—The model features agents who know their own stochastic productivity level each period before they make their labor allocation decision but are unaware of the stochastic productivity levels or the labor allocation decisions of their potential bartering partners.
3. Decentralized markets—The model decentralizes markets and does away with the construct of the Walrasian Auctioneer by featuring agents who engage in barter.
4. Implicit price stickiness—The fixed one-for-one exchange rate of one variety for another is equivalent to price stickiness. Alternatively, this fixed exchange rate can also consider an outcome of the Leontief production function for the final good that is then split equally between each pair of intermediate goods providers.
5. Markets that could fail to clear—Although in one sense markets clear trivially in all barter economies, this model exhibits disequilibria in a deeper sense where agents are unable to satisfy their desired demands and supplies and may regret that they have produced too little or too much.

---

[2]In addition, the RBC model described in [20] features capital in the production function as well as the associated consumption-investment trade-off. Adding capital to the model in this paper would not affect its central findings but would complicate the analysis considerably since agents would need to base their own decisions not just on their expectations of others' stochastic productivity levels and labor allocation decisions but also on their expectations of others' capital holdings. This increase in dimensionality renders the problem computationally impractical if REH is assumed along with heterogeneous shocks and a finite number (i.e., not a continuum) of distinct individuals.

Rational Vs. Adaptive Expectations in a Barter Economy

Table 2 All possible outcomes for a pair of bartering agents

Situation #	Productivity		Effort		Output		Consumption		Utility		Overproduction	Underproduction	Market clearing?
	Agent 1	Agent 2	Agent 1	Agent 2	Agent 1	Agent 2	Agent 1	Agent 2	Agent 1	Agent 2			
1	1	1	1	1	1	1	1	1	$2^{(1-\alpha)}$	$2^{(1-\alpha)}$	0	0	Yes
2	1	1	1	2	1	1	1	1	$2^{(1-\alpha)}$	1	0	0	Yes
3	1	1	2	1	1	1	1	1	1	$2^{(1-\alpha)}$	0	0	Yes
4	1	1	2	2	1	1	1	1	1	1	0	0	Yes
5	1	2	1	1	1	1	1	1	$2^{(1-\alpha)}$	$2^{(1-\alpha)}$	0	0	Yes
6	1	2	1	2	1	2	1	1	$2^{(1-\alpha)}$	1	1	0	No
7	1	2	2	1	1	1	1	1	1	$2^{(1-\alpha)}$	0	0	Yes
8	1	2	2	2	1	2	1	1	1	1	1	0	No
9	2	1	1	1	1	1	1	1	$2^{(1-\alpha)}$	$2^{(1-\alpha)}$	0	0	Yes
10	2	1	1	2	1	1	1	1	$2^{(1-\alpha)}$	1	0	0	Yes
11	2	1	2	1	2	1	1	1	1	$2^{(1-\alpha)}$	1	0	No
12	2	1	2	2	2	1	1	1	1	1	1	0	No
13	2	2	1	1	1	1	1	1	$2^{(1-\alpha)}$	$2^{(1-\alpha)}$	0	2	Yes
14	2	2	1	2	1	2	1	1	$2^{(1-\alpha)}$	1	1	1	No
15	2	2	2	1	2	1	1	1	1	$2^{(1-\alpha)}$	1	1	No
16	2	2	2	2	2	2	2	2	2^α	2^α	0	0	Yes

Note: $2^\alpha > 2^{(1-\alpha)} > 1$ because $0.5 < \alpha < 1$. In particular, since $\alpha = 0.8$, $1.74 > 1.15 > 1$

The presence of these features allows this model to occupy a space in-between standard DSGE models and various agent-based models. The next section analyzes the results of this model under various alternative assumptions.

3 Results and Analyses

Many different versions of the model were analyzed in order to capture the role of various assumptions. Each version with a discrete number of agents was simulated over multiple periods multiple times using Monte Carlo techniques in order to obtain statistically reliable interpretations. For the Monte Carlo simulations, 100 different random seeds (one for each simulation) were applied uniformly across all heterogeneous agent models over 50 periods (the initial period was treated as the zeroth period and dropped from the analysis). The version with a continuum of agents could not be simulated and so theoretical results from that version are included instead. The time series and ensemble means and standard deviations of the various outcome measures for the various versions analyzed are presented in Table 3.

3.1 The Benchmark Version: Homogeneous Agents

In this version of the model, all agents are assumed to be perpetually identical in terms of their realized stochastic productivity levels. Every agent finds herself in either situation 1 (20% of the time, on average) or situation 16 (80% of the time) of Table 2. If we add the assumption of Pareto-optimality in the benchmark case, the REH where every agent expends high effort when productivity levels are high and low effort when productivity levels are low can be sustained.[3] Consequently, there is no overproduction, no underproduction, markets always clear, and the level of utility is always the highest attainable in each and every period. Another important feature of the homogeneous case is that outcomes are scale invariant and results remain identical regardless of the number of agents (as long as the number of red variety producers is equal to the number of blue variety producers, an assumption that is held true throughout this paper).

---

[3]In theory, non-Pareto optimal outcomes are possible with REH here. For instance, it would be rational for all agents to expend low effort regardless of productivity levels with the expectation that everyone else too expends low effort in every state.

Table 3 A comparison of time series and ensemble means and standard deviations for various model versions

						Output per capita		Utility per capita		Over production per capita		Under production per capita	
									Over 50 periods				
						Mean	Std. Dev.	Mean	Std. Dev.	Mean	Std. Dev.	Mean	Std. Dev.
Homogeneous agents					Mean	1.81	0.38	1.63	0.22	0.00	0.00	0.00	0.00
					Std. Dev.	0.09	0.08	0.05	0.05	0.00	0.00	0.00	0.00
Heterogeneous agents	Symmetric rational expectations equilibrium	4 agents		Strategy 1 (0)	Mean	1.00	0.00	1.15	0.00	0.00	0.00	0.63	0.33
					Std. Dev.	0.00	0.00	0.00	0.00	0.00	0.00	0.08	0.04
				Strategy 2 (65,535)	Mean	1.64	0.33	1.50	0.22	0.16	0.16	0.00	0.00
					Std. Dev.	0.07	0.03	0.02	0.01	0.03	0.01	0.00	0.00
				Strategy 3 (32,489)	Mean	1.40	0.42	1.38	0.26	0.08	0.14	0.24	0.35
					Std. Dev.	0.10	0.04	0.06	0.02	0.03	0.02	0.05	0.03
		Continuum (theoretical)		Strategy 1	Mean	1.00	0.00	1.15	0.00	0.00	0.00	0.64	0.00
					Std. Dev.	0.00	0.00	0.00	0.00	0.00	0.00	0.00	0.00
				Strategy 2	Mean	1.64	0.00	1.50	0.00	0.16	0.00	0.00	0.00
					Std. Dev.	0.00	0.00	0.00	0.00	0.00	0.00	0.00	0.00
	Adaptive expectations based on individual	4 agents			Mean	1.01	0.03	1.15	0.02	0.01	0.03	0.62	0.33
					Std. Dev.	0.02	0.06	0.01	0.03	0.01	0.03	0.08	0.03
		100 agents			Mean	1.01	0.03	1.15	0.01	0.01	0.02	0.61	0.09
					Std. Dev.	0.00	0.01	0.00	0.00	0.00	0.00	0.01	0.01
	Adaptive expectations based on population	4 agents			Mean	1.14	0.17	1.21	0.11	0.06	0.10	0.38	0.28
					Std. Dev.	0.20	0.17	0.12	0.10	0.04	0.04	0.17	0.11
		100 agents			Mean	1.64	0.07	1.31	0.04	0.16	0.03	0.00	0.00
					Std. Dev.	0.02	0.01	0.01	0.01	0.01	0.00	0.00	0.00

(Across 100 Simulations)

3.2 The REH Version with 4 Agents

The next model analyzed assumes REH and includes four agents.[4] Though other information sets compatible with REH are possible, in the version of REH assumed here, the information set of each agent is assumed to consist of the actual past productivities of all agents. With this assumption in place, all four agents are fully aware of all four productivity levels in the previous period, the fixed transition matrix Π determining the probabilities of current period stochastic productivity level outcomes, and their own current period realized stochastic productivity level. With REH, each agent can come up with a strategy that involves a certain action for each possible state of affairs. For any agent, the most parsimonious description for the current state of affairs from his own perspective (his state vector, in other words) involves one unit of information for the agent's own current period shock and four units of information for the four individual shocks in the previous period. Each unit of information involves one of two possible values (high or low). Therefore, an agent can find himself in any one of 2^5 or 32 different states.[5]

A full description of an agent's strategy requires specifying an action for each possible state. Since an agent can choose an action of either high effort or low effort for each state, the total number of possible strategies for an agent is 2^{32} or 4,294,967,296. However, from the structure of the model we know that no agent would find it optimal to put in high effort when their own current stochastic productivity level is low and so in 16 of those 32 states, the optimal action is necessarily that of low effort. This leads to each agent effectively having 2^{16} or 65,536 potentially optimal strategies. If all four distinctly identifiable agents are allowed to pursue different strategies, then the overall number of possible strategies to be considered in the model is $(65,536)^4 = 1.8447e + 19$ if ordering of strategies matters or $\frac{65,539 \times 65,538 \times 65,537 \times 65,536}{4!} = 7.6868e + 17$ if the ordering of strategies is irrelevant. Since either of these is computationally intractable for the purposes of this research paper, the analysis here is restricted to symmetric REE ("SREE") which can be thought of as REE with the additional requirement of symmetry across agents. It bears repeating that despite the considerable computing power of the microcomputer at my disposal, tractability in the model requires the imposition of symmetry as an assumption. Methodologically, this calls into question

---

[4] A two-agent version was also analyzed but this version is qualitatively different since it does not include the randomness associated with the pairing for the barter process. Results for this version are available upon request.

[5] An implicit assumption here is that all agents are distinctly identifiable. If such distinctions are assumed away, then the most parsimonious description for the current state involves one unit of information for the agent's own current period shock (which can take on one of the two values), one unit of information for the number of red producers who received high productivity levels in the last period (which can take on three values—none, one, and both) and one unit of information for the number of blue producers who received high productivity levels in the last period (which can also take on three values). Thus if, we assume away distinctiveness, an agent can find himself in any one of $2 \times 3 \times 3 = 18$ different states.

Rational Vs. Adaptive Expectations in a Barter Economy

the reasonableness of assuming strict REH in the real world where symmetry is not commonly observed and agents rely primarily on mental calculations rather than supercomputers.

With the assumption of symmetry in place, the 65,536 strategies (numbered 0 through 65,535) were examined one-at-a-time using a simple programming approach that tested whether an individual agent could expect a higher potential utility if she alone were to deviate from the strategy under consideration. 64 of the 65,536 potentially optimal strategies were found to be SREE, i.e., strategies such that no individual agent could gain from deviating away from that strategy if everyone else in the economy were following that same strategy.[6,7] The full set of 1.8447e+19 possible combinations of strategies was not explored, but the examination of a random sample suggested the existence of numerous asymmetric REE as well.[8] As stated previously, the existence of multiple equilibria is a common outcome in models with complementarities and coordination failures. Although the various equilibria can be Pareto-ranked, if REH is followed strictly, choosing between these equilibria is theoretically impossible, because each REE, by definition, satisfies all REH requirements. Cooper [5] provides some suggestions on how this problem could potentially be resolved through the use of history as a focal point or through best response dynamics to out-of-equilibrium scenarios. Within the SREE framework adopted here however, the latter is ruled out by definition while the former is helpful to the extent that it suggests that any chosen SREE is likely to continue unchanged across periods.

Table 4 contains the means and standard deviations of the various outcomes of interest across the 64 different SREE strategies as well as the minimum and maximum values of each outcome variable and the associated strategy numbers. These results were obtained through Monte Carlo simulations where 100 different random seeds (one for each simulation) were applied uniformly across all 64 SREE strategies over 50 periods (the initial period was treated as the zeroth period and dropped from the analysis). There were no readily discernible common properties of these 64 different SREE strategies.[9] On the other hand, results suggest that outcomes differ greatly based on which of the multiple SREEs are assumed to be in place. In particular, it can be seen that two contrasting strategies, Strategy #0 and Strategy #65,535, are both SREEs and result in significantly different outcomes. Strategy #0 corresponds to all four agents always choosing low effort regardless of stochastic

---

[6]Technically, since distinctiveness is unnecessary if SREE is assumed, the more parsimonious method mentioned in footnote 5 can be adopted resulting in the need to explore only 2^9 or 512 strategies. As one of the exercises in this paper allowed for a sampling of asymmetric strategies, all 65,535 possibilities were treated as distinct for comparability.

[7]The number of SREEs obtained was robust to small changes in parameter values. However, the number of SREEs obtained was affected by large changes in parameter values and even dropped to as low as one (e.g., when the low productivity level was a near certainty, the only SREE was to always provide low effort).

[8]In the 2 agent version of the model, 16 REEs were found of which 8 were also SREEs.

[9]Detailed results for all SREE strategies are available from the author upon request.

Table 4 Summary of results across the 64 SREE strategies

	Output per capita		Utility per capita		Overproduction per capita		Underproduction per capita	
	Value	Strategy code #	Value	Strategy code #	Value	Strategy code #	Value	Strategy code #
Mean	1.32		1.33		0.08		0.32	
Std. Dev.	0.21		0.12		0.05		0.21	
Min.	1.00	0	1.15	0	0.00	0	0.00	65,535
Max.	1.64	65,535	1.50	65,535	0.16	65,535	0.64	0

productivity levels, while Strategy #65,535 corresponds to all 4 agents choosing low effort only when their own idiosyncratic stochastic productivity level is low.[10] In order to allow at a glance comparisons across models, only three SREEs are included in Table 3: Strategy #0, Strategy #65,535, and a randomly chosen SREE, Strategy #32,489.

Using the terminology employed by Page [19] and developed further by Jackson and Ken [13], the properties for the various versions of the model in this paper can be studied in terms of state dependence, path dependence, and equilibrium dependence.[11] Based on this, the SREE with Strategy #0 displays independence (i.e., outcomes and strategies are independent of the current period shocks and past period shocks). In all the SREE cases other than Strategy #0, state dependence exists in a trivial sense since the stochastic productivity level evolves according to a Markov process and the decisions by the agents themselves depend only on their own current state vectors. Consequently, both outcomes and decisions in these SREE cases are only state dependent rather than path dependent. Also, unless an existing SREE is perturbed, there is no change in the strategy pursued. Therefore, the SREE version of the model can also be considered equilibrium independent since the long-term evolution of the system does not change regardless of the evolution of stochastic productivity levels.

3.3 The REH Version with a Continuum of Agents

In terms of scale, the REH model above becomes significantly less tractable with additional agents. For instance, with six potentially asymmetric distinctly identifiable agents, even after the simplification that no agent would put in high

---

[10] It can be shown mathematically that regardless of how many agents are included in the model, both of these contrasting strategies (i.e., 1. always choosing low and 2. choosing low only when individual productivity level is low) will be equilibria for the chosen values of α and Π in this model.

[11] The concept of equilibrium here is the one adopted by Page [19] and is completely unrelated to economic equilibrium or market clearing.

effort in a situation where she has received a low productivity level, the number of potential optimal strategies that need to be searched is $(2^{2^6})^6$ which approximately equals 3.9402 e+115 and is intractable. In general, the number of strategies that would need to be examined with N distinct agents equals $(2^{2^N})^N$ if asymmetric equilibriums are allowed and order matters (i.e., agents are distinctly identifiable) or $\frac{(2^{2^N}+N-1)!}{N!(2^{2^N}-1)!}$ if asymmetric equilibriums are allowed but the order of strategies is irrelevant (i.e., the identity of individual agents does not matter). Even if attention is restricted to SREE and agents are considered to be indistinct from each other, under the most parsimonious formulation, the number of strategies that would need to be examined equals $2^{(\frac{N}{2}+1)^2}$.

This combinatorial explosion of strategy space occurs because the information set for each agent increases as the number of agents increases. With a larger information set, each agent can fine tune her strategy based on many more possible combinations of prior productivity levels across all the agents. Consequently, as the number of agents increases, the number of possible SREE strategies also increases explosively.

Interestingly, one possible way to escape this combinatorial explosion is to assume a continuum of agents. With a continuum of agents, if we force SREE (i.e., require all agents to follow the exact same strategy), the information set in effect collapses completely. The probability of encountering a high or low productivity level agent with a set strategy no longer depends on the precise combinations of prior stochastic productivity levels and remains constant across all periods. Each agent now only needs to consider what to do when her own productivity is low and what to do when her own productivity level is high since the state of the aggregate economy is unchanged across periods due to the averaging over a continuum. Thus, only four strategies are possible (expend high effort always, expend low effort always, expend effort to match own productivity level, and expend effort inversely to own productivity level) only two of which are reasonable (i.e., do not involve expending high effort during a low productivity period). These two equilibria remain for the values of α and Π chosen in this paper, and as can be predicted on the basis of footnote 10 (page 152), these equilibria correspond to always choosing low regardless of productivity levels and choosing low only when individual productivity level is low. However, note that if we assume a continuum of agents, entire functions of asymmetric equilibria are also possible.

Theoretical results with this assumption of a continuum of agents are available in Table 3. It is easy to see on the basis of the discussion above why homogeneity and continuums are often assumed in the REH literature for the purposes of tractability even though neither specification accurately reflects real world economies where a finite number of distinct agents interact. In fact, as Table 3 suggests, the homogeneous case and the continuum case are very different from those of an SREE such as Strategy # 32,489. In other words, assuming homogeneity or a simplified continuum ignores many more equilibrium possibilities than merely assuming SREE.

Table 3 also highlights the difference between the homogeneous case and the continuum case with Strategy #2. Intuitively, this difference arises because there are no mismatched productivities in the homogeneous case, while they occur in the latter. Also, note that the standard deviations are zero for the continuum case because of averaging over a continuum of agents.[12]

3.4 The Adaptive Expectations ("AE") Versions

In the AE model, agents are assumed to not know the underlying stochastic model. In particular, it is assumed that agents do not know the value of Π and, furthermore, do not know the values of the past and current stochastic productivity levels of other agents. The only information that an agent in the AE model is allowed to retain is information about the output produced in the previous period. Two versions of the AE model are considered based loosely on the long-standing literature that distinguishes between individual and social learning; for instance [23], argues that there are fundamental distinctions between the two modeling approaches with significant implications in terms of greatly differing outcomes based on whether individual or social learning is assumed. Some summary comparative results for individual and social adaptive learners are provided in Table 3.

1. Adaptive Expectations—Individual ("AE–I"): In this version, an agent naïvely predicts that the individual she will be paired with in the current period would be someone who has produced the same quantity as the individual she encountered in the previous period. In other words, agents in this version form their expectations solely on the basis of their most recent individual experience.
2. Adaptive Expectations—Population ("AE–P"): In this version, an agent who produces a particular variety is aware of the economy-wide distribution of the other variety produced in the previous period and predicts that distribution will remain unchanged in the current period. In other words, agents in this version form their expectations on the basis of the most recent population-wide outcome.

It must be noted that the agents in both these versions are goal-directed expected utility maximizers who use information about their observable idiosyncratic shocks and their naïve expectations about others actions to determine their optimal level of effort. It is also assumed that all adaptive agents start with the high effort strategy (in the first period, each agent expends high effort if her own stochastic productivity level is high). A few interesting results emerge that highlight the contrast with the SREE version. Although some of these differences can be found in Table 3, it is easier to visualize these differences through graphs rather than summary statistics.

---

[12] I would like to thank an anonymous referee for highlighting this point.

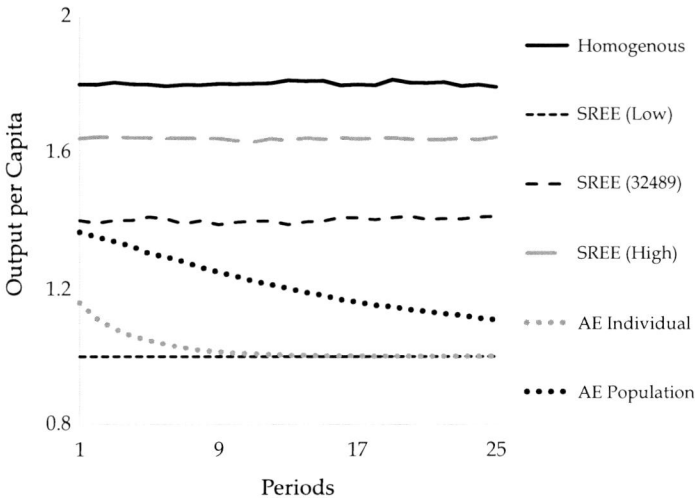

Fig. 1 Ensemble averages of output per capita over 1000 simulations in a four-agent model

Figure 1 depicts ensemble averages of per capita output over time for AE–I, AE–P, the high SREE, the low SREE, and the randomly chosen SREE. As can be seen, the high SREE and the randomly chosen SREE bounce around throughout, while AE–I and AE–P bounce around for a while before settling at 1 which equals the low SREE. In other words, it appears that the low SREE is the eventual outcome of the adaptive expectations process. Notice that none of the SREEs have any perceptible trend, while both AE–I and AE–P have a perceptible trend that eventually settles down at the low SREE.

In the AE–I case, the convergence to the low equilibrium occurs because each time an agent trades with another agent who has produced a low level of output, she decides to produce low output herself in the next period regardless of her productivity level. This action of hers in turn causes the agent she interacts with in the next period to produce low output in the period subsequent to the next. On the other hand, if an agent encounters another agent who has produced a high level of output, she will only produce a high level herself in the next period if she herself also has a high outcome for her stochastic productivity level. The asymmetry in this model is such that with adaptive expectations, the outcome is inexorably drawn towards the low equilibrium due to the asymmetry embedded in effort choice: high effort is optimal only when one's own productivity level is high *and* one reasonably expects to meet someone else who has produced a high level of output; low effort is optimal both when one's own productivity is low and *or* one expects to have a trading partner who has produced low output.

The reason for the convergence of the AE–P case to the low equilibrium is similar. As long as the population-wide distribution of outputs is above a certain threshold, agents continue to adopt the high effort strategy (in which they expend high effort whenever their own stochastic productivity level is high) because they reasonably expect to meet a high output agent the next period. However, as soon as the population-wide distribution of outputs falls below a certain threshold, each agent assumes that the probability of meeting a high output agent is too low to justify expending high effort even when their own stochastic productivity outcome is high the next period. As soon as this threshold is breached, all agents expend low effort in the next period, leading to low effort remaining the adopted strategy for all periods thereafter. When the population size is large, the likelihood of the population-wide distribution of productivity levels being extreme enough to cause the distribution of outputs to fall below the threshold mentioned previously is low. Consequently, as the population size grows larger, the results of AE–P converge towards the continuum case.

Various papers in the multiple equilibrium literature have a similar result where one or more specific equilibria tend to be achieved based on best response dynamics [6], genetic algorithms [1, 3, 15], various forms of learning [8, 10, 16], etc.

As can be seen from the discussion on AE–P above, the effect of scale depends crucially on the modeling assumptions. While issues of dimensionality preclude the analysis of all SREE in models with more than four agents, as discussed earlier, the low SREE and the high SREE can be shown to exist regardless of the number of agents. In order to examine the effects of scale more explicitly, consider Fig. 2 which compares one sample run each (with different random evolutions of the stochastic processes across the different models) for 4 agents, 12 agents, and 100 agents. From the one run graphs in Fig. 2, it appears that scale matters a great deal for AE–P, whereas scale does not matter for SREE or AE–I, except in the sense that variance of outputs is predictably lower with more agents. This effect of scale is more clearly visible if we look at the ensemble averages presented in Fig. 3.[13] Scale matters to a very limited extent for AE–I (being visible only when zoomed into the short run), and a great deal for AE–P as we go from 4 agents to 100 agents. AE–P most closely resembles AE–I in a model with 4 agents but by the time the economy is comprised of 100 agents, AE–P starts to resemble SREE high. For the models with for 4 agents and 8 agents, AE–P starts with a high output per capita that settles eventually to the low SREE.

Based on the above results as well as the underlying theory, it can be seen that unlike the SREE versions, the AE models are path dependent; it is not just the productivity level in the last period that matters but also the specific sequence of past stochastic productivity levels and past random pairings. However, continuing with the terminology employed by Page [19], the AE models are equilibrium independent in the sense that the long-run equilibrium converges with SREE low. This happens fairly rapidly for AE–P with a small number of agents and AE–I in all situations. For AE–P with a large number of agents, this could take a very long time indeed but

---

[13] For all figures, the initial (zeroth) period is dropped from consideration.

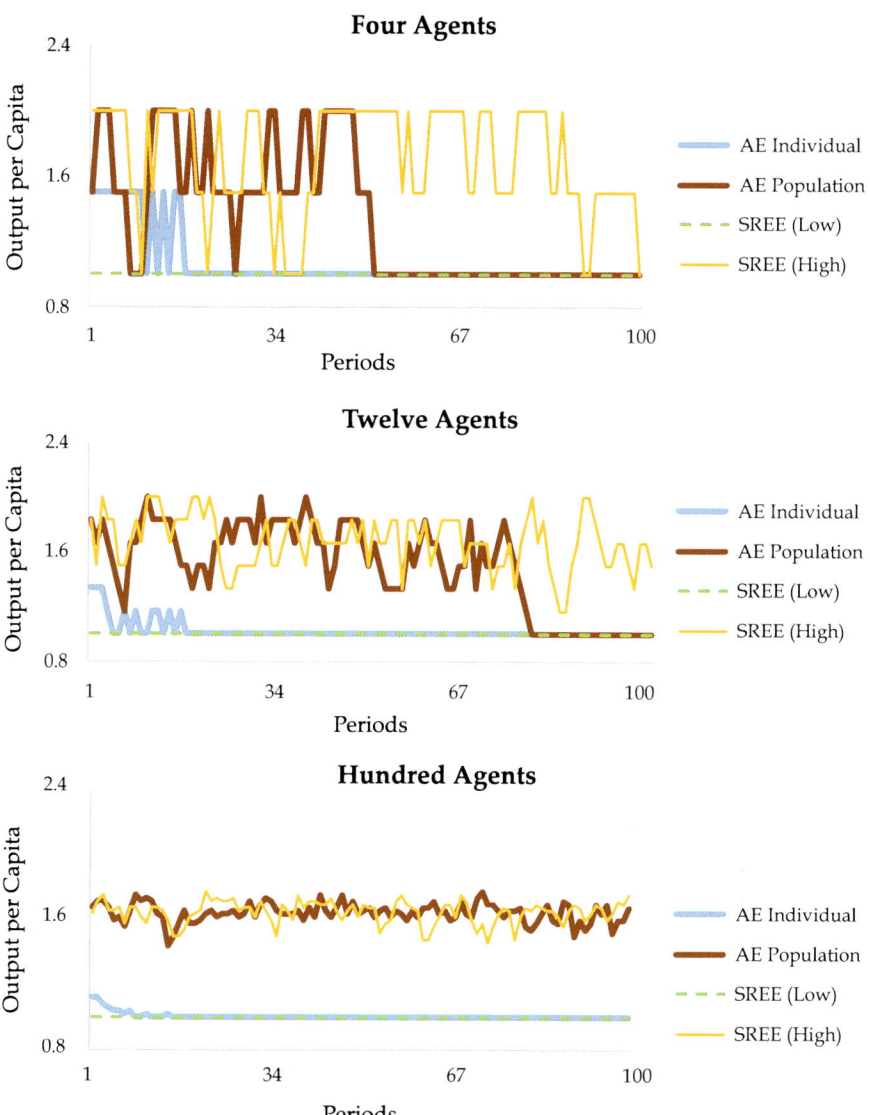

Fig. 2 Scale effects—single runs of models with 4 agents, 12 agents, and 100 agents

will indeed eventually occur.[14] Assuming SREE as a simplification is equivalent to ignoring short and intermediate run dynamics and those might matter in the real world where the long-run equilibrium could indeed be a long way away, as in the AE–P case with a large number of agents.

---

[14]It can be shown, for instance, that if all agents were to receive a low productivity level in any period (a scenario with an extremely low but still nonzero probability in a model with many agents) then everyone would switch to the low strategy in the next period.

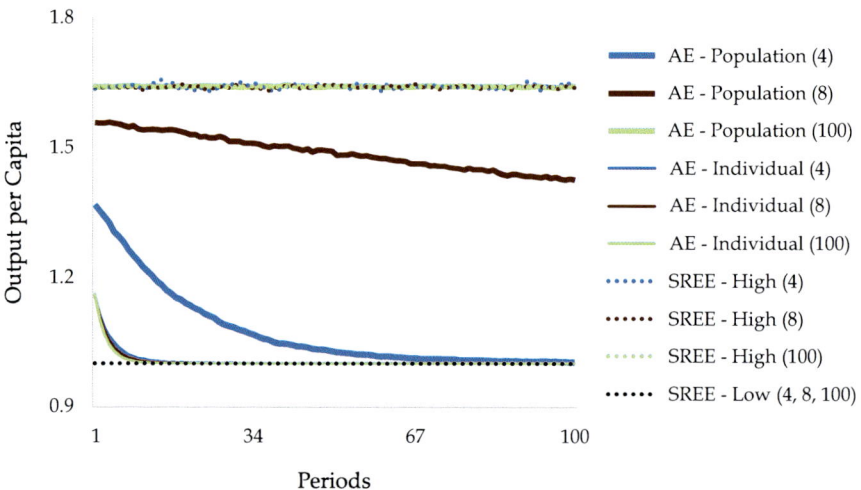

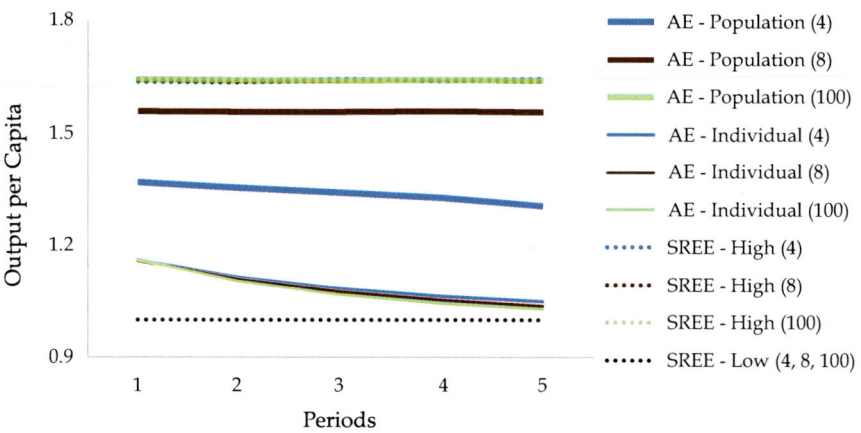

Fig. 3 Scale effects—ensemble means of models with 4 agents, 12 agents, and 100 agents

4 Further Discussion

In keeping with much of the literature surveyed in the introduction, the results from this paper suggest that the assumptions of homogeneity, REH, and Walrasian equilibria are far from innocuous. Even within the REH framework, assuming continuums of agents and implementing SREE restrictions could lead to focusing on a small subset of interesting results, even though those assumptions and restrictions are extremely helpful in terms of providing tractability. Furthermore, results obtained through REH are not path dependent and could feature multiple equilibria without a robust theoretical basis for choosing among them although various alternatives have been proposed in the literature. AE–I and AE–P, on the other hand, are path dependent and still eventually converge to certain SREE, thereby allowing more precise and testable analyses of trajectories in the economy through simulation-based event studies and Monte Carlo approaches. It should be noted that the AE models employed here have a limitation. These AE models assume that memory lasts only one period; changes in results caused by increasing memory could be addressed in further work.[15] Finally, SREE is scale free (although not in terms of variance of output) whereas AE–I and AE–P are not (AE–P, in particular, shows much greater sensitivity to scale) thereby allowing for richer models.

Overall, the exercises undertaken in this paper suggest that REE can indeed be implemented in agent-based models as a benchmark. However, even a small increase in the state space or the number of agents renders a fully specified REE computationally intractable. ABM researchers interested in realism may find REE benchmarks of limited applicability since modifying models to allow for tractable REE could involve imposing severe and debilitating limitations on the model in terms of the number of agents, their heterogeneity, the types of equilibria allowed, and other forms of complexities.

Acknowledgements I would like to thank anonymous reviewers for their invaluable suggestions. I would also like to acknowledge my appreciation for the helpful comments I received from the participants of the 20th Annual Workshop on the Economic Science with Heterogeneous Interacting Agents (WEHIA) and the 21st Computing in Economics and Finance Conference.

References

1. Arifovic, J. (1994). Genetic algorithm learning and the cobweb model. *Journal of Economic Dynamics and Control, 18*(1), 3–28.
2. Bryant, J. (1983). A simple rational expectations Keynes-type model. *Quarterly Journal of Economics, 98*(3), 525–528.
3. Chen, S.-H, John, D., & Yeh, C.-H. (2005). Equilibrium selection via adaptation: Using genetic programming to model learning in a coordination game. In A. Nowak & S. Krzysztof (Eds.), *Advances in dynamic games*. Annals of the international society of dynamic games (Vol. 7, pp. 571–598). Boston: Birkhäuser.
4. Colander, D. (2006). *Post Walrasian economics: Beyond the dynamic stochastic general equilibrium model*. New York: Cambridge University Press.
5. Cooper, R. (1994). Equilibrium selection in imperfectly competitive economies with multiple equilibria. *Economic Journal, 104*(426), 1106–1122.

---

[15]Intuitively, with more memory, AE–I and AE–P may sustain the high effort equilibrium for longer because individuals would take into consideration more periods with high outputs while making decisions in subsequent periods. For instance, even in the period right after an economy-wide recession, if agents possessed longer memories, they could revert to expending high effort the next period if their own individual stochastic productivity level was high.

6. Cooper, R., & Andrew, J. (1988). Coordinating coordination failures in Keynesian models. *Quarterly Journal of Economics, 103*(3), 441–463.
7. Diamond, P. (1982). Aggregate demand management in search equilibrium. *Journal of Political Economy, 90*(5), 881–894.
8. Evans, G., & Seppo, H. (2013). Learning as a rational foundation for macroeconomics and finance. In R. Frydman & P. Edmund (Eds.), *Rethinking expectations: The way forward for macroeconomics* (pp. 68–111). Princeton and Oxford: Princeton University Press.
9. Fagiolo, G., & Andrea, R. (2012). Macroeconomic policy in DSGE and agent-based models. *Rev l'OFCE, 124*(5), 67–116.
10. Frydman, R. (1982). Towards an understanding of market processes: Individual expectations, learning, and convergence to rational expectations equilibrium. *The American Economic Review, 72*(4), 652–668.
11. Frydman, R., & Edmund, P. (2013). Which way forward for macroeconomics and policy analysis? In R. Frydman & P. Edmund (Eds.), *Rethinking expectations: the way forward for macroeconomics* (pp. 1–46). Princeton and Oxford: Princeton University Press.
12. Howitt, P., & Robert, C. (2000). The emergence of economic organization. *Journal of Economic Behavior and Organization, 41*(1), 55–84
13. Jackson, J., & Ken, K. (2012). Modeling, measuring, and distinguishing path dependence, outcome dependence, and outcome independence. *Political Analysis, 20*(2), 157–174.
14. Lengnick, M. (2013). Agent-based macroeconomics: A baseline model. *Journal of Economic Behavior and Organization, 86*, 102–120.
15. Marimon, R., Ellen, M., & Thomas, S. (1990). Money as a medium of exchange in an economy with artificially intelligent agents. *Journal of Economic Dynamics and Control, 14*(2), 329–373.
16. Milgrom, P., & John, R. (1990). Rationalizability, learning, and equilibrium in games with strategic complementarities. *Econometrica, 58*(6), 1255–1277.
17. Napoletano, M., Jean Luc, G., & Zakaria, B. (2012). *Agent Based Models: A New Tool for Economic and Policy Analysis*. Briefing Paper 3 OFCE Sciences Po, Paris.
18. Oeffner, M. (2008). *Agent-Based Keynesian Macroeconomics: An Evolutionary Model Embedded in an Agent-Based Computer Simulation*. PhD Thesis, Julius-Maximilians-Universität, Würzburg.
19. Page, S. (2006). Path dependence. *Quarterly Journal of Political Science, 1*(1), 87–115.
20. Plosser, C. (1989). Understanding real business cycles. *The Journal of Economic Perspectives, 3*(3), 51–77.
21. Stiglitz, J., & Mauro, G. (2011). Heterogeneous interacting agent models for understanding monetary economies. *Eastern Economic Journal, 37*(1), 6–12.
22. Tesfatsion, L. (2006). Agent-based computational modeling and macroeconomics. In D. Colander (Ed.), *Post Walrasian economics: Beyond the dynamic stochastic general equilibrium model* (pp. 175–202). New York: Cambridge University Press.
23. Vriend, N. (2000). An illustration of the essential difference between individual and social learning, and its consequences for computational analyses. *Journal of Economic Dynamics and Control, 14*(1), 1–19.

Does Persistent Learning or Limited Information Matter in Forward Premium Puzzle?

Ya-Chi Lin

Abstract Some literature explains the forward premium puzzle by the learning process of agents' behavior parameters, and this work argues that their conclusions may not be convincing. This study extends their model to the limited information case, resulting in several interesting findings: First, the puzzle happens when the proportion of full information agents is small, even people make expectation near rationally. Second, allowing the proportion of full information agents to be endogenous and highly relied on the performance of forecasting, agents turn to become full information immediately, and the puzzle disappears. These results are similar in different learning gain parameters. Our finding shows that limited information would be more important than learning, when explaining forward premium puzzle. Third, the multi-period test of Fama equation is also examined by the exchange rate simulated by learning in the limited information case. The Fama coefficients are positive, and the puzzle will not remain. It is consistent with the stylized facts in the multi-period version of Fama regression, which is found in McCallum (J Monet Econ 33(1):105–132, 1994). Finally, we also find that if agents rely on the recent data too much when forecasting, they tend to overreact on their arbitrage behavior. The Fama coefficient deviates further from unity. People might not benefit from having more information.

Keywords Limited information · Persistent learning · Market efficiency · Forward premium puzzle · Dual learning

---

Y.-C. Lin (✉)
Hubei University of Economics, Wuhan, China
e-mail: yachi.lin@hbue.edu.cn

© Springer Nature Switzerland AG 2018
S.-H. Chen et al. (eds.), *Complex Systems Modeling and Simulation in Economics and Finance*, Springer Proceedings in Complexity,
https://doi.org/10.1007/978-3-319-99624-0_8

1 Introduction

The forward premium puzzle is a long-standing paradox in international finance. Foreign exchange market efficiency is in a status that exchange rates fully reflect all available information. Unexploited excess profit opportunity does not exist when market efficiency is hold. Therefore, the forward exchange rate should be the best predictor of the future spot exchange rate. However, most empirical researches have the opposite findings. The slope coefficient in a regression of the future spot rate change on the foreign premium is significantly negative, which is expected to be unity if market is efficient. If s_t is the natural log of the current spot exchange rate (defined as the domestic price of foreign exchange), Δs_{t+1} is the depreciation of the natural log of the spot exchange rate from period t to $t+1$, i.e., $\Delta s_{t+1} = s_{t+1} - s_t$, and f_t is the natural log of the one-period forward rate at period t. The forward premium puzzle is examined by the following Fama equation:

$$\Delta s_{t+1} = \hat{\beta}(f_t - s_t) + \hat{u}_{t+1} \tag{1}$$

The Fama coefficient $\hat{\beta}$ is unity if the efficient market hypothesis holds. However, in the majority of researches, $\hat{\beta}$ is negative. It is what we called "forward premium puzzle."

Examining market efficiency by regressing future spot rate change on the forward premium is based on the assumption that agents are rational and risk neutral. The rejection of market efficiency has the following explanations. First, if agents are risk averse, the forward exchange rate contains a risk premium. Hodrick [13] and Engel [9] apply Lucas asset pricing model to price forward foreign exchange risk premium, which shows future spot rate deviating from forward rate. Second, agents may have incomplete knowledge about the underlying economic environment. During the transitional period, they can only guess the forward exchange rates by averaging several spot exchange rates that may possibly happen. This makes systematic prediction errors even though they are behaving rational. It is what we called "peso problem," which is originally studied by Krasker [15]. Lewis [16] assumes that agents update their beliefs about the regime shifts in fundamentals by Bayesian learning. During the learning period, the forecast errors are systematic and serially correlated. Motivated by the fact that even today only a tiny fraction of foreign currency holdings are actively managed, [1] assume that agents may not incorporate all information in their portfolio decisions. Most investors do not find it in their interest to actively manage their foreign exchange positions since the resulting welfare gain does not outweigh the information processing cost. It leads to a negative correlation between exchange rate change and forward premium for five to ten quarters. The puzzle disappears over longer horizons. Scholl and Uhlig [23] consider agents are Bayesian investors, who trades contingent on a monetary policy shock, and uses Bayesian VAR to assess the posterior uncertainty regarding the resulting forward discount premium. Forward discount premium diverges for several countries even without delayed overshooting. Forward discount puzzle seems to be robust.

The third explanation is that agents behave far from rationality. Agents have difficulty to distinguish between pseudo-signals and news, and then they take actions on these pseudo-signals. The resulting trading dynamics produce transitory deviations of the exchange rate from fundamental values [20]. Agents' prospective spot exchange rates are distorted by waves of excessive optimism and pessimism. Black [3] calls these kinds of agents "noise traders." Literatures explain forward premium puzzle by irrational traders are as below. Gourinchas and Tornell [12] suppose that investors are adaptive expectations and misperceive interest rate shocks as transitory. In the following period, they find that interest rate turns out to be higher than they first expected, which leads them to revise upward their beliefs about the persistence of the original interest rate shock and triggers a further appreciation of the dollar. Therefore, gradual appreciation is accompanied with increases in interest rates, and also forward premium. Chakraborty and Evans [8] replace rational expectations by perpetual learning, which assumes that agents in the economy face some limitations on knowledge in the economy. Agents do not themselves know the parameter values and must estimate them econometrically. They adjust their forecast for the parameter values as new data become available over time. The model generates a negative bias, and the bias becomes stronger when fundamentals are strongly persistent or the learning gain parameter is larger.

Lewis [16] and Gourinchas and Tornell [12] assume that the unexpected shift in fundamentals is initially unsure for agents, or is misperceived as temporary. Agents learn about the changes in fundamentals but learning takes time. Based on such model setting, the $\hat{\beta}$ in Eq. (1) could be less than unity. On the other hand, [8] assume that the fundamentals are observable by agents. The thing that unknown by agents is the behavior parameters, which is adaptively learned by agents over time.

In this paper, we investigate whether forward premium puzzle remains under the interaction of limited information flows and adaptive learning. There are three reasons to examine forward premium puzzle under limited information flows. First, although [8] provide the numerical evidences of forward premium puzzle under adaptive learning, the puzzle is not discovered in the theoretical model. Refer to Fig. 1, the Fama coefficient decreases in gain parameter, but it never turns to be negative in the reasonable gain parameter range from 0 to 1, despite the value of fundamental persistence. Second, the negative bias provided in the numerical exercise decreases as the sample size increases, refer to Table 3 in [8]. The numerical results are correlated to the sample size. Third, the analysis of [8] is based on full information, and the simulated exchange rate always reflects full market information. Therefore, they purely explain the impact of learning process on forward premium puzzle, instead of the foreign exchange market efficiency. The full information in [8] does not mean complete information since the adaptive learning model implies model uncertainty, the imperceptibility of behavior parameters to agents.

The limited information flows and learning model derive several interesting findings. First, the simulated $\hat{\beta}$ is negative when the proportion of full information agents is small, even in small learning parameter and rational expectation. It implies that the effect of limited information on asymptotic bias dominates that of persistent

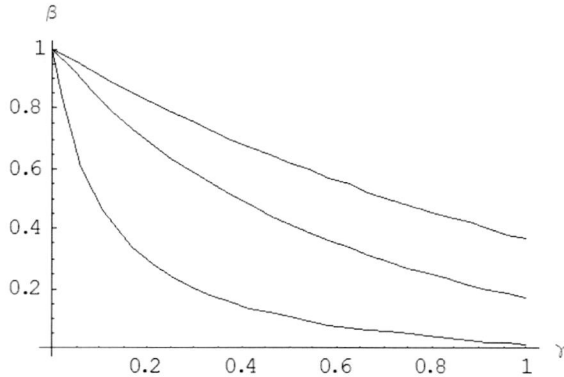

Fig. 1 Theoretical plim$\hat{\beta}$ in [8] for $\theta = 0.9$ and $\rho = 0.9, 0.95, 0.99$

learning. Second, the limited information could not only explain forward premium puzzle, it is consistent with the stylized facts, which is found in [21]. The puzzle will not remain in the multi-period test of Fama equation. Third, agents are free to choose the full or limited information when forecasting. Then the proportion of full information agents is endogenous, which is influenced by the information or switching costs. The learning in forecasting rule is accompanied with the learning in behavior parameters, which is called "dual learning."İ When agents react strongly to the relative performance of the forecasting rule by using full or limited information, all agents switch to apply full information quickly. The proportion of full information agents approaches to unity in a short period, and market efficiency is hold. This result is similar for different gain parameters. Forward premium puzzle happens when agents are uninformed. Finally, it is possible for agents to choose the limited information forecasting rule rather than full information one. If agents rely on the recent data too much, they tend to overreact on their arbitrage behavior. Then agents might not benefit from too much information.

Our study is organized as follows. In Sect. 2 we present the impact of limited information and learning on forward premium puzzle. The simulation is applied by assuming the proportion of full information agents is exogenous. Section 3 introduces the dual learning into the model and investigates the forward premium puzzle by simulation. Finally, Sect. 4 concludes.

2 The Theoretical Model

2.1 Monetary Model

To show the impact of limited information and adaptive learning on forward exchange market efficiency, we use a simple monetary exchange rate model based on purchasing power parity and uncovered interest parity.

$$s_t = p_t - p_t^* \tag{2}$$

$$i_t = i_t^* + E_t s_{t+1} - s_t \tag{3}$$

where p_t is the log price level, and $E_t s_{t+1}$ is the log of the forward rate at time t for foreign currency at $t+1$. The variables with a star represent foreign variables. The money market equilibrium in the home country is

$$m_t - p_t = c_1 y_t + c_2 i_t \tag{4}$$

where m_t is the log money stock and y_t is log real output. The money market equilibrium is also held in the foreign country with identical parameters.

Combining the equations above, the reduced form of exchange rate can be solved.

$$s_t = \theta E_t s_{t+1} + v_t \tag{5}$$

where $\theta = c_2/(1-c_2)$, and $v_t = (1-\theta)((m_t - m_t^*) - c_1(y_t - y_t^*))$. Suppose that v_t follows a stationary first-order autoregressive process,

$$v_t = \rho v_{t-1} + \varepsilon_t \tag{6}$$

where $\varepsilon_t \sim N(0, \sigma_\varepsilon^2)$, and $0 \leq \rho < 1$. By method of undetermined coefficients, the rational expectation solution to this model is

$$s_t = \bar{b} v_{t-1} + \bar{c} \varepsilon_t \tag{7}$$

where $\bar{b} = \rho(1-\rho\theta)^{-1}$ and $\bar{c} = (1-\rho\theta)^{-1}$. Here risk neutrality is assumed, thus the forward exchange rate equals to the market expectation of s_{t+1} held at time t.

$$f_t = E_t s_{t+1} \tag{8}$$

Examining forward premium puzzle by using the exchange rate generated from the rational expectation solution above, the puzzle does not exist.[1] However, [8] show that when replacing rational expectations by constant gain learning, a negative bias exists in $\hat{\beta}$. The result coincides with the concept of adaptive market hypothesis, as proposed by Lo [17]. Even in an ideal world of frictionless markets and costless trading, market efficiency still fails because of the violation of rationality in the agents' behavior, such as overconfidence, overreaction, and other behavioral biases. Overconfident individuals overreact to their information about future inflation. The forward rate will have larger response than the spot rate because that the forward rate is more driven by speculation and the spot rate is the consequence of transaction

---

[1] From the rational expectation solution, we could derive the following relationships, $cov(\Delta s_{t+1}, f_t - s_t) = var(f_t - s_t) = (1-\rho)\sigma_\varepsilon^2/[(1+\rho)(1-\rho\theta)^2]$. Therefore, $\hat{\beta} = 1$, and the puzzle does not exist.

demand for money [7]. Therefore the slope coefficient of the spot return on the forward premium turns to below unity but still positive, as shown in Fig. 1.

Based on the result of [8], $\hat{\beta}$ are negative only in the numerical simulation, but it is always positive in the theoretical model. The forward premium puzzle is not discovered theoretically, which represents that the puzzle may not be explained thoroughly by simply adaptive learning. This study combines adaptive learning with limited information flows, and we attempt to explain the puzzle by these two factors.

2.2 Limited Information Flow Model

The limited information flow model in our study is constructed by simplifying the sticky information model in [18].[2]

Define I_{t-j} as the information set consists of all explanatory variables dated $t-j$ or earlier. We first assume that the latest information is not available for all agents, but the explanatory variables dated $t-1$ or earlier are public information. This assumption conforms to the economic reality that there may be information costs or barriers when receiving the latest information. Once the latest information becomes history and is announced, it is available to all agents. We also assume that there exist two groups of agents, $i = 1, 2$. The first group make forecast with full information, i.e., $j = 0$. The second group is blocked from the latest information, and they could only make forecast by the past one, i.e., $j = 1$.[3]

$$s^e_{1,t+1} = E_1(s_{t+1}|I_t) = b_{t-1}v_t \tag{9}$$

$$s^e_{2,t+1} = E_2(s_{t+1}|I_t) = b_{t-2}E_2(v_t|I_{t-1}) = b_{t-2}\rho_{t-2}v_{t-1} \tag{10}$$

where b_{t-j} and ρ_{t-j} are estimates based on information up to time t-j.

At the beginning of time t, agents with full information have estimates b_{t-1} of the coefficients b, based on information through $t-1$. Together with the observed fundamentals v_t, agents make the forecasting exchange rates for the next periods. For agents not updating the information sets at time t, the forecasting exchange rate is equivalent to the two-period forecasting at time $t-1$. The forecasting formula is referred to [4] and [19]. The forecasting exchange rate for the next period is formed by the behavior parameter b_{t-2} multiplying the expected fundamentals based on I_{t-1}.

---

[2]Mankiw and Reis [18] assume that an exogenous proportion κ of agents will update the information sets in each period. Then, at each t there are κ percent of agents with $E(s_{t+1}|I_t)$, $\kappa(1-\kappa)$ percent of agents with $E(s_{t+1}|I_{t-1})$, $\kappa(1-\kappa)^2$ percent of agents with $E(s_{t+1}|I_{t-2})$, and so on. Thus the mean forecast is $\kappa \sum_{j=0}^{\infty}(1-\kappa)^j E(s_{t+1}|I_{t-j})$.

[3]Full information in this setting is not complete information since the behavioral parameters are unknown to agents.

The market forecast $E_t(s_{t+1})$ is a weighted average of forecasts for these two kinds of agents. For simplicity, we assume that the proportion of agents using full information is an exogenous n in every period.

$$E_t s_{t+1} = n s^e_{1,t+1} + (1-n) s^e_{2,t+1} \tag{11}$$

The exchange rate is obtained by substituting the market forecast in Eq. (5).

$$s_t = \theta(n b_{t-1} v_t + (1-n) b_{t-2} \rho_{t-2} v_{t-1}) + v_t \tag{12}$$

Although agents know the structural form relationship between the fundamentals and the exchange rate, the parameter values are imperceptible to them. The parameter values b should be surmised by regressing s_t on v_{t-1} from the observed data. Likewise, the parameter values ρ should be surmised by regressing v_t on v_{t-1} from the observed data. We introduce constant gain learning into the model. The learning process obeys

$$b_t = b_{t-1} + \gamma (s_t - b_{t-1} v_{t-1}) v_{t-1} R^{-1}_{t-1} = b_{t-1} + \gamma Q(t, b_{t-1}, v_{t-1})$$
$$\rho_t = \rho_{t-1} + \gamma (v_t - \rho_{t-1} v_{t-1}) v_{t-1} R^{-1}_{t-1}$$
$$R_t = R_{t-1} + \gamma (v_t^2 - R_{t-1}) \tag{13}$$

where $\gamma > 0$ is the learning gain parameter. R_t could be viewed as an estimate of the second moment of the fundamentals. The gain parameter reflects the trade-off between tracking and filtering, and it is set to be $1/t$ under decreasing gain learning. The decreasing gain learning b will converge to $\bar{b}$ with probability one. Under constant gain learning, agents weight recent data more heavily than past data and the gain parameter is therefore set to be a constant. The parameter b_t is allowed to vary over time, which in turn allows for potential structural change taking an unknown form. As suggested by Orphanides and Williams [22], the model with the gain parameter in a range of $\gamma = 0.02 \sim 0.05$ will provide a better forecasting performance for GDP growth and inflation. Based on the above model setting, we discuss the impact of persistent learning and information flow on market efficiency.

2.3 Forward Exchange Market Efficiency

In this subsection we would like to show how learning process and limited information model affect the relationship between forward exchange rates and future spot rates. The tests of market efficiency are usually based on Eq. (1), corresponding to Fama's definition. Fama [11] summarizes the idea saying that a market is called efficiency, when prices fully reflect all available information. Therefore under market efficiency, the only reason for the price changes is the arrival of news or unanticipated events. It also implies that forecast errors are unpredictable on the basis of any information that is available at the time forecast is made. According to the above definition, market efficiency could be represented as below if agents are rational expectation and risk neutrality.

$$s_{t+1} = E_t s_{t+1} + u_{t+1} \tag{14}$$

u_{t+1} is the news in exchange rate markets, randomly distributed and expected to be zero at time t. In other words, it is impossible to make superior profits by trading on the basis of information that is available at time t if market is efficiency. From Eqs. (5) to (11), the time $t+1$ adaptive learning forecast error is

$$\begin{aligned} s_{t+1} - E_t s_{t+1} &= \theta E_{t+1} s_{t+2} + v_{t+1} - [n s^e_{1,t+1} + (1-n) s^e_{2,t+1}] \\ &= (1 + \theta n b_t)\epsilon_{t+1} - (1-n) b_{t-2} \rho_{t-2} v_{t-1} \\ &\quad + \{\rho - (1 - \rho\theta) b_t + (1-n)[(1 - \rho\theta) b_t \\ &\quad + \theta b_{t-1} \rho_t] + n\gamma Q(t, b_{t-1}, v_{t-1})\} v_t \end{aligned} \tag{15}$$

According to Fama's definition, efficient market hypothesis is supported if the expected forecast error based on information available at time t is equal to zero. In the case that all agents have full information ($n = 1$), Eq. (15) is rewritten as below.

$$s_{t+1} - E_t s_{t+1} = [\rho - (1 - \rho\theta) b_t + \gamma Q(t, b_{t-1}, v_{t-1})] v_t + (\theta b_t + 1)\varepsilon_{t+1} \tag{16}$$

If agents follow rational expectations, b_t is equal to $\bar{b} = \rho/(1 - \rho\theta)$ in every period, and the gain parameter $\gamma \to 0$. The coefficient of v_t is asymptotically equal to zero. With rational expectation, agents' forecast error cannot be predicted by fundamental before current period, $v_{t-1}, v_{t-2}, \ldots$. However, agents in real lives usually make forecast by adaptive learning ($\gamma > 0$), which will induce systematic forecast errors. The failure of market efficiency could be ascribed to adaptive learning under full information. In the case of limited information flows ($n < 1$), the forecast error has systematic components because the coefficients on v_t and v_{t-1} are not zero. There are two channels that could affect the result of market efficiency, the proportion of agents with limited information, $1 - n$, and adaptive learning, γ. The effect of adaptive learning on the expected forecast error depends on parameters ρ, θ, n. Therefore, the effect of adaptive learning on the expected forecast error is not monotonic under limited information flows, which is contradict with the findings in [8] (as shown in Appendix 1). We also find that the expected forecast error is not equal to zero even when agents are rational and information flow is limited.

2.4 Numerical Calibration

In this section, we simulate the exchange rate under limited information flow model with adaptive learning, given $\bar{b}$ as the initial parameter. To find out the impact of limited information flows on the results of forward premium puzzle, the proportion of agents using full information, n, is assumed to be exogenous and constant. The exchange rates are simulated based on various n tested by Eq. (1), Fama regression without intercept.

In the beginning of the simulation, we adopt the parameter setting in [8] and compare the results with them. The results are shown in Table 1. In the case of full information, $n = 1$, the simulated $\hat{\beta}$ are positive in small gain parameters. The simulated $\hat{\beta}$ are decreasing to negative as the gain parameter increases. This result coincides with the findings in [8]. When the spot and forward exchange rates derived from near unit root process leads the simulated slope coefficient of the spot return on the forward premium to be negative. The statistical explanation is that there exists long memory behavior of the conditional variance in forward premium. Although $\hat{\beta}$ is centered around unity, but it is widely dispersed and converge very slowly to its true value of unity. A relatively large number of observations is required to provide reliable estimates [2, 14].

Table 1 Simulated $\hat{\beta}$ and $t_{\hat{\beta}}$ under persistent learning and limited information by using the parameter settings in [8]

γ	$n = 0.1$	$n = 0.4$	$n = 0.7$	$n = 1$
(a) $\theta = 0.9$, $\rho = 0.99$				
RE	−5.45 (−23.17)	−3.57 (−10.24)	−0.95 (−1.76)	1.29 (1.56)
0.02	−6.22 (−27.70)	−4.59 (−12.50)	−1.69 (−2.73)	1.03 (1.15)
0.03	−6.65 (−30.72)	−4.84 (−13.13)	−2.12 (−3.23)	0.77 (0.81)
0.04	−6.75 (−31.95)	−5.09 (−13.82)	−2.35 (−3.68)	0.39 (0.45)
0.05	−6.71 (−31.67)	−5.27 (−14.37)	−2.79 (−4.03)	0.15 (0.18)
0.10	−6.21 (−26.87)	−4.64 (−13.62)	−2.44 (−4.56)	−0.81 (−1.02)
(b) $\theta = 0.6$, $\rho = 0.99$				
γ	$n = 0.1$	$n = 0.4$	$n = 0.7$	$n = 1$
RE	−1.35 (−19.26)	−1.23 (−8.96)	−0.81 (−2.67)	1.20 (1.48)
0.02	−1.40 (−20.61)	−1.31 (−9.75)	−1.08 (−3.55)	0.23 (0.24)
0.03	−1.39 (−20.67)	−1.30 (−9.86)	−1.08 (−3.70)	−0.14 (−0.18)
0.04	−1.38 (−20.59)	−1.31 (−9.96)	−1.05 (−3.74)	−0.34 (−0.52)
0.05	−1.36 (−20.21)	−1.27 (−9.94)	−1.05 (−3.92)	−0.32 (−0.56)
0.10	−1.28 (−19.24)	−1.15 (−9.67)	−0.85 (−3.91)	−0.39 (−1.01)

Note: Results from 1000 simulations with sample size equal to 360 after discarding the first 20,000 data points. Table gives medians of $\hat{\beta}$ and $t_{\hat{\beta}}$ for testing $H_0: \beta = 0$, without intercept in the Fama regression. The medians of $t_{\hat{\beta}}$ are shown in parentheses

Table 2 Simulated Fama coefficient in multi-period version under persistent learning and limited information

γ	$n = 0.1$	$n = 0.4$	$n = 0.7$	$n = 1$
(a) $\theta = 0.9$, $\rho = 0.99$				
RE	1.14 (19.02)	1.53 (28.92)	1.36 (26.10)	1.00 (18.95)
0.02	1.14 (19.12)	1.54 (28.88)	1.38 (26.42)	1.02 (19.25)
0.03	1.14 (19.08)	1.54 (29.01)	1.39 (26.68)	1.03 (19.44)
0.04	1.14 (19.09)	1.54 (29.08)	1.39 (26.68)	1.03 (19.50)
0.05	1.14 (19.06)	1.54 (28.95)	1.40 (26.91)	1.05 (19.74)
0.10	1.12 (18.81)	1.52 (28.47)	1.42 (27.48)	1.08 (20.35)
(b) $\theta = 0.6$, $\rho = 0.99$				
RE	1.30 (13.44)	1.81 (24.39)	1.39 (23.21)	1.00 (18.98)
0.02	1.29 (13.26)	1.83 (24.22)	1.41 (23.15)	1.02 (19.08)
0.03	1.28 (13.17)	1.82 (23.93)	1.41 (23.06)	1.02 (19.10)
0.04	1.27 (13.18)	1.80 (23.61)	1.41 (22.97)	1.03 (19.22)
0.05	1.25 (13.06)	1.77 (23.15)	1.41 (23.01)	1.03 (19.23)
0.10	1.16 (12.61)	1.66 (21.51)	1.40 (22.51)	1.05 (19.40)

$s_{t+1} - s_{t-1} = \hat{\beta}(f_t - s_{t-1}) + \hat{u}_{t+1}$

Note: Results from 1000 simulations with sample size equal to 360 after discarding the first 20,000 data points. Table gives medians of $\hat{\beta}$ and $t_{\hat{\beta}}$ for testing $H_0: \beta = 0$, without intercept in the Fama regression. The medians of $t_{\hat{\beta}}$ are shown in parentheses

In the case of limited information, $n < 1$, the simulated $\hat{\beta}$ are negative even when gain parameters are small. Therefore, it may not be possible to explain forward premium puzzle by gain parameter when information flow is limited. In the limited information case of $\theta = 0.6$ and $\rho = 0.99$, the simulated $\hat{\beta}$ is increasing in the gain parameter γ, contrary to the case of full information. The impact of gain parameter on the simulated $\hat{\beta}$ is not monotonic when considering limited information.

McCallum [21] argues that the forward premium puzzle does not exist in another form of the test that in a multi-period version. A multi-period version for Eq. (1) is as below:

$$s_{t+1} - s_{t+1-j} = \hat{\beta}(f_t - s_{t+1-j}) + \hat{u}_{t+1}, j = 2, 3, \ldots. \quad (17)$$

McCallum [21] finds that $\hat{\beta}$ is close to one if $j \geq 2$, which coincides with the theory. As shown in Table 2, our simulated estimates support McCallum's findings. The simulated $\hat{\beta}$ are close to unity and significantly greater than zero. Although the forward premium puzzle resulted from persistent learning or limited information in a single-period version, the puzzle will disappear when testing in a multi-period version.

This section integrates the concept of limited information flow with the model of [8]. We find that when information flow is restricted, the impact of gain parameter on Fama coefficient is small relative to the impact of the proportion of full information agents on Fama coefficient. The innocence of behavior parameter may not be the critical reason under limited information for forward premium puzzle, while the innocence of the current information about fundamentals might be the main point.

3 Dual Learning

The proportion of agents with full information n is assumed to be exogenous in prior section. Here n is allowed to vary over time. This section introduces dual learning into the model: parameter learning and dynamic predictor selection. The market forecast is a weighted average of the two forecasts, and the proportion of agents is time varying.

$$E_t s_{t+1} = n_{t-1} s^e_{1,t+1} + (1 - n_{t-1}) s^e_{2,t+1} \tag{18}$$

Each period, agents could choose to use the first or second forecasting rule. This choice is based on the models' forecasting performance in each period. Following [5], agents update their mean square error (MSE) estimates according to a weighted least squares procedure with geometrically decreasing weights on past observations.

$$\text{MSE}_{i,t} = \text{MSE}_{i,t-1} + \gamma[(s_t - s^e_{i,t}) - \text{MSE}_{i,t-1}] \tag{19}$$

Refer to [6], predictor proportions are determined according to the following discrete choice formula:

$$n_t = \frac{exp(-\alpha \text{MSE}_{1,t})}{exp(-\alpha \text{MSE}_{1,t}) + exp(-\alpha \text{MSE}_{2,t})} \tag{20}$$

where α measures the sensitivity of agents reacting to the relative performance of the two forecasting rules. The value of α is correlated to the switching barriers on choosing forecasting rules, such as information costs or switching costs. When the switching parameter α equals to zero, agents do not react to any changes of the relative performance and their weight are sticky at 0.5. As the switching parameter approaches to infinity, agents are sensitive and switch the forecasting rule instantaneously. The proportion of full information agents then become either 0 or 1.

The dual learning algorithm is given by a loop over Eqs. (5), (9), (10), (13), (18), (19), and (20). In period t, agents in first group make exchange rate forecast $s^e_{1,t+1}$ by predetermined parameter estimates b_{t-1} and the current observed fundamentals v_t. Since agents in the second group are restricted to the latest information, the exchange rate forecast $s^e_{2,t+1}$ is made by predetermined parameter estimates b_{t-2}, ρ_{t-2} and the fundamentals v_{t-1}, which is observed in the last period. The market expectation is formed as a weighted average of the forecast for group 1 and 2, with predetermined weights n_{t-1} and $(1 - n_{t-1})$. When agent observe s_t, they begin to update their belief parameters b_t, ρ_t, and $\text{MSE}_{i,t}$. The relative performance of

Table 3 Simulated $\hat{\beta}$ and $t_{\hat{\beta}}$ by dual learning

		$\alpha = 0.01$	$\alpha = 0.05$	$\alpha = 0.1$	$\alpha = 500$
(a) $\gamma = 0.02$					
$\bar{n}$		0.67	0.94	0.98	1
$\hat{\beta}$	$j=1$	−1.94 (−3.59)	0.07 (0.10)	0.19 (0.28)	0.35 (0.47)
$\hat{\beta}$	$j=2$	1.41 (26.75)	1.07 (19.99)	1.04 (19.46)	1.02 (19.18)
(b) $\gamma = 0.03$					
$\bar{n}$		0.68	0.95	0.98	1
$\hat{\beta}$	$j=1$	−1.95 (−3.53)	0.02 (0.03)	0.10 (0.15)	0.18 (0.22)
$\hat{\beta}$	$j=2$	1.40 (26.46)	1.07 (19.99)	1.05 (19.52)	1.03 (19.32)
(c) $\gamma = 0.04$					
$\bar{n}$		0.69	0.95	0.98	1
$\hat{\beta}$	$j=1$	−1.87 (−3.58)	−0.24 (−0.32)	−0.14 (−0.17)	−0.09 (−0.15)
$\hat{\beta}$	$j=2$	1.39 (26.27)	1.08 (20.01)	1.05 (19.62)	1.04 (19.43)
(d) $\gamma = 0.05$					
$\bar{n}$		0.69	0.95	0.98	1
$\hat{\beta}$	$j=1$	−1.79 (−3.52)	−0.32 (−0.46)	−0.28 (−0.37)	−0.23 (−0.29)
$\hat{\beta}$	$j=2$	1.40 (26.22)	1.09 (20.14)	1.06 (19.7)	1.05 (19.58)
(e) $\gamma = 0.1$					
$\bar{n}$		0.69	0.93	0.98	0.98
$\hat{\beta}$	$j=1$	−1.60 (−3.7)	−0.74 (−1.30)	−0.75 (−1.17)	−0.69 (−1.07)
$\hat{\beta}$	$j=2$	1.40 (26.16)	1.13 (20.87)	1.10 (20.38)	1.09 (20.23)

$s_{t+1} - s_{t+1-j} = \hat{\beta}(f_t - s_{t+1-j}) + \hat{u}_{t+1}, j = 1, 2$
Note: Results from 1000 simulations with sample size equal to 360. The simulation is based on the parameter assumption of $\theta = 0.9$ and $\rho = 0.99$. Table gives medians of $\hat{\beta}$ and $t_{\hat{\beta}}$ for testing $H_0: \beta = 0$, with intercept in the Fama regression. The medians of $t_{\hat{\beta}}$ are shown in parentheses

forecasting rule is used to determine the new predictor proportions n_t for the next period. All initial values of the parameters are set by the rational expectation values. The initial weight on model 1 is drawn from a standard uniform distribution $U(0, 1)$.

The median Fama coefficients estimated by simulated exchange rates are shown in Table 3. The result in Table 3 is simulated by the parameter assumptions of $\theta = 0.9$ and $\rho = 0.99$. For given γ, the average proportion for the first forecasting rule over the period, $\bar{n}$, increases in α, the sensitivity of the relative performance to the predictors. If agents react more strongly to the relative performance of the forecasting rules, they tend to forecast by using full information. Figure 2 plots the cross-section median of the 1000 simulations for the proportion of the first forecast rule n_t. When α equals to 0.01, n_t increases in relatively slow speed. While it switches to unity immediately when α is larger than 500. The gain parameter γ also influences the time path of n_t. Agents rely more on recent data to update their belief parameter when γ is larger. The time path of n_t will converge to unity in shorter time in larger gain parameter. Highly dependent on the recent data to forecast may also lead to overreaction to the market news, the time path of n_t is more volatile in

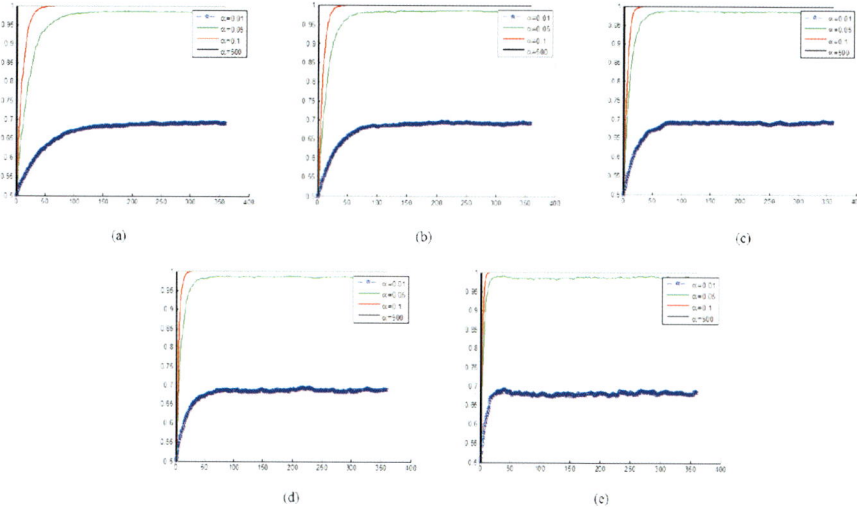

Fig. 2 Weights on full information under dual learning for $\theta = 0.9$ and $\rho = 0.99$. (**a**) $\gamma = 0.02$. (**b**) $\gamma = 0.03$. (**c**) $\gamma = 0.04$. (**d**) $\gamma = 0.05$. (**e**) $\gamma = 0.1$

larger gain parameters. The situation might happen when agents are overconfident on the recent data, they tend to overreact on their arbitrage behavior. Then having information might lead to market inefficiency. Agents might not benefit from too much information and would rather drop some of information when forecasting.

In the case of single-period version of Fama regression ($j = 1$), when α is equal to 0.01, the average proportion of full information agents is about 0.7, while it increases to unity when α is equal to 500. It also shows that when α is equal to 0.01, the simulated $\hat{\beta}$ are significantly negative despite the value of gain parameter. Both $\bar{n}$ and $\hat{\beta}$ increase with α. The puzzle tends to disappear as the switching parameter is increasing. $\hat{\beta}$ are positive in small gain parameter, and turn to be negative when the gain parameter is larger than 0.03. However, $\hat{\beta}$ are still not significantly negative as the gain parameter increases to 0.1. Here we find that forward premium puzzle is better explained by the switching parameter, instead of gain parameter.

The Fama regression in a multi-period version is also examined under dual learning. The results are reported in the case of $j = 1$ in Table 3. The simulated $\hat{\beta}$ are close to unity and significantly greater than zero. The value of simulated $\hat{\beta}$ is uncorrelated to the average proportion for the first forecast rule over the period, $\bar{n}$, and also the sensitivity to the relative performance of the predictors, α. The puzzle will disappear when testing in a multi-period version.

4 Conclusion

The forward premium puzzle was explained in several directions, such as irrational agents, incomplete knowledge, or risk neutral agents. Chakraborty and Evans [8] assume that agents are misperceive of behavior parameters and have to update their knowledge of behavior parameters by persistent learning. They show that the puzzle coexists with larger gain parameters. However, the Fama coefficient is always positive in their theoretical model and the simulated Fama coefficient is also highly correlated with sample size. This paper continues the discussion of [8]. The persistent learning model of [8] was enlarged to a limited information case, that the current fundamentals are not perceivable by some agents.

Based on the assumptions above, we find that: First, the simulated $\hat{\beta}$ is negative when the proportion of full information agents is small, even in small learning parameter and rational expectation. It implies that the effect of limited information on asymptotic bias dominates that of persistent learning. Second, the limited information could not only explain forward premium puzzle, it is consistent with the stylized facts, which is found in [21]. The puzzle will not remain in the multi-period test of Fama equation. Third, when agents react strongly to the relative performance of the forecasting rule by using full or limited information, all agents switch to apply full information quickly. The proportion of full information agents approaches to unity. Therefore, market efficiency hold when the sensitivity of the relative performance of the forecasting rule is high. This result is similar for different gain parameters. Finally, the same as the case that the proportion of full information agents is exogenous, we find that the Fama coefficients are positive when examining a multi-period Fama regression. These concepts will help for further understanding of the reasons for forward premium puzzle.

Appendix 1

Following [8], we could derive the distribution of b_t for small $\gamma > 0$, which is approximately normal with mean $\bar{b}$ and variance γC, where $C = (1 - \rho^2)(1 - (1 - n)\rho\theta)^2/2(1 - \rho\theta)^3$. Using this property, we could derive the least-square estimates $\hat{\beta}$ under the null hypothesis of $H_0 : \beta = 1$ in Eq. (1). We have the following findings:

First, under $H_0 : \beta = 1$ and sufficiently small $\gamma > 0$, the asymptotic bias plim $B(\gamma, \theta, \rho, n) = plim \hat{\beta} - 1$ is approximately equal to

$$B(\gamma, \theta, \rho, n) = -\frac{\gamma(1 - \rho\theta)(1 + \rho)(1 - \theta)((1 - n)\rho^2 + n) - (1 - n)A}{(\gamma(1 - \theta)^2(1 + \rho) + 2(1 - \rho)(1 - \rho\theta)((1 - n^2)\rho^2 + n^2) + (1 - n)F} \tag{21}$$

where $A \equiv (1+\rho)(\gamma n(1-\theta)(1-\rho^2) + 2\rho(1-\rho\theta)(n\rho(1-\theta) - (1-\rho\theta)))$, and $F \equiv 2(1-\rho\theta)^2(1+\rho)((1-n)(1-\rho\theta) + 2n(1-\rho))$. When all agents in the society update their information sets freely, i.e., $n = 1$, $B(\gamma, \theta, \rho, 1)$ will be equal to that provided by Chakraborty and Evans [8].

Plim$\hat{\beta}$ is below unity when the parameters satisfy $0 \le \theta < 1$, $0 \le \rho < 1$ and $0 \le n < 1$. As shown in Fig. 3, when the fundamentals are highly persistent, $\hat{\beta}$ turns

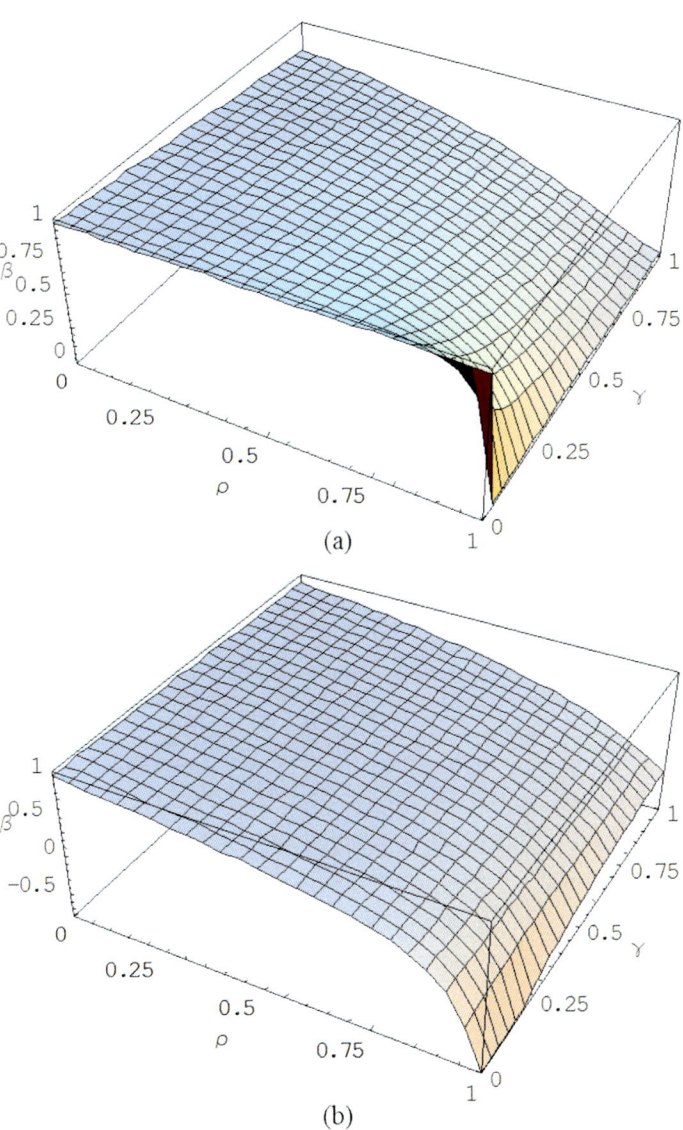

Fig. 3 Theoretical plim $\hat{\beta}$ for (**a**) $\theta = 0.6$, $n = 1$ and (**b**) $\theta = 0.6$, $n = 0.6$

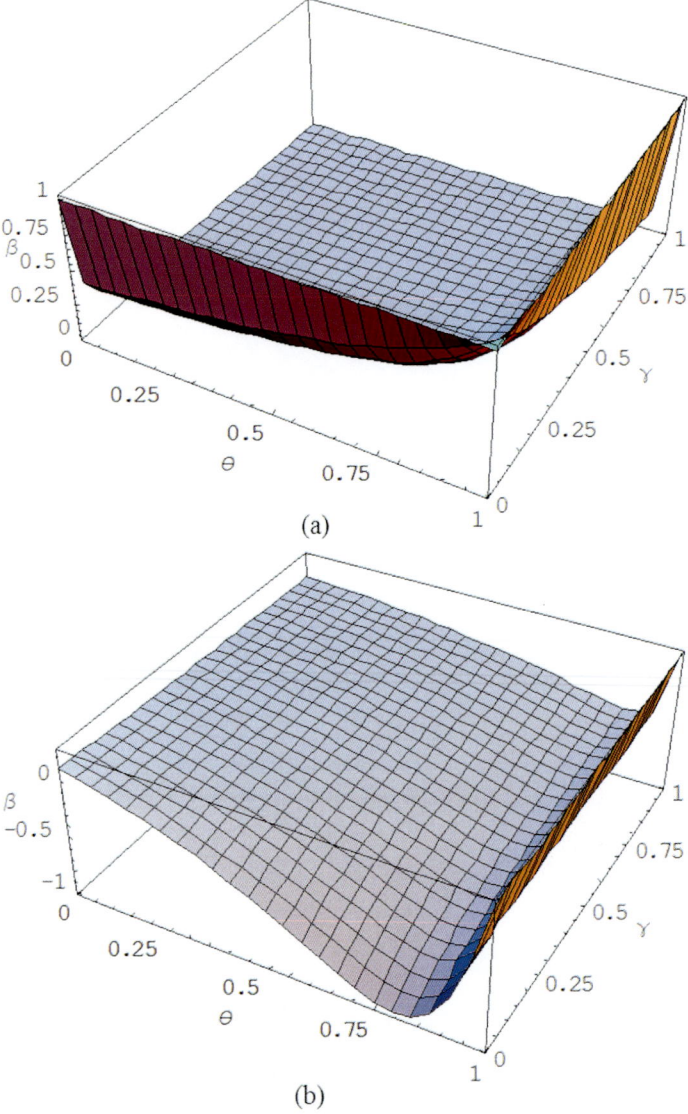

Fig. 4 Theoretical plim $\hat{\beta}$ for (**a**) $\rho = 0.98, n = 1$ and (**b**) $\rho = 0.98, n = 0.6$

to negative under limited information ($n = 0.6$) and is always in the range between 0 and 1 under full information.

Second, as $\rho \longrightarrow 1$, the asymptotic bias $B(\gamma, \theta, \rho, n) = plim\hat{\beta} - 1$ approximate to below -1, and the OLS estimates $\hat{\beta}$ turns to be negative, i.e., forward premium puzzle exists.

$$\lim B(\gamma, \theta, \rho, n) = -\frac{\gamma + 2(1-n)^2}{\gamma + 2(1-n)^2(1-\theta)} \leq 1 \qquad (22)$$

In the case of full information ($n = 1$), the asymptotic bias is equal to -1, which means the lower bound of theoretical $\hat{\beta}$ is 0.

Third, contrary to [8], our study finds that the learning process might either magnify or minify the asymptotic bias when information flows are limited. The impact of gain parameter on $plim\hat{\beta}$ is not monotonic, depending on the persistence of fundamental ρ and the dependence of that current exchange rate on the expectation of future exchange rate θ. $\hat{\beta}$ may either decrease or increase in learning parameter under limit information, as shown in the lower panel of Figs. 3 and 4.

Appendix 2

Under constant gain learning, there is a non-zero correction for the estimated coefficients through the forecasting error, even in the limit as t approaches to infinity. The constant gain algorithm converges to a distribution rather than the constant value of rational expectation. Evans [10] show that under the two conditions below, a constant gain algorithm converges to a normal distribution centered the rational expectation equilibrium. First, the gain parameter should be in the range between 0 and 1. Second, the parameter values should coincide with the E-stability conditions.

The learning process obeys

$$b_t = b_{t-1} + \gamma(s_t - b_{t-1}v_{t-1})v_{t-1}R_{t-1}^{-1} = b_{t-1} + \gamma Q(t, b_{t-1}, v_{t-1})$$

$$\rho_t = \rho_{t-1} + \gamma(v_t - \rho_{t-1}v_{t-1})v_{t-1}R_{t-1}^{-1}$$

$$R_t = R_{t-1} + \gamma(v_t^2 - R_{t-1}) \qquad (23)$$

The learning system is E-stable if it is a locally asymptotically stable equilibrium of the ordinary differential equation.

$$\frac{db}{d\tau} = (\theta\rho - 1)b$$

$$\frac{d\rho}{d\tau} = 0 \qquad (24)$$

Then an rational expectation equilibrium is E-stable if the Jacobian of $db/d\tau$ evaluated at the rational expectation equilibrium is a stable matrix, implying that it has eigenvalues with strictly negative real parts.

$$J(\bar{b}) = \frac{db/d\tau}{db}|_{b=\bar{b}} = \theta\rho - 1 \tag{25}$$

Since the parameters θ and ρ are in the range of 0 and 1, the unique stationary rational expectation equilibrium is E-stable.

References

1. Bacchetta, P., & Wincoop, E. V. (2009). *On the Unstable Relationship between Exchange Rates and Macroeconomic Fundamentals.* NBER Working Papers 15008, National Bureau of Economic Research, Inc.
2. Baillie, R. T., & Bollerslev, T. (2000). The forward premium anomaly is not as bad as you think. *Journal of International Money and Finance, 19*, 471–488.
3. Black, F. (1986). Noise. *Journal of Finance, 41*, 529–543.
4. Branch, W. A. (2007). Sticky information and model uncertainty in survey data on inflation expectations. *Journal of Economic Dynamics and Control, 31*, 245–276.
5. Branch, W. A., & Evans, G. W. (2007). Model uncertainty and endogenous volatility. *Review of Economic Dynamics, 10*(2), 207–237.
6. Brock, W. A., & Hommes, C. H. (1997). A rational route to randomness. *Econometrica, 65*(5), 1059–1096.
7. Burnside, C., Han, B., Hirshleifer, D., & Wang, T. Y. (2011). Investor overconfidence and the forward premium puzzle. *Review of Economic Studies, 78*, 523–558.
8. Chakraborty, A., & Evans, G. W. (2008). Can perpetual learning explain the forward-premium puzzle? *Journal of Monetary Economics, 55*(3), 477–490.
9. Engel, C. (1992). On the foreign exchange risk premium in a general equilibrium model. *Journal of International Economics, 32*(3–4), 305–319.
10. Evans, G. W., & Honkapohja, S. (2001). *Learning and expectations in macroeconomics.* Princeton, NJ: Princeton University Press.
11. Fama, E. F. (1970). Efficient capital markets: A review of theory and empirical work. *Journal of Finance, 25*, 383–417.
12. Gourinchas, P. O., & Tornell, A. (2004). Exchange rate puzzles and distorted beliefs. *Journal of International Economics, 64*(2), 303–333.
13. Hodrick, R. J. (1987). *The empirical evidence on the efficiency of forward and futures foreign exchange markets.* Chur: Harwood Academic Publishers.
14. Kellard, N., & Sarantis, N. (2008). Can exchange rate volatility explain persistence in the forward premium? *Journal of Empirical Finance, 15*(4), 714–728.
15. Krasker, W. S. (1980). The 'peso problem' in testing the efficiency of forward exchange markets. *Journal of Monetary Economics, 6*(2), 269–276.
16. Lewis, K. (1989). Changing beliefs and systematic rational forecast errors with evidence from foreign exchange. *American Economic Review, 79*, 621–36.
17. Lo, A. (2004). The adaptive markets hypothesis: Market efficiency from an evolutionary perspective. *Journal of Portfolio Management, 30*, 15–29.
18. Mankiw, N. G., & Reis, R. (2002). Sticky information versus sticky prices: A proposal to replace the New Keynesian Phillips curve. *The Quarterly Journal of Economics, 117*(4), 1295–1328.

19. Mark, N. C. (2009). Changing monetary policy rules, learning, and real exchange rate dynamics. *Journal of Money, Credit and Banking, 41*(6), 1047–1070.
20. Mark, N. C., & Wu, Y. (1998). Rethinking deviations from uncovered interest parity: The role of covariance risk and noise. *Economic Journal, 108*(451), 1686–1706.
21. McCallum, B. T. (1994). A reconsideration of the uncovered interest parity relationship. *Journal of Monetary Economics, 33*(1), 105–132.
22. Orphanides, A., & Williams, J. C. (2005). The decline of activist stabilization policy: Natural rate misperceptions, learning, and expectations. *Journal of Economic Dynamics and Control, 29*(11), 1927–1950.
23. Scholl, A., & Uhlig, H. (2008). New evidence on the puzzles: Results from agnostic identification on monetary policy and exchange rates. *Journal of International Economics, 76*(1), 1–13.

Price Volatility on Investor's Social Network

Yangrui Zhang and Honggang Li

Abstract Based on an Ising model proposed by Harras and Sornette, we established an artificial stock market model to describe the interactions among diverse agents. We regard these participants as network nodes and link them with their correlation. Then, we analyze the financial market based on the social network of market participants. We take the random network, scale-free network, and small-world network into consideration, and then build the stock market evolution model according to the characteristics of the investors' trading behavior under the different network systems. This allows us to macroscopically study the effects of herd behavior on the rate of stock return and price volatility under different network structures. The numerical simulation results show that herd behavior will lead to excessive market volatility. Specifically, the greater the degree of investor's trust in neighbors and their exchange, the greater the volatility of stock price will be. With different network synchronization capabilities, price fluctuations based on the small-world network are larger than those based on the regular network. Similarly, price fluctuations based on the random network are larger than those based on the small-world network. On the other hand, price fluctuations based on both the random network and the small-world network firstly increase and then decrease with the increase of the average node degree. All of these results illustrate the network topology that has an important impact on the stock market's price behavior.

Keywords Artificial stock market · Social network · Price volatility · Herd behavior · Network structures

---

Y. Zhang · H. Li (✉)
School of Systems Science, Beijing Normal University, Beijing, People's Republic of China
e-mail: hli@bnu.edu.cn

1 Introduction

For decades, a multitude of complex and macroscopic behavioral characteristics have constantly sprung up in financial market. But it is difficult to accurately estimate the nonlinear relationships between variables based on the traditional probability theory and econometrics. Establishing artificial financial markets, using objective bounded rationality assumptions to simulate the interaction between investor behaviors and studying the evolution and dynamics process, enable us to have a deeper understanding of the mechanism driving macroscopic behavior characteristics.

Establishing the artificial stock market model to study the interactions of investor behaviors and macroscopic behavior characteristics has become an important field of computational finance. Economists have proposed different mechanisms to describe the diverse agents, which are used to simulate macroscopic behavior and dynamic evolution in the market, and the interactions among them. Some scholars such as Arifovic [1], Lettau [2], and Kirman [3] analyzed the influence of information on investors' decisions through the observation of real participants' trading behavior. In the model of Johnson [4], when the value of the best strategy was below a certain threshold value, traders remained inactive, so the number of traders in the system is time-varying. Based on the Ising model, Iori [5] established the feedback of price and threshold value, and concluded that price and volatility are related. Johansen and Sornette [6] pointed out that all traders can be regarded as interactive sources of opinion, they found that the feedback effects and the memory function play an important role in the market. Basically, these researchers proposed a variety of artificial financial market models with heterogeneous interacting agents in order to account for the observed scaling laws, most of these models simply assume either a fully connected network or a regular structure. More reasonable network structures are needed to be taken into account with the improvement of the research. In the framework of the imitative conformity mechanism, Kim [7] researched the correlation of stock price with the scale-free network structure. Liang [8] established an artificial stock market model on the basis of the small-world network and explored some factors that lead to the complexity of financial markets. Alfarano and Milakovic [9] analyzed the correlation of herd behavior and the systemic size based on the network topology.

In the financial market, the trading behavior of investors, who have bounded rationality, leads to the complex nonlinear mechanism of the whole financial system. Investors are more prone to being influenced by their peers, the media, and other channels that combine to build a self-reflexive climate of optimism. Particularly, social network communication from investors may greatly affect investment opinion. At the same time, this communication may lead to significant imitation, herding, and collective behaviors [10]. Therefore, it is necessary to establish a reasonable social network to study interactions between investors and herd behavior from the microscopic aspect. Based on the complex network concept, we treat these participants as network nodes and link them according to their correlation. Then, we analyze the financial market with the social network.

The preceding research has adopted the agent-based model to study the complexity of the financial market. However, most of agent-based models do not consider the social network between traders as a key element. Our model focuses on the interaction among traders by a comparative analysis of different complex network structures. We take the random network, the scale-free network [11], and the small-world network [12] into consideration, and build the dynamic model according to the characteristics of the traders' investing behavior. Technically, inspired by Harras and Sornette model [6, 13], we establish the artificial stock market model to study the effect of herd behavior on the rate of return and price volatility under different network structures, from a kind of macroscopic aspect.

2 The Model

2.1 Investor's Social Network

The relationship between investors in the stock market can be described through the network structure, in which the connection may lead to the spread of trading behavior.

It is necessary to establish reasonable interpersonal network topology to research the interactions between investors and herd behavior. Due to the real stock market network scale and the complex interactions between nodes, its topology structure is still unexplored. Therefore, we mainly take the random network, the scale-free network, and the small-world network into consideration, and build the stock market evolution model so that we can macroscopically study the effects of herd behavior on the rate of stock return and price volatility under different network structures.

In addition, in our opinion, the social network structure itself evolves very slowly. Therefore, in this paper we assume that the network topology remains unchanged during the process of simulation.

2.2 Market

We focus on the interactive influence among traders and set an artificial stock market model to describe the interaction based on a prototypical interaction-based herding model proposed by Harras and Sornette [6].

We consider a fixed environment is composed of N agents, who are trading a single asset, as a stock market, and at each time phase, agents have the possibility to either trade (buy or sell) or to remain passive. The trading decision $s_i(t)$ of agent i is opinion-based on the future price development and the opinion of agent i at time t,

$w_i(t)$. It consists of three different sources: idiosyncratic opinion, global news, and acquaintance network.

$$w_i(t) = c_{1i} \sum_{j=1}^{J} k_{ij}(t-1) E_i[s_j(t)] + c_{2i} u(t-1) n(t) + c_{3i} \varepsilon_i(t)$$

Where $\varepsilon_i(t)$ represents the private information of agent i, $n(t)$ is the public information. J is the number of neighbors that agent i polls for their opinion and $E_i[s_j(t)]$ is the action of the neighbor j at time $t-1$; (c_{1i}, c_{2i}, c_{3i}) is the form of the weights the agent attributes to each of the three pieces of information.

It is assumed that each agent is characterized by a fixed threshold $\underline{w_i}$ to control the triggering $s_i(t)$ of an investment action. An agent i uses a fixed fraction g of his cash to buy a stock if his conviction $w_i(t)$ is sufficiently positive so as to reach the threshold: $w_i(t) \geq \underline{w_i}$. On the contrary, he sells the same fixed fraction g of the value of his stocks if $w_i(t) \leq -\underline{w_i}$.

$$w_i(t) \geq \underline{w_i}: \quad s_i(t) = +1 (\text{buying})$$
$$v_i(t) = g \cdot \frac{cash_i(t)}{p(t-1)}$$
$$w_i(t) \leq -\underline{w_i}: \quad s_i(t) = -1 (\text{selling})$$
$$v_i(t) = g \cdot stocks_i(t)$$

Once all of the agents have determined their orders, the new price of the asset is determined by the following equations:

$$r(t) = \frac{1}{\lambda \cdot N} \sum_{i=1}^{N} s_i(t) \cdot v_i(t)$$

$$\log[p(t)] = \log[p(t-1)] + r(t)$$

Here $r(t)$ is the return and $v_i(t)$ is the volume at time t; λ represents the relative impact of the excess demand upon the price, i.e., the market depth.

When the return and the new price are determined, the cash and number of stocks held by each agent i are updated according to

$$cash_i(t) = cash_i(t-1) - s_i(t) v_i(t) p(t)$$
$$stocks_i(t) = stocks_i(t-1) + s_i(t) v_i(t)$$

2.3 Adaption

At each step traders change the trust factor according to the related traders' decisions and public information from the previous period. We assume that agents adapt their beliefs concerning the credibility of the news $n(t)$ and their trust in the advice $E_i[s_j(t)]$ of their social contacts, according to time-dependent weights $u(t)$ and $k_{ij}(t)$, which take into account their recent performance. The implementation is achieved by a standard auto-regressive update:

$$u(t) = \alpha u(t-1) + (1-\alpha) n(t-1) \frac{r(t)}{\sigma_r}$$

$$k_{ij}(t) = \alpha k_{ij}(t-1) + (1-\alpha) E_i[s_j(t-1)] \frac{r(t)}{\sigma_r}$$

where α represents traders' memory in the process, σ_r is the volatility of returns.

3 Simulations and Results

3.1 Parameter Setting

In our simulations, we fix the network parameters to $N = 2500$, the market parameters to $\lambda = 0.25$, $g = 2\%$, $w_i = 2$, the initial amount of cash and stocks held by each agent to $cash_i(0) = 1$ and $stocks_i(0) = 1$, and the memory discount factor to $\alpha = 0.90$.

3.2 Market Price

Firstly, we assume that all traders lack communication with others. The numerical experimental results in Fig. 1 show that, when traders make independent decisions, fluctuations in market price sequence (red) are equal to fluctuations in fundamental value (blue), which is determined by the market information. We consider that this market is efficient in this case.

We then allow the communication between traders with the random network structure. In Fig. 2, we can observe the excessive volatility and price bubbles; the volatility of return has significant cluster characteristics.

Figures 3 and 4 show that the distribution of return has the characteristics of a higher peak and a fat tail compared with the normal distribution The right panel also shows the slow attenuation of the autocorrelation function; the absolute value of the returns is related for a long time, which also suggests that the price fluctuation has the volatility clustering property. All of these results concur with the empirical research.

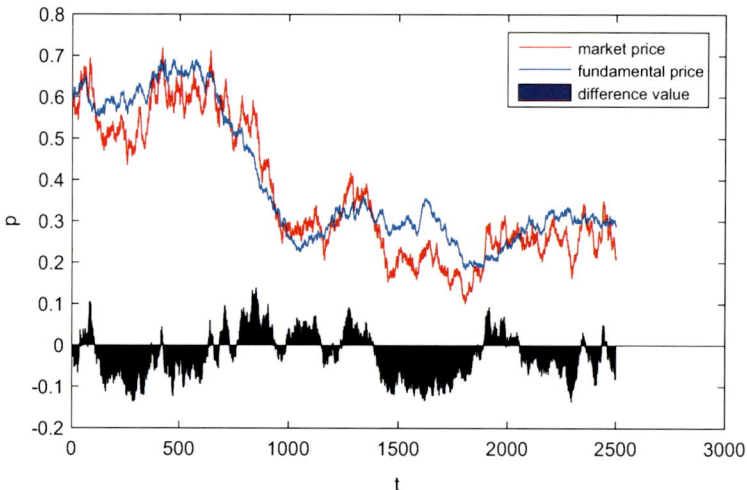

Fig. 1 Series of market price and fundamental value

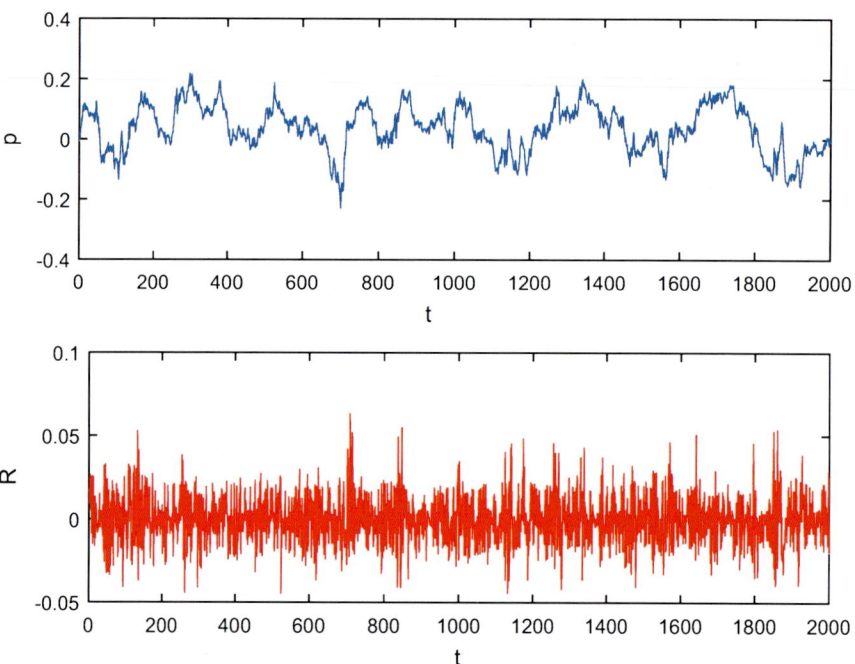

Fig. 2 Market price and returns

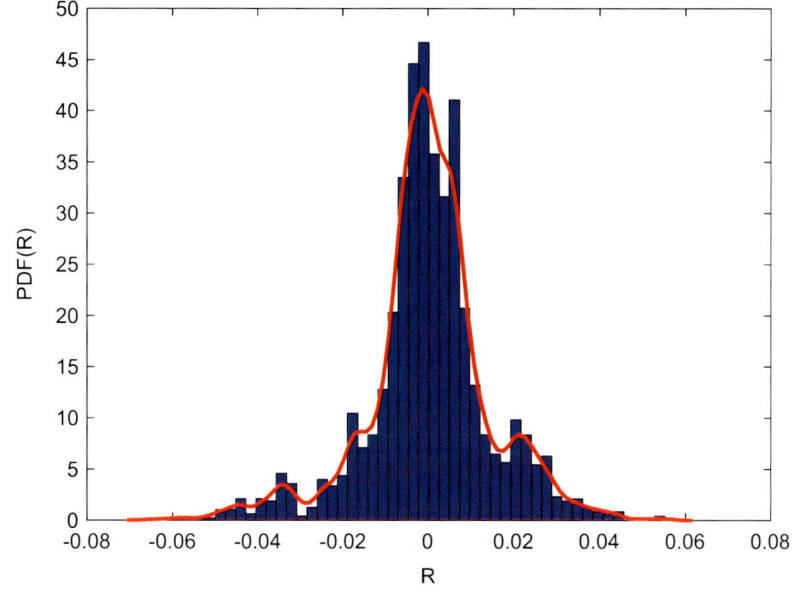

	Mean	Maximum	Minimum	Std.Dev	Skewness	Kurtosis	Jarque-Bera	Probability
Data	1.15e-004	0.146000	-0.11000	0.027559	0.189764	6.497000	548.6494	0

Fig. 3 Distribution of return

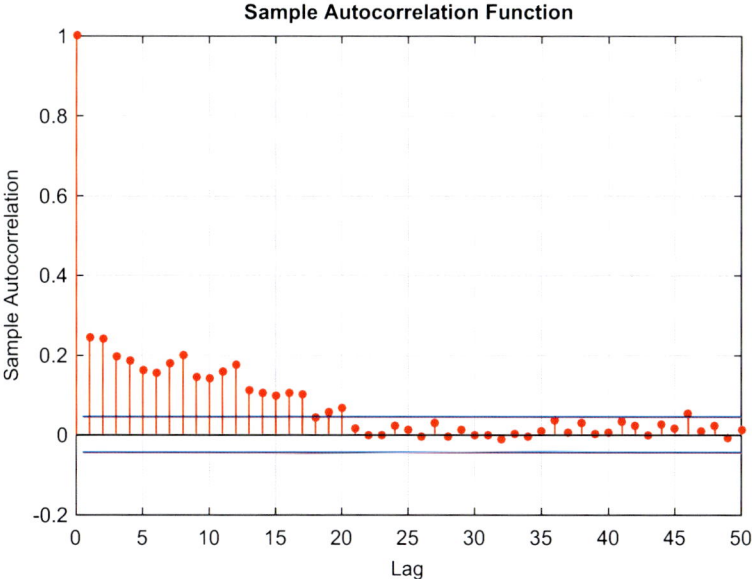

Fig. 4 Correlation of volatility (measured as the absolute value of returns)

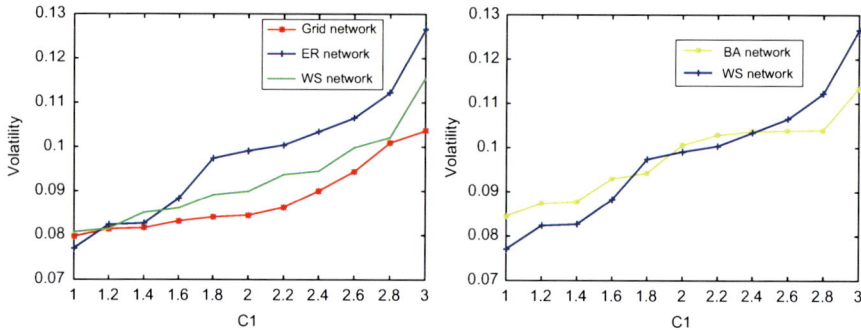

Fig. 5 Volatility changes with c_1 under different network structures

3.3 Market Volatility

To research the relationship between investors' trust in neighbors and market volatility under different network structures with c_1 values of 1 to 3, we set the average node of three network types to 8. Taking averaged value of the return volatility from 200 times simulations, as shown in Fig. 5, we can conclude that in the same network structure, the greater degree of investors trust in neighbors and their exchange, the greater the volatility of stock price will be. With the different network synchronization capabilities, the difference between volatility under the small-world (SW) network and the scale-free (BA) network is small. Furthermore, the price fluctuations based on the small-world network are larger than those based on the regular grid network. Finally, equally, the price fluctuations based on the random (ER) network are larger than those based on the small-world network.

For a larger system size, it can be seen as a collection of small networks, the excessive volatility still exists since the influence of the macro- information. So the model is robust with respect to an enlargement of system size (Fig. 6).

3.4 Network Characteristics

We refer to the construction algorithm of the WS small-world network to change the values of rewiring probability p to transfer the network structure from the regular network ($p = 0$) to the random network ($p = 1$). Figure 7 shows the impact of p on the market volatility. During the transition from the regular network to the random network, the volatility increases with the increase of the rewiring probability, under the same transmission sensitivity. We also find that the market volatility has a fast declining trend at first, and then leveled off when the clustering coefficient increases (Fig. 8).

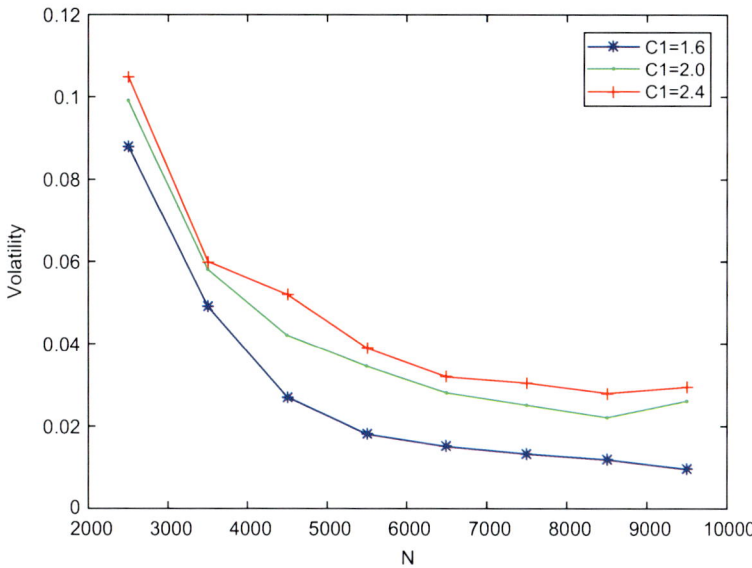

Fig. 6 Volatility changes with N (the total number of agents) for $C_1 = 1.6, 2.0, 2.4$ (ER network)

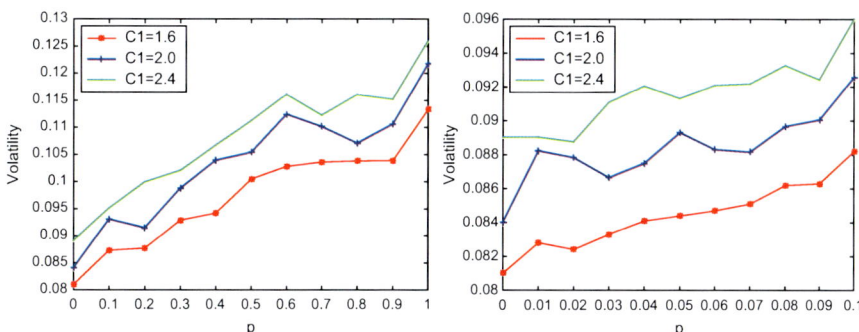

Fig. 7 Volatility changes with p (rewiring probability) for $C_1 = 1.6, 2.0, 2.4$

Figure 9 displays the volatility changes with the average node degree under the different network structures. Where the volatility firstly increases and then slowly rests at a lower level with the increase of the average node degree. The cause of this phenomenon may be attributed to the long-range connection and the higher synchronizability of the small-world network. On the other hand, when the network node degree increases to a certain degree, the connection of the activity nodes will be wider. Finally, the behavior of traders in the market will tend to make an overall decision, resulting in a loss of market volatility.

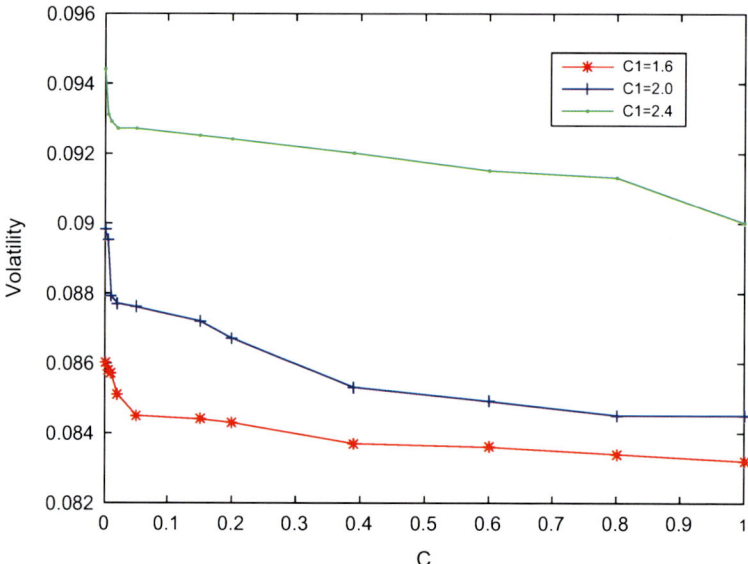

Fig. 8 Volatility changes with C (Clustering coefficient) for $C_1 = 1.6, 2.0, 2.4$

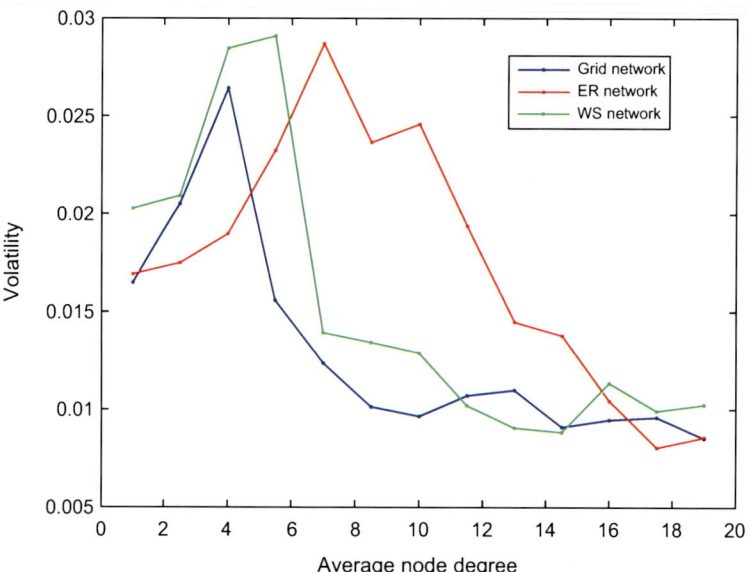

Fig. 9 Volatility changes with average node degree changes under different network structures

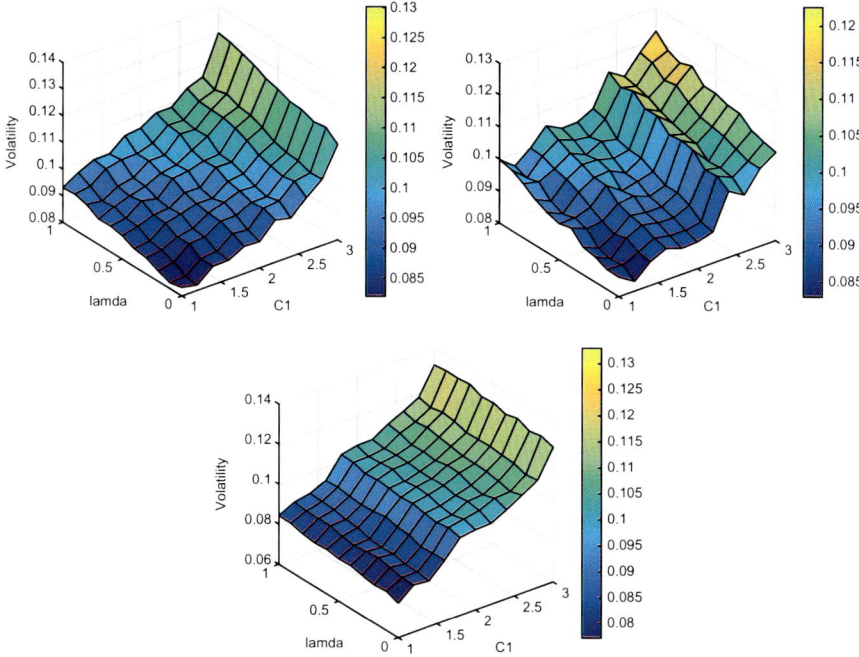

Fig. 10 Volatility with C_1 and λ changes under different network structures (upper left panel: ER network. upper right panel: SW network; lower panel: BA network)

3.5 Guru

In the communication network, we add a given special agent named "guru." It attracts the largest number of in-coming links. In our settings, each agent i chooses to link a probability λ to the fixed agent guru, and with probability $1 - \lambda$, to keep the original connection.

Figure 10 shows the volatility changes with C_1 and λ under different network structures. It can be seen that the volatility increases with the increase of λ under the same transmission sensitivity. This means that the connection of the guru changes the network structure and increases the market volatility.

4 Conclusion

Based on the artificial stock market model, we research the effect of the network topology on market volatility. When investors establish contact with others, the herd behavior can lead to excessive volatility of the market. The greater the degree of investor trust in neighbors and their exchange, the greater the volatility of stock price will be. On the other hand, the volatility firstly increases and then slowly rests at a lower level with the increase of the average node degree, under three kinds of network structure. Furthermore, the results show that the topology of the traders' network has an important effect on the herding behavior and the market price.

It goes without saying that, in this paper, we only analyze one type of market model. However, the market volatility based on different types of market models and different mechanisms of information transmission still requires further research.

Acknowledgements We wish to thank Xuezhong He for useful comments and discussions, and anonymous referees for helpful comments. None of the above is responsible for any of the errors in this paper. This work was supported by the Fundamental Research Funds for the Central Universities under Grant No. 2012LZD01 and the National Natural Science Foundation of China under Grant No. 71671017.

References

1. Arifovic, J. (1996). The behavior of the exchange rate in the genetic algorithm and experimental economies. *Journal of Political Economy, 104*, 510–541.
2. Lettau, M. (1997). Explaining the facts with adaptive agents: The case of mutual fund flows. *Journal of Economic Dynamics and Control, 21*(7), 1117–1147.
3. Kirman, A. (1991). Epidemics of opinion and speculative bubbles in financial markets. In *Money and financial markets* (pp. 354–368). Cambridge: Blackwell.
4. Johnson, N. F., Hart, M., Hui, P. M., & Zheng, D. (2000). Trader dynamics in a model market. *International Journal of Theoretical and Applied Finance, 3*(03), 443–450.
5. Iori, G. (2002). A microsimulation of traders activity in the stock market: The role of heterogeneity, agents' interactions and trade frictions. *Journal of Economic Behavior & Organization, 49*(2), 269–285.
6. Harras, G., & Sornette, D. (2011). How to grow a bubble: A model of myopic adapting agents. *Journal of Economic Behavior & Organization, 80*(1), 137–152.
7. Kim, H. J., Kim, I. M., Lee, Y., & Kahng, B. (2002). Scale-free network in stock markets. *Journal-Korean Physical Society, 40*, 1105–1108.
8. Liang, Z. Z., & Han, Q. L. (2009). Coherent artificial stock market model based on small world networks. *Complex Systems and Complexity Science, 2*, 70–76.
9. Alfarano, S., & Milaković, M. (2009). Network structure and N-dependence in agent-based herding models. *Journal of Economic Dynamics and Control, 33*(1), 78–92.
10. Tedeschi, G., Iori, G., & Gallegati, M. (2009). The role of communication and imitation in limit order markets. *The European Physical Journal B, 71*(4), 489–497.
11. Barabási, A. L., & Albert, R. (1999). Emergence of scaling in random networks. *Science, 286*(5439), 509–512.
12. Watts, D. J., & Strogatz, S. H. (1998). Collective dynamics of small-world networks. *Nature, 393*(6684), 440–442.
13. Sornette, D., & Zhou, W. X. (2006). Importance of positive feedbacks and overconfidence in a self-fulfilling Ising model of financial markets. *Physica A: Statistical Mechanics and Its Applications, 370*(2), 704–726.

The Transition from Brownian Motion to Boom-and-Bust Dynamics in Financial and Economic Systems

Harbir Lamba

Abstract Quasi-equilibrium models for aggregate or emergent variables over long periods of time are widely used throughout finance and economics. The validity of such models depends crucially upon assuming that the system participants act both independently and without memory. However important real-world effects such as herding, imitation, perverse incentives, and many of the key findings of behavioral economics violate one or both of these key assumptions.

We present a very simple, yet realistic, agent-based modeling framework that is capable of simultaneously incorporating many of these effects. In this paper we use such a model in the context of a financial market to demonstrate that herding can cause a transition to multi-year boom-and-bust dynamics at levels far below a plausible estimate of the herding strength in actual financial markets. In other words, the stability of the standard (Brownian motion) equilibrium solution badly fails a "stress test" in the presence of a realistic weakening of the underlying modeling assumptions.

The model contains a small number of fundamental parameters that can be easily estimated and require no fine-tuning. It also gives rise to a novel stochastic particle system with switching and re-injection that is of independent mathematical interest and may also be applicable to other areas of social dynamics.

Keywords Financial instability · Far-from-equilibrium · Endogenous dynamics · Behavioral economics · Herding · Boom-and-bust

---

H. Lamba (✉)
Department of Mathematical Sciences, George Mason University, Fairfax, VA, USA
e-mail: hlamba@gmu.edu

1 Introduction

In the physical sciences using a stochastic differential equation (SDE) to model the effect of exogenous noise upon an underlying ODE system is often straightforward. The noise consists of many uncorrelated effects whose cumulative impact is well-approximated by a Brownian process B_s, $s \geq 0$ and the ODE $df = a(f, t) \, dt$ is replaced by an SDE $df = a(f, t) \, dt + b(f, t) \, dB_t$.

However, in financial and socio-economic systems the inclusion of exogenous noise (i.e., new information entering the system) is more problematic—even if the noise itself can be legitimately modeled as a Brownian process. This is because such systems are themselves the aggregation of many individuals or trading entities (referred to as *agents*) who typically

(a) interpret and act differently to new information,
(b) may act differently depending upon the recent system history (i.e., non-Markovian behavior), and
(c) may not act independently of each other.

The standard approach in neoclassical economics and modern finance is simply to "average away" these awkward effects by assuming the existence of a single representative agent as in macroeconomics [7], or by assuming that the averaged reaction to new information is correct/rational, as in microeconomics and finance [4, 13]. In both cases, the possibility of significant endogenous dynamics is removed from the models resulting in unique, Markovian (memoryless), (quasi)-equilibrium solutions. This procedure is illustrated in Fig. 1 where the complicated "human filter" that lies between the new information and the aggregate variables (such as price) does not alter its Brownian nature. This then justifies the use of SDEs upon aggregate variables directly.

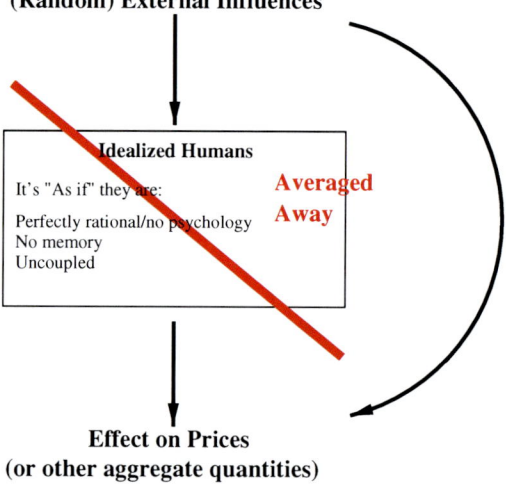

Fig. 1 In standard financial and economic models exogenous effects act directly upon the model variables. This is achieved by assuming that the system participants behave independently and are (at least when averaged) perfectly rational and memoryless

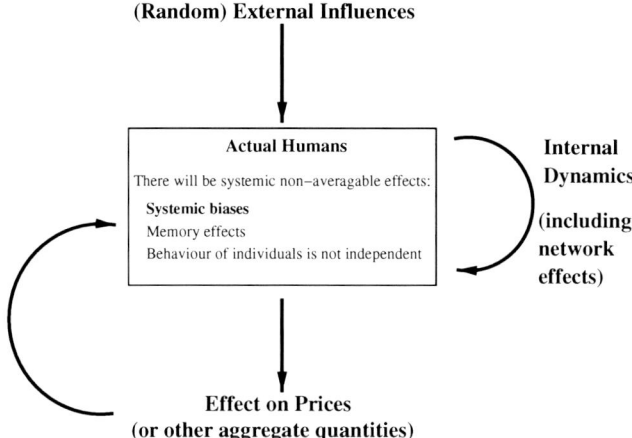

Fig. 2 A more realistic "human filter" is capable of introducing endogenous, far-from-equilibrium, dynamics into the system and economic history suggests that these are often of the boom-and-bust type. Unfortunately, the difference between such dynamics and the instantly-equilibrating models represented in Fig. 1 may only be apparent over long timescales or during extreme events corresponding to endogenous cascading processes

However, the reality is far more complicated, as shown in Fig. 2. Important human characteristics such as psychology, memory, systemic cognitive or emotional biases, adaptive heuristics, group influences, and perverse incentives will be present, as well as various possible positive feedbacks caused by endogenous dynamics or interactions with the aggregate variables.

In an attempt to incorporate some of these effects, many heterogeneous agent models (HAMs) have been developed [6] that simulate agents directly rather than operate on the aggregate variables. These have demonstrated that it is relatively easy to generate aggregate output data, such as the price of a traded asset, that approximate reality better than the standard averaging-type models. In particular the seemingly universal "stylized facts" [2, 11] of financial markets such as *heteroskedasticity* (volatility clustering) and *leptokurtosis* (fat-tailed price-return distributions resulting from booms-and-busts) have been frequently reproduced. However, the effects of such research upon mainstream modeling have been minimal perhaps, in part, because some HAMs require fine-tuning of important parameters, others are too complicated to analyze, and the plethora of different HAMs means that many are mutually incompatible.

The purpose of this paper is rather different from other studies involving HAMs. We are not proposing a stand-alone asset pricing model to replace extant ones (although it could indeed be used that way). Rather we are proposing an entire framework in which the stability of equilibrium models can be tested in the presence of a wide class of non-standard perturbations that weaken various aspects of the efficiency and rationality assumptions—in particular those related to (a), (b), and (c) above.

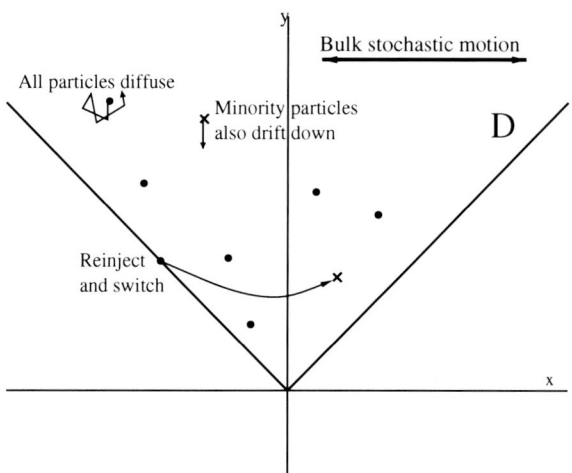

Fig. 3 The M signed particles are subject to a horizontal stochastic forcing and they also diffuse independently. Minority particles also drift downwards at a rate proportional to the imbalance. When a particle hits the boundary it is re-injected with the opposite sign and a kick is added to the bulk forcing that can trigger a cascade (see text)

In this paper we focus upon the effects of herding as it is an easily understood phenomenon that has multiple causes (rational, psychological, or due to perverse incentives) and is an obvious source of lack of independence between agents' actions. We start by introducing a simplified version of the modeling framework introduced in [9, 10] that can also be described as a particle system in two dimensions (Fig. 3). A web-based interactive simulation of the model can be found at http://math.gmu.edu/~harbir/market.html. It provides useful intuition as to how endogenous multi-year boom-and-bust dynamics naturally arise from the competition between equilibrating and disequilibrating forces that are usually only considered important over much shorter timescales.

2 A Stochastic Particle System with Re-injection and Switching

We define the open set $D \subset \Re^2$ by $D = \{(x, y) : -y < x < y, \ y > 0\}$. There are M signed particles (with states $+1$ or -1) that move within D subject to four different motions. Firstly, there is a bulk Brownian forcing B_t in the x-direction that acts upon every particle. Secondly, each particle has its own independent two-dimensional diffusion process. Thirdly, for agents *in the minority state only*, there is a downward (negative y-direction) drift that is proportional to the imbalance.

Finally, when a particle hits the boundary ∂D it is re-injected into D *with the opposite sign* according to some predefined probability measure. When this happens, the position of the other particles is kicked in the x-direction by a (small) amount $\pm \frac{2\kappa}{M}$, $\kappa > 0$, where the kick is positive if the switching particle goes from the -1 state to $+1$ and negative if the switch is in the opposite direction. Note that the particles do not interact locally or collide with one another.

2.1 Financial Market Interpretation

We take as our starting point the standard geometric Brownian motion (gBm) model of an asset price p_t at time t with $p_0 = 1$. It is more convenient to use the log-price $r_t = \ln p_t$ which for constant drift a and volatility b is given by the solution $r_t = at + bB_t$ to the SDE

$$dr_t = a\, dt + b\, dB_t. \tag{1}$$

Note that the solution r_t depends only upon the value of the exogenous Brownian process B_t at time t and not upon $\{B_s\}_{s=0}^{t}$. This seemingly trivial observation implies that r_t is Markovian and consistent with various notions of market efficiency. Thus gBm can be considered a paradigm for economic and financial models in which the aggregate variables are assumed to be in a quasi-equilibrium reacting instantaneously and reversibly to new information.

The model involves two types of agent and a separation of timescales. "Fast" agents react near instantaneously to the arrival of new information B_t. Their effect upon the asset price is close to the standard models and they will not be modeled directly. However, we posit the existence of M "slow" agents who are primarily motivated by price changes rather than new information and act over much longer timescales (weeks or months). At time t the ith slow agent is either in state $s_i(t) = +1$ (owning the asset) or $s_i(t) = -1$ (not owning the asset) and the *sentiment* $\sigma(t) \in [-1, 1]$ is defined as $\sigma(t) = \frac{1}{M}\sum_{i=1}^{M} s_i(t)$. The ith slow agent is deemed to have an evolving strategy that at time t consists of an open interval $(L_i(t), U_i(t))$ containing the current log-price r_t (see Fig. 4). The ith agent switches state whenever the price crosses either threshold, i.e., $r_t = L_i(t)$ or $U_i(t)$, and a new strategy interval is generated straddling the current price. Note that slow agents wishing to trade do not need to be matched with a trading partner—it is assumed that the fast agents provide sufficient liquidity.

We assume in addition that each threshold for every slow agent has its own independent diffusion with rate α_i (corresponding to slow agents' independently evolving strategies) and those in the minority (i.e., whose state differs from $|\sigma|$) also have their lower and upper thresholds drift inwards each at a rate $C_i|\sigma|$, $C_i \geq 0$.

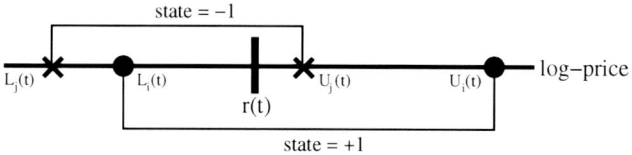

Fig. 4 A representation of the model showing two agents in opposite states at time t. Agent i is in the $+1$ state and is represented by the two circles at $L_i(t)$ and $U_i(t)$ while agent j is in the -1 state and is represented by the two crosses

These *herding constants* C_i are crucial as they provide the only (global) coupling between agents. The inward drift of the minority agents' strategies makes them more likely to switch to join the majority—they are being pressured/squeezed out of their minority view. Herding and other mimetic effects appear to be a common feature of financial and economic systems. Some causes are irrationally human while others may be rational responses by, for example, fund managers not wishing to deviate too far from the majority opinion and thereby risk severely under-performing their average peer performance. The reader is directed to [9] for a more detailed discussion of these and other modeling issues.

Finally, changes in the sentiment σ feed back into the asset price so that gBm (1) is replaced with

$$dr_t = a\,dt + b\,dB_t + \kappa\Delta\sigma \qquad (2)$$

where $\kappa > 0$ and the ratio κ/b is a measure of the relative impact upon r_t of exogenous information versus endogenous dynamics. Without loss of generality we let $a = 0$ and $b = 1$ by setting the risk-free interest rate to zero and rescaling time.

One does not need to assume that all the slow agents are of equal size, have equal strategy-diffusion, and equal herding propensities. But if one does set $\alpha_i = \alpha$ and $C_i = C\ \forall i$, then the particle system above is obtained by representing the ith agent, not as an interval on $\mathfrak{R}$, but as a point in $D \subset \mathfrak{R}^2$ with position $(x_i, y_i) = (\frac{U_i+L_i}{2}-r_t, \frac{U_i-L_i}{2})$. To make the correspondence explicit: the bulk stochastic motion is due to the exogenous information stream, the individual diffusions are caused by strategy-shifting of the slow agents; the downward drift of minority agents is due to herding; the re-injection and switching are the agents changing investment position; and the kicks that occur at switches are due to the change in sentiment affecting the asset price via the linear supply/demand price assumption.

2.2 Limiting Values of the Parameters

There are different parameter limits that are of interest.

1. $\underline{M \to \infty}$ In the continuum limit the particles are replaced by a pair of evolving density functions $\rho^+(x, y, t)$ and $\rho^-(x, y, t)$ representing the density of each agent state on D—such a mesoscopic Fokker–Planck description of a related, but simpler, market model can be found in [5]. The presence of nonstandard boundary conditions, global coupling, and bulk stochastic motion present formidable analytic challenges for even the most basic questions of existence and uniqueness of solutions. However, numerical simulations strongly suggest that, minor discretization effects aside, the behavior of the system is independent of M for $M > 1000$.

2. $B_t \to 0$ As the external information stream is reduced the system settles into a state where σ is close to either ± 1. Therefore this potentially useful simplification is not available to us.
3. $\alpha \to 0$ or ∞ In the limit $\alpha \to 0$ the particles do not diffuse, i.e., the agents do not alter their thresholds between trades/switches. This case was examined in [3] and the lack of diffusion does not significantly change the boom–bust behavior shown below. On the other hand, for $\alpha \gg \max(1, C)$ the diffusion dominates both the exogenous forcing and the herding/drifting and equilibrium-type dynamics is re-established. This case is unlikely in practice since slow agents will alter their strategies more slowly than changes in the price of the asset.
4. $C \to 0$ This limit motivates the next section. When $C = 0$ the particles are uncoupled and if the system is started with approximately equal distributions of ± 1 states, then σ remains close to 0. Thus (2) reduces to (1) and the particle system becomes a standard equilibrium model—agents have differing expectations about the future which causes them to trade but on average the price remains "correct." In Sect. 3 we shall observe that endogenous dynamics arise as C is increased from 0 and the equilibrium gBm solution loses stability in the presence of even small amounts of herding.
5. $\kappa \to 0$ For $\kappa > 0$ even one agent switching can cause an avalanche of similar switches, especially when the system is highly one-sided with $|\sigma|$ close to 1. When $\kappa = 0$ the particles no longer provide kicks (or affect the price) when they switch although they are still coupled via $C > 0$. The sentiment σ can still drift between -1 and $+1$ over long timescales but switching avalanches and large, sudden, price changes do not occur.

3 Parameter Estimation, Numerical Simulations, and the Instability of Geometric Brownian Pricing

In all the simulations below we use $M = 10{,}000$ and discretize using a timestep $h = 0.000004$ which corresponds to approximately 1/10 of a trading day if one assumes a daily standard deviation in prices of $\approx 0.6\%$ due to new information. The price changes of ten consecutive timesteps are then summed to give daily price return data making the difference between synchronous vs asynchronous updating relatively unimportant.

We choose $\alpha = 0.2$ so that slow agents' strategies diffuse less strongly than the price does. A conservative choice of $\kappa = 0.2$ means that the difference in price between neutral ($\sigma = 0$) and polarized markets $\sigma = \pm 1$ is, from (2), $\exp(0.2) \approx 22\%$.

After switching, an agent's thresholds are chosen randomly from a uniform distribution to be within 5% and 25% higher and lower than the current price. This allows us to estimate C by supposing that in a moderately polarized market with $|\sigma| = 0.5$ a typical minority agent (outnumbered 3–1) would switch due to herding

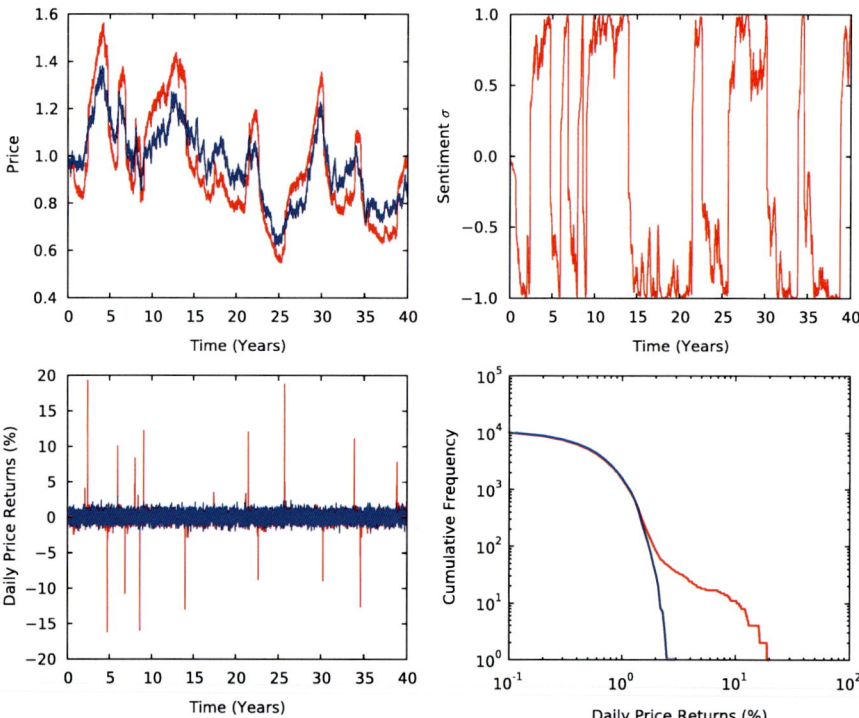

Fig. 5 The top left figure shows the prices $p(t)$ for both our model and the gBm pricing model (1) with the same exogenous information stream B_s. The herding model is clearly more volatile but the other pictures demonstrate the difference more clearly. In the bottom left, the same data is plotted in terms of daily price changes. Over the 40-year period not a single instance of a daily price change greater than 2% occurred in the gBm model. All the large price fluctuations are due to endogenous dynamics in the herding model. This is shown even more clearly in the top right picture where sentiment vs time is plotted for the herding model—the very sudden large switches in sentiment are due to cascading changes amongst the agents' states. It should be noted that the sentiment can remain polarized close to ±1 for unpredictable and sometimes surprisingly long periods of time. Finally, the bottom right picture shows the cumulative log–log plot of daily price changes that exceed a given percentage for each model. The fat-tailed distribution for the herding model shows that the likelihood of very large price moves is increased by orders of magnitude over the gBm model

pressure after approximately 80 trading days (or 3 months, a typical reporting period for investment performance) [14]. The calculation $80C|\sigma| = |\ln(0.85)|/0.00004$ gives $C \approx 100$. Finally, we note that no fine-tuning of the parameters is required for the observations below.

Figure 5 shows the results of a typical simulation, started close to equilibrium with agents' states equally mixed and run for 40 years. The difference in price history between the above parameters and the equilibrium gBm solution is shown in the top left. The sudden market reversals and over-reactions can be seen more clearly in the top right plot where the market sentiment undergoes sudden shifts due to

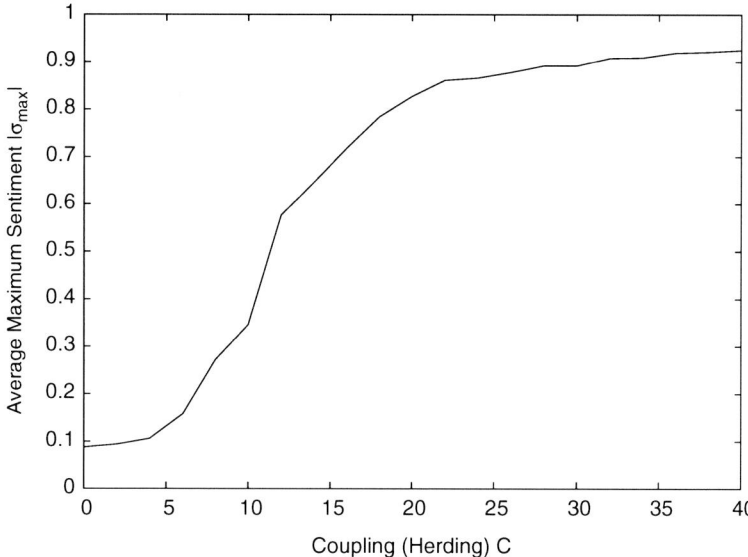

Fig. 6 A measure of disequilibrium $|\sigma|_{max}$ averaged over 20 runs as the herding parameter C changes

switching cascades. These result in price returns (bottom left) that could quite easily bankrupt anyone using excessive financial leverage and gBm as an asset pricing model! Finally in the bottom right the number of days on which the magnitude of the price change exceeds a given percentage is plotted on log–log axes. It should be emphasized that this is a simplified version of the market model in [9] and an extra parameter that improves the statistical agreement with real price data (by inducing volatility clustering) has been ignored.

To conclude we examine the stability of the equilibrium gBm solution using the herding level C as a bifurcation parameter. In order to quantify the level of disequilibrium in the system we record the maximum value of $|\sigma|$ ignoring the first 10 years of the simulation (to remove any possible transient effects caused by the initial conditions) and average over 20 runs each for values of $0 \leq C \leq 40$. All the other parameters and the initial conditions are kept unchanged.

The results in Fig. 6 show that even for values of C as low as 20 the deviations from the equilibrium solution are close to being as large as the system will allow with $|\sigma|$ usually getting close to ± 1 at some point during the simulations. To reiterate, this occurs at a herding strength C which is a factor of 5 lower than the value of $C = 100$ estimated above for real markets! It should also be noted that there are other significant phenomena that have not been included, such as new investors and money entering the asset market after a bubble has started, and localized interactions between certain subsets of agents. These can be included in the model by allowing κ to vary (increasing at times of high market sentiment, for example) and, as expected, they cause the equilibrium solution to destabilize even more rapidly.

4 Conclusions

Financial and economic systems are subject to many different kinds of interdependence between agents and potential positive feedbacks. However, even those mainstream models that attempt to quantify such effects [1, 14] assume that the result will merely be a shift of the equilibria to nearby values without qualitatively changing the nature of the system. We have demonstrated that at least one such form of coupling (incremental herding pressure) results in the loss of stability of the equilibrium. Furthermore the new dynamics occurs at realistic parameters and is clearly recognizable as "boom-and-bust." It is characterized by multi-year periods of low-level endogenous activity (long enough, certainly, to convince equilibrium-believers that the system is indeed in an equilibrium with slowly varying parameters) followed by large, sudden, reversals involving cascades of switching agents triggered by price changes.

A similar model was studied in [8] where momentum-traders replaced the slow agents introduced above. The results replicated the simulations above in the sense that the equilibrium solution was replaced with multi-year boom-and-bust dynamics but with the added benefit that analytic solutions can be derived, even when agents are considered as nodes on an arbitrary network rather than being coupled globally.

The model presented here is compatible with existing (non-mathematized) critiques of equilibrium theory by Minsky and Soros [12, 15]. Furthermore, work on related models to appear elsewhere shows that positive feedbacks can result in similar non-equilibrium dynamics in more general micro- and macro-economic situations.

Acknowledgements The author thanks Michael Grinfeld, Dmitri Rachinskii, and Rod Cross for numerous enlightening conversations and Julian Todd for writing the browser-accessible simulation of the particle system.

References

1. Akerlof, G., & Yellen, J. (1985). Can small deviations from rationality make significant differences to economic equilibria? *The American Economic Review, 75*(4), 708–720.
2. Cont, R. (2001). Empirical properties of asset returns: Stylized facts and statistical issues. *Quantitative Finance, 1*, 223–236.
3. Cross, R., Grinfeld, M., Lamba, H., & Seaman, T. (2005). A threshold model of investor psychology. *Physica A, 354*, 463–478.
4. Fama, E. F. (1965). The behavior of stock market prices. *Journal of Business, 38*, 34–105.
5. Grinfeld, M., Lamba, H., & Cross, R. (2013). A mesoscopic market model with hysteretic agents. *Discrete and Continuous Dynamical Systems B, 18*, 403–415.
6. Hommes, C. H. (2006). *Handbook of computational economics* (Vol. 2, pp. 1109–1186). Amsterdam: Elsevier.
7. Kirman, A. (1992). Who or what does the representative agent represent? *Journal of Economic Perspectives, 6*, 117–136.

8. Krejčí, P., Melnik, S., Lamba, H., & Rachinskii, D. (2014). Analytical solution for a class of network dynamics with mechanical and financial applications. *Physical Review E, 90*, 032822.
9. Lamba, H. (2010). A queueing theory description of fat-tailed price returns in imperfect financial markets. *The European Physical Journal B, 77*, 297–304.
10. Lamba, H., & Seaman, T. (2008). Rational expectations, psychology and learning via moving thresholds. *Physica A: Statistical Mechanics and its Applications, 387*, 3904–3909.
11. Mantegna, R., & Stanley, H. (2000). *An introduction to econophysics*. Cambridge: Cambridge University Press.
12. Minsky, H. P. (1992). *The Financial Instability Hypothesis*. The Jerome Levy Institute Working Paper, 74.
13. Muth, J. A. (1961). Rational expectations and the theory of price movements. *Econometrica*, 6. https://doi.org/10.2307/1909635
14. Scharfstein, D., & Stein, J. (1990). Herd behavior and investment. *The American Economic Review, 80*(3), 465–479.
15. Soros, G. (1987). *The alchemy of finance*. New York: Wiley.

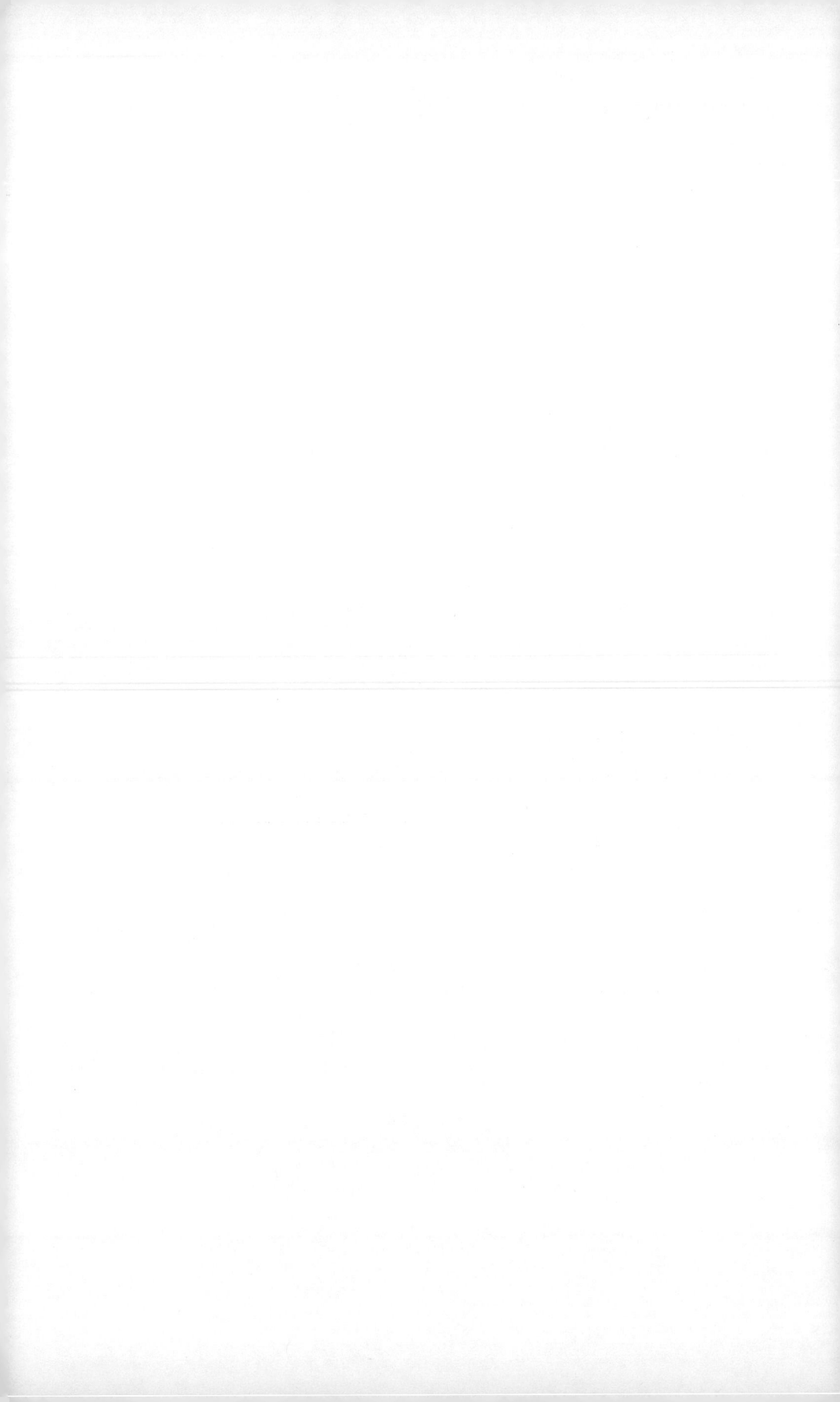

Product Innovation and Macroeconomic Dynamics

Christophre Georges

Abstract We develop an agent-based macroeconomic model in which product innovation is the fundamental driver of growth and business cycle fluctuations. The model builds on a hedonic approach to the product space and product innovation developed in Georges (A hedonic approach to product innovation for agent-based macroeconomic modeling, 2011).

Keywords Innovation · Growth · Business cycles · Agent-based modeling · Agent-based macroeconomics

1 Introduction

Recent evidence points to the importance of product quality and product innovation in explaining firm level dynamics. In this paper we develop an agent-based macroeconomic model in which both growth and business cycle dynamics are grounded in product innovation. We take a hedonic approach to the product space developed in [23] that is both simple and flexible enough to be suitable for modeling product innovation in the context of a large-scale, many-agent macroeconomic model.

In the model, product innovation alters the qualities of existing goods and introduces new goods into the product mix. This novelty leads to further adaptation by consumers and firms. In turn, both the innovation and adaptation contribute to complex market dynamics. Quality adjusted aggregate output exhibits both secular endogenous growth and irregular higher frequency cycles. There is ongoing churning of firms and product market shares, and the emerging distribution of these shares depends on opportunities for niching in the market space.

C. Georges (✉)
Hamilton College, Department of Economics, Clinton, NY, USA
e-mail: cgeorges@hamilton.edu

2 Background

Recent research suggests that product innovation is a pervasive force in modern advanced economies. For example, Hottman et al. [28] provide evidence that product innovation is a central driver of firm performance. They offer an accounting decomposition that suggests that 50–70% of the variance in firm size at the aggregate level can be attributed to differences in product quality, whereas less than 25% can be attributed to differences in costs of production. Further, in their analysis, individual firm growth is driven predominantly by improvements in product quality. Similarly, Foster et al. [21] found that firm level demand is a more powerful driver of firm survival than is firm level productivity. Broda and Weinstein [7] and Bernard et al. [6] further document the substantial pace of churning in product markets with high rates of both product creation and destruction and changes of product scope at the firm level.

We explore the implications of product innovation for growth and business cycle fluctuations in an agent-based macroeconomic model. While there is a literature (e.g., [24]) that attributes economic growth to growth in product variety, variety is only one expression of product innovation. Both the product turnover and skewed distributions of firm sizes and product market shares that we observe indicate that it is conventional for some products to drive out other products and develop outsized market shares due to superiority in perceived quality. Our agent-based approach is well suited to model the types of heterogeneity and churning dynamics that we observe empirically.

Our approach revisits [31, 32]. Preferences are defined over a set of product characteristics, and products offer various bundles of these characteristics. As Lancaster notes, the characteristics approach allows for new products to be introduced seamlessly, as new products simply offer new possibilities for the consumption of an unchanging set of characteristics.

Of course there is a large literature on product innovation, and there are a number of existing models that bear some relation to the one developed here. See, for example, [2, 4, 11–16, 25, 30, 33, 34, 36–38]. For comparisons with the current approach, see [23].

The current paper is in the recent tradition of agent-based macroeconomics. This literature builds macroeconomic models from microfoundations, but in contrast to standard macroeconomic practice, treats the underlying agents as highly heterogeneous, boundedly rational, and adaptive, and does not assume a priori that markets clear. See, e.g., [17–20, 26].

3 The Macroeconomic Environment

Our goal is to understand the role of product innovation in driving growth and fluctuations in a very simple macroeconomic environment. Here are the fundamental features of the model.

Product Innovation and Macroeconomic Dynamics

- There are n firms, each of which produces one type of good at any time.
- There are m characteristics of goods that consumers care about. Each good embodies distinct quantities of these characteristics at any given time. Product innovation affects these quantities.
- The probability that a firm experiences a product innovation at any time depends on its recent investments in R&D, which in turn is the outcome of a discrete choice rule.
- Each firm produces with overhead and variable labor. It forecasts the final demand for its product by extrapolating from recent experience, sets its price as a constant mark-up over marginal cost, and plans to produce enough of its good to meet its expected final demand given this price.
- There is a single representative consumer who spends all of her labor income each period on consumption goods and searches for better combinations of products to buy within her budget.
- If a firm becomes insolvent, it exits the market and is replaced by a new entrant.

4 Consumer Preferences

The representative consumer's preferences are defined on the characteristics space. Specifically, the momentary utility from consuming the vector $z \in R_m$ of hedonic characteristics is $u(z)$.[1]

In addition to this utility function, the consumer has a home production function $g(q)$ that maps bundles $q \in R_n$ of products into perceived bundles $z \in R_m$ of the characteristics.[2]

More specifically, we assume that the representative consumer associates with each good i a set of *base characteristic* magnitudes z-base$_i \in R_m$ per unit of the good, as well as a set of *complementarities* with other goods. If the consumer associates good k as complementary to good i, then she associates with the goods pair (i,k) an additional set of characteristic magnitudes z-comp$_{i,k} \in R_m$, per composite unit $q_{i,k} = \theta(q_i, q_k)$ of the two goods (defined below).[3]

Intuitively, a box of spinach may offer a consumer certain quantities of subjectively valued characteristics like nutrition, flavor, and crunchiness. However, the flavor characteristic might also be enhanced by consuming the spinach in combination with a salad dressing, so that the total quantity of flavor achieved by

---

[1] While we are working with a representative consumer in the present paper for convenience, it is a simple step in the agent-based modeling framework to relax that assumption and allow for idiosyncratic variation of consumer preferences.

[2] This is essentially the approach taken by Lancaster, and shares some similarities with others such as [5, 35]. The primary deviation of our approach from that of Lancaster is the construction of our home production function $g(q)$.

[3] This vector is associated with good i, and it is convenient to assume that the complementarities are independent across goods (i.e., that the vectors z-comp$_{i,k}$ and z-comp$_{k,i}$ are independent).

eating these in combination is greater than the sum of the flavor quantities from consuming each separately.

Similarly, in isolation, an Apple iPad may provide a consumer some modest degree of entertainment, but this entertainment value is dramatically enhanced by consuming it along with a personal computer, an internet access subscription, electricity, apps, and so on.

We assume that both base characteristic magnitudes and complementary characteristic magnitudes are additive at the level of the individual good. Thus, for good i and hedonic characteristic j, the consumer perceives

$$z_{i,j} = \text{z-base}_{i,j} \cdot q_i + \sum_k \text{z-comp}_{i,j,k} \cdot q_{i,k}. \tag{1}$$

These characteristic magnitudes are then aggregated over products by a CES aggregator:

$$z_j = \left[\sum_{i=1}^{n} z_{i,j}^{\rho_1}\right]^{1/\rho_1} \tag{2}$$

with $\rho_1 < 1$. Equations (1) and (2) define the mapping $g(q)$ introduced above. The CES form of (2) introduces some taste for variety across products.[4]

We assume that the utility function u for the representative consumer over hedonic characteristics is also CES, so that

$$u = \left[\sum_{j=1}^{m} (z_j + \bar{z}_j)^{\rho_2}\right]^{1/\rho_2} \tag{3}$$

where $\bar{z}_j$ is a shifter for characteristic j (see [29]), and $\rho_2 < 1$. Thus, utility takes a nested CES form. Consumers value variety in both hedonic characteristics and in products.

Finally, we specify the aggregator for complements $\theta(q_i, q_k)$ as floor(min($q_i \cdot \frac{1}{\lambda}, q_k \cdot \frac{1}{\lambda}$)) $\cdot \lambda$. I.e., complementarities are defined per common (fractional) unit λ consumed.[5]

---

[4]Note that if $\rho_1 = 1$, the number of viable products in the economy would be strongly limited by the number of hedonic elements, as in Lancaster, who employs a linear activity analysis to link goods and characteristics.

[5]Note that this introduces a (fractional) integer constraint on the consumer's optimization and search problem. $\lambda > 0$ but need not be less than one.

5 Product Innovation

A product innovation takes the form of the creation of a new or improved product that, from the point of view of the consumer, combines a new set of characteristics, or enhances an existing set of characteristics, when consumed individually or jointly with other products. The new product will be successful if it is perceived as offering utility (in combination with other goods) at lower cost than current alternatives. The product may fail due to high cost, poor search by consumers, or poor timing in terms of the availability or desirability of other (complementary and non-complementary) goods.

In the present paper, at any time the base and complementary set of hedonic characteristic magnitudes (z-base$_i$ and z-comp$_{i,k}$) associated by the consumer with good i are coded as m dimensional vectors of integers. These characteristics vectors are randomly initialized at the beginning of the simulation.

A product innovation is then a set of random (integer) increments (positive or negative) to one or more elements of z-base$_i$ or z-comp$_{i,k}$. Product innovation for continuing firms is strictly by mutation. Product innovations can be positive or negative. I.e., firms can mistakenly make changes to their products that consumers do not like. However, there is a floor of zero on characteristic values. Further, innovations operate through preferential attachment; for a firm that experiences a product innovation, there is a greater likelihood of mutation of non-zero hedonic elements.[6]

The probability that a firm experiences product innovation in any given period t is increasing in its recent *R&D* activity.

6 R&D

The R&D investment choice is binary—in a given period a firm either does or does not engage in a fixed amount of R&D. If a firm engages in R&D in a given period, it incurs additional overhead labor costs R in that period.

In making its R&D investment decision at any time, the firm compares the recent profit and R&D experiences of other firms and acts according to a discrete choice rule. Specifically, firms observe the average recent profits π_H and π_L of other firms with relatively high and low recent R&D activity.[7] Firms in the lower profit group switch their R&D behavior with a probability related to the profitability differential between the two groups. Specifically, they switch behavior with probability $2\Phi - 1$, where

---

[6]This weak form of preferential attachment supports specialization in the hedonic quality space.

[7]Each firm's recent profits and recent R&D activity are (respectively) measured as exponentially weighted moving averages of its past profits and R&D activity.

$$\Phi = \frac{e^{\gamma \pi_1}}{e^{\gamma \pi_1} + e^{\gamma \pi_2}}$$

$\gamma > 0$ measures the intensity of choice and π_1 and π_2 are measures of the average recent profits of the high and low profit R&D groups.[8] There is additionally some purely random variation in R&D choice.

7 Production and Employment

Each firm i produces its good with labor subject to a fixed labor productivity A_i and hires enough production labor to meet its production goals period by period. Each firm also incurs a fixed overhead labor cost H in each period and additional overhead R&D labor cost R in any period in which it is engaged in R&D. In this paper, our focus is on product innovation rather than process innovation. Consequently, we suppress process innovation and hold A_i, H, and R constant over time.[9]

8 Consumer Search

The consumer spends her entire income each period and selects the shares of her income to spend on each good. Each period, she experiments with random variations on her current set of shares. Specifically, she considers randomly shifting consumption shares between some number of goods, over some number of trials, and selects among those trials the set of shares that yields the highest utility with her current income. While the consumer engages in limited undirected search and is not able to globally maximize utility each period, she can always stick to her current share mix, and so never selects new share mixes that reduce utility below the utility afforded by the current one. I.e., the experimentation is a thought exercise not an act of physical trial and error.

9 Entry and Exit

When a firm does not have enough working capital to finance production, it shuts down and is replaced. The new firm adopts (imitates) the product characteristics of a randomly selected existing firm, retains the exiting firm's current share of consumer demand, and is seeded with startup capital.

---

[8]I.e., if $\pi_H > \pi_L$, then $\pi_1 = \pi_H$ and $\pi_2 = \pi_L$, and firms with relatively low recent R&D activity switch R&D on with a probability that is greater the larger is the difference between π_1 and π_2.
[9]Process innovation that affects these parameters across firms and over time can easily be introduced.

10 Timing

The timing of events within each period t is as follows:

- R&D: firms chose their current R&D investment levels (0 or 1).
- Innovation: firms experience product innovation with probabilities related to their recent R&D investments.
- Production: firms forecast sales, hire production labor, and produce to meet forecasted demand.
- Incomes: wages and salaries are paid to all labor employed (production, standard overhead, R&D overhead).
- Consumer search: the representative consumer searches and updates the consumption basket.
- Sales: the consumer spends all of the labor income (above) on the consumption basket.
- Entry and Exit: firms with insufficient working capital are replaced.

11 Features of the Model

If we eliminate all heterogeneity in the model, there is a steady state equilibrium which is stable under the simple dynamics described above. Assuming that all firms engage in R&D investment, steady state aggregate output is a multiple of the combined overhead labor cost $H + R$ and is independent of product quality. See the Appendix for details. Quality adjusted output and consumer utility, on the other hand, will grow over time in equilibrium as product quality improves.

With firm heterogeneity, the evolution of the hedonic characteristics of firms' products will influence product market shares as well as aggregate activity and living standards.

The stochastic evolution of product quality with zero reflective lower bounds on individual characteristic magnitudes will tend to generate skewness in the distribution of individual product qualities and thus in the distribution of market shares.

Working against the skewness of the distribution of product shares is the variety-loving aspect of consumer preferences and the number of product characteristics and complementarities. The greater the number of characteristics, the greater the opportunities for individual firms to find niches in the characteristics space. A small number of characteristics relative to the number of products creates a "winner take all" environment in which firms compete head to head in a small number of dimensions, whereas a large number of characteristics may create opportunity for the development of a (so-called) "long tail" of niche products.[10]

---

[10] A recent literature argues that the growth of internet retail has allowed niche products that better suit existing consumer preferences to become profitable, eroding the market shares of more broadly popular "superstar" products [8, 9].

Turning to aggregate activity, with heterogeneity, there are several channels through which product innovation may be a source of output and utility fluctuations.[11]

First, product innovation produces ongoing changes in the pattern of demand across goods. This may affect aggregate output for several reasons. For one, it takes time for production to adjust to the changing demand pattern resulting in both direct and indirect effects on output. Additionally, as demand patterns change, demand may shift between goods with different production technologies.[12] Further, if network effects are large or product innovation results in highly skewed distributions of firm sizes and product shares, then idiosyncratic firm level fluctuations may not wash out in the aggregate, even as the number of firms and products grows large.[13]

A second mechanism is that R&D investment has a direct impact on aggregate demand by increasing labor income and thus consumer spending on all goods with positive demand shares. Since R&D investment decisions are conditioned by social learning, there can be cyclical herding effects in this model.

Finally, quality adjusted output and consumer utility will fluctuate and grow due to the evolution of both physical output and product qualities.

12 Some Simulation Results

Runs from a representative agent baseline version of the model behave as expected. If all firms have the same parameters and the same productivities and hedonic qualities and all engage in R&D, then output converges to the analytical equilibrium discussed in the Appendix.

For the heterogeneous firms model, ongoing endogenous innovation tends to generate stochastic growth in consumer utility and ongoing fluctuations in total output. For the simulation experiments discussed here, as the number of firms increases, the output fluctuations become increasingly dominated by variations in R%D investment spending through the herding effect noted above.

---

[11] We exclude other sources of business cycle fluctuations and growth from the model to focus on the role of product innovation.

[12] Below, we suppress the latter effect, standardizing productivity across firms in order to focus more directly on product innovation.

[13] Intuitively, if some firms or sectors remain large and/or central to the economy, even under highly disaggregated measurement, then idiosyncratic shocks will have aggregate effects. For formalizations, see, for example, [1, 3, 10, 18, 22, 27].

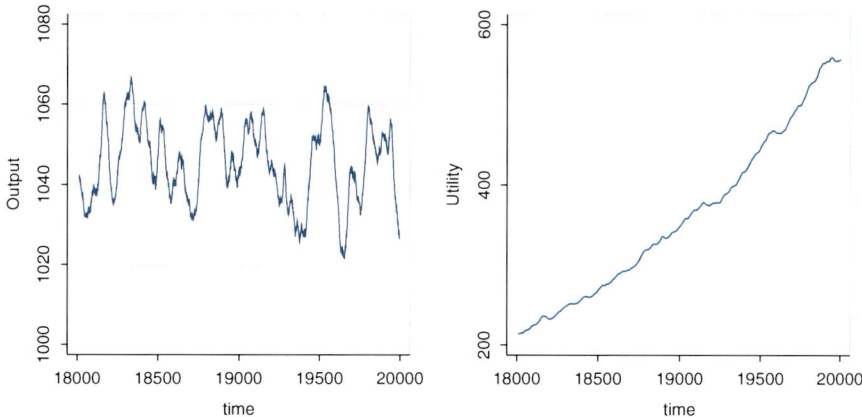

Fig. 1 Representative run. Output and utility, for rounds 18,000–20,000. Output is measured weekly as average output for the last quarter. The number of hedonic characteristics is $m = 50$ and the intensity of choice for firms' R&D decisions is $\gamma = 0.2$

Figure 1 is produced from a representative run with 1000 firms. In this case the number of product characteristics is 50, $\rho_1 = \rho_2 = 0.8$, and mutation is multiplicative.[14] There is no variation in labor productivity A_i over time or across firms i.

The time period is considered to be a week, so we are showing approximately 40 years of simulated data well after transitory growth dynamics have died out. Aggregate output exhibits irregular fluctuations and cycles; peak to trough swings in GDP are on the order of 1–3%. Utility also fluctuates with both output and the evolution of the quality of products produced and consumed, but also grows due to long-term net improvements in product quality. These net improvements are driven both directly and indirectly by innovation. Existing firms' product qualities may rise or fall in response to innovations, but also as consumers shift toward higher quality product mixes, less successful firms are driven from the market, and new firms enter.

The evolution of output in Fig. 1 follows the evolution of R&D investment spending fairly closely (with correlation close to 0.9). There is substantial additional variation of firm level output due to the churning of demand shares, with much of that variation washing out in the aggregate.

When the number of firms engaging in R&D increases, spending on R&D increases, driving up consumption, output, and utility.[15] A second effect is that more firms innovate over time, driving the rate of growth of utility upward. Figure 2 illustrates these level and growth effects on utility (equivalently, quality adjusted

---

[14]The number of firm in this simulation is small (1000), but can easily be scaled up to several million. Similarly the number of hedonic characteristics (50) can be increased easily (though in both cases, of course, at some computational cost).

[15]In the representative agent case, the multiplier for output is $\frac{1}{\eta-1}$, where η is the markup.

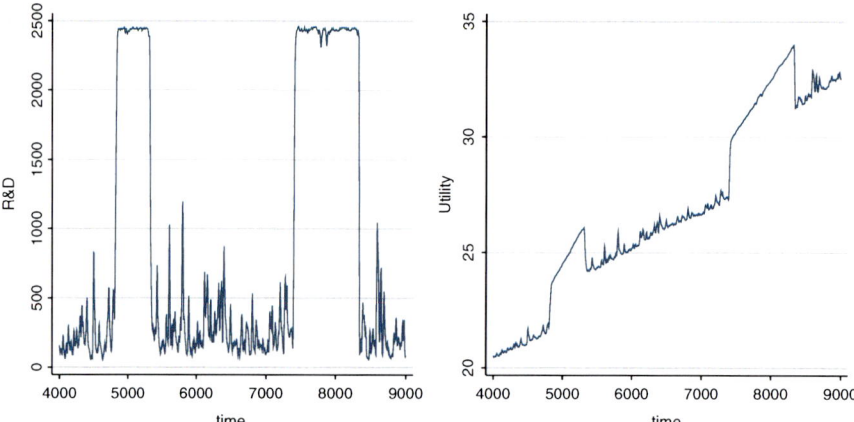

Fig. 2 Run similar to that in Fig. 1, but with high intensity of choice $\gamma = 10$ for R&D decisions. The number of firms n is 2500

output) in a run with the intensity of choice for firms' R&D decisions increased dramatically (from $\gamma = 0.2$ above, to $\gamma = 10$).

Here, as the relative profits of firms engaging in *R&D* evolves, there is strong herding of the population of firms toward and away from engaging in *R&D*. These shifts lead to shifts in both the level and growth rate of overall consumer utility.

Note that for the individual firm, R&D investment involves a tradeoff between its current profit and its future potential profits. R&D spending reduces the firm's profit today (by R), but potentially raises its profit in the future via increased demand for its product. Firms with relatively high demand shares can spread their overhead costs over more output and make positive profit, while firms with chronically low demand shares tend to run losses and ultimately fail due to their inability to cover even their basic overhead costs (H). Nevertheless, current R&D is a direct drag on current profit, with the potential benefits via product innovation and increased demand accruing in the future. This tension between the short and medium term costs and benefits of *R&D* investment supports the complex patterns of R&D investment behavior in the model.

Now consider the distribution of firm sizes as measured by final demand shares. We start the simulations with characteristic magnitudes (z-base$_i$ and z-comp$_{i,k}$) initialized to random sequences of zeros and ones, so that there is idiosyncratic variation in product quality across firms. However, the representative consumer initially sets equal shares across firms. As each simulation proceeds, the consumer searches for better bundles of goods.

If there is neither product nor process innovation, then the optimum bundle for the consumer is static, and demand is redistributed across firms over time, leading to a skewed distribution of demand shares. The degree of skewness is influenced by the CES utility elasticities. Lower values of ρ_1 and ρ_2 indicate a greater taste for variety in characteristics and products, limiting the impact of product quality

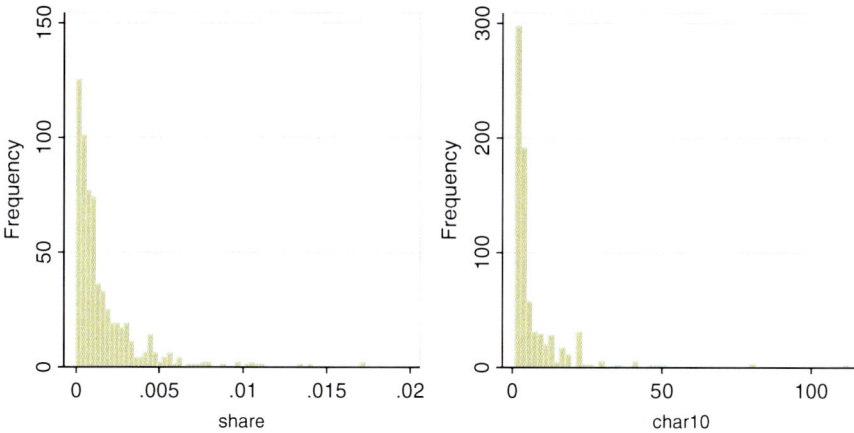

Fig. 3 Distributions of product shares and values of hedonic characteristic 10 by firm at time 20,000 in the run in Fig. 1

on market shares. The shape of the distribution is also affected by the ratio $\frac{m}{n}$ of product characteristics to goods, as well as the distribution of complementarities and the entry and exit process. For example, if $\frac{m}{n}$ is small, then there is less room for firms to have independent niches in the product space, and so the market tends to become more concentrated.

Including ongoing endogenous product innovation leads to ongoing changes in the relative qualities of goods and so ongoing churning in demand shares. The share distribution follows a similar pattern to that above in early rounds. Starting from a degenerate distribution with all the mass on $1/n$, it first spreads out and then becomes skewed as shares are progressively reallocated among products.

As innovation proceeds, and product churning emerges, the degree of skewness continues to evolve. The ultimate limiting distribution depends on the stochastic process for product innovation in addition to the factors above (ρ_1, ρ_2, $\frac{m}{n}$, the distribution of complementarities, and the entry and exit process).

Figure 3 shows the distribution of product shares and the values of one of the 50 hedonic characteristics (characteristic 10) over the firms at time 20,000 in the run in Fig. 1. Here we exclude zero shares and zero characteristic values (held by 374 and 262 of the 1000 firms respectively) and display the distributions of non-zero values.

13 Conclusion

We have developed an agent-based macroeconomic model in which product innovation is the fundamental driver of growth and business cycle fluctuations. The model builds on a hedonic approach to the product space and product innovation proposed in [23]. Ongoing R&D activities, product innovation, and consumer search yield ongoing firm level and aggregate dynamics. Holding productivity constant, output fluctuates but does not grow in the long run. However, utility, or equivalently quality adjusted output, does exhibit long run endogenous growth as a result of product innovation. The distribution of product market shares tends to become skewed, with the degree of skewness depending on the opportunities for niching in the characteristics space. As the number of firms grows large, business cycle dynamics tend to become dominated by an innovation driven investment cycle.

Acknowledgements I am grateful to participants at CEF 2015 and a referee for useful comments, suggestions, and discussions. All errors are mine.

Appendix: Representative Agent Benchmark

Consider the case in which all firms are identical and each starts with a 1/n share of the total market. Suppose further that all firms engage in R&D in every period and experience identical innovations over time.

In this case, the equal individual market shares will persist, since there is no reason for consumers to switch between firms with identical product qualities and identical prices. Further, there is a unique equilibrium for the real production and sales of consumer goods Y at which demand and supply are in balance. At this equilibrium, aggregate real activity depends on the markup η, the per firm overhead labor costs H and R, the wage rate W (for production workers), labor productivity A (for production workers), and the number of firms n. Specifically, at this equilibrium $Y = (\frac{1}{\eta-1}) \cdot \frac{A}{W} \cdot (H+R) \cdot n$. See below for details. Further, this equilibrium is a steady state of the agent dynamics in the model and is locally stable under those dynamics. If, for example, firms all start with production less than steady state production, then since the markup $\eta > 1$, demand will be greater than production for each firm, and production will converge over time to the steady state equilibrium.

We can see this as follows. Since all firms are identical, they produce identical quantities q of their goods. Then total labor income is:

$$E = n \cdot \left[\frac{W}{A} \cdot q + H + R \right] \qquad (4)$$

Each firm also charges an identical price p for its good, which is a markup η on marginal cost

$$p = \eta \cdot \text{MC}$$
$$\text{MC} = \frac{W}{A} \qquad (5)$$

Product Innovation and Macroeconomic Dynamics

and so produces and sells

$$q = \frac{E}{n \cdot p} \qquad (6)$$

units of its good.

These relationships yield the following steady state equilibrium value for (per firm) production.

$$q^* = \frac{(H+R) \cdot \frac{A}{W}}{\eta - 1} \qquad (7)$$

We can see that $\partial q^*/\partial(H+R) > 0$, $\partial q^*/\partial A > 0$, $\partial q^*/\partial W < 0$, and $\partial q^*/\partial \eta < 0$. These are all demand driven. An increase in the cost of overhead labor (H or R) raises the incomes of overhead workers, raising AD and thus equilibrium output. An increase in labor productivity A will cause firms to lower their prices, raising aggregate demand and equilibrium output. Similarly an increase in the wage rate W of production workers or in the mark-up η will cause firms to raise their prices, lowering AD and equilibrium output.

Further, if in this representative agent case firms follow simple one period adaptive expectations for demand, then firm level output dynamics are given by:

$$q_t = \frac{1}{\eta} \cdot q_{t-1} + \frac{1}{\eta} \cdot (H+R) \cdot \frac{A}{W} \qquad (8)$$

Thus, given $\eta > 1$, the steady state equilibrium q^* is asymptotically stable.

Total market output in the steady state equilibrium is just

$$Y^* = n \cdot q^*$$
$$= \frac{n \cdot (H+R) \cdot \frac{A}{W}}{\eta - 1} \qquad (9)$$

Clearly, this will be constant as long as there is no change in the parameters H, R, A, W, η and n. If we were to allow them to change, growth in the number of firms n or the productivity of production labor A would cause equilibrium total production Y^* to grow over time, while balanced growth in production wages W and labor productivity A would have no impact on equilibrium production. Note that, in the full heterogeneous agent model, while all of the parameters above are fixed, the fraction of firms adopting R&D investment varies endogenously, contributing to the disequilibrium dynamics of the model.

Importantly for the present paper, note that the representative agent steady state equilibrium production Y^* above is entirely independent of product quality. Improvements in product quality will, however, increase consumer utility, or equivalently, the quality adjusted value of total production at this equilibrium.

Specifically, given the nested CES formulation of utility, if the magnitudes of all product characteristics grow at rate g due to innovation, then the growth rate of consumer utility will converge in the long run to g. If the magnitudes of different characteristics grow at different rates, the growth rate of utility will converge to the rate of growth of the fastest growing characteristic. All else equal, the long run growth path of utility will be lower in the latter case than in the former case.

It is also worth noting that, at the representative agent steady state equilibrium above, firms make zero profit. Revenues net of the cost of production labor are just great enough to cover all overhead labor costs. Once we move to the heterogeneous firm case, in the comparable steady state equilibrium, profits are distributed around zero across firms. Firms will vary as to R&D investment status, with firms who are not engaging in R&D investment saving overhead cost R per period. Firms will also vary with respect to demand shares (driven by product qualities which are themselves related to past investments in R&D), and firms with relatively high demand shares are able to spread overhead cost over greater sales. Thus, for two firms with the same overhead cost (i.e., the same current R&D investment status), the firm with greater demand for its product will have higher profit, while for two firms with the same product demand shares, the one with the lower overhead cost (lower current R&D investment) will have higher profit. Firms that face chronic losses will eventually fail and be replaced.

References

1. Acemoglu, D., Carvalho, V. M., Ozdaglar, A., & Tahbaz-Selehi, A. (2012). The network origins of aggregate fluctutations. *Econometrica, 80*(5), 1977–1016.
2. Acemoglu, D., Akcigit, U., Bloom, N., & Kerr, W. R. (2013). *Innovation, reallocation and growth*. In NBER WP 18933.
3. Acemoglu, D., Akcigit, U., & Kerr, W. R. (2015). Networks and the macroeconomy: An empirical exploration. In *NBER macroeconomic annual* (Vol. 30).
4. Akcigit, U., & Kerr, W. R. (2018). Growth through heterogeneous innovations. *Journal of Political Economy, 126*(4), 1374–1443.
   Akcigit, U., & Kerr, W. R. (2018). *Growth through heterogeneous innovations*. Journal of Political Economy (Vol. 136) No.4
5. Becker, G. S. (1965). A theory of the allocation of time. *Economic Journal, 75*(299), 493–517.
6. Bernard, A. B., Redding, S. J., & Schott, P. K. (2010). Multi-product firms and product switching. *American Economic Review, 100*(1), 70–97.
7. Broda, C., & Weinstein, D. E. (2010). Product creation and destruction: Evidence and price implications. *American Economic Review, 100*(3), 691–723.
8. Brynjolfsson, E., Hu, Y., & Smith, M. D. (2010). *The longer tail: The changing shape of Amazon's sales distribution curve*. Manuscript, MIT Sloan School.
9. Brynjolfsson, E., Hu, Y., & Simester, D. (2011). Goodbye Pareto principle, hello long tail: The effect of search costs on the concentration of product sales. *Management Science, 57*(8), 1373–1386.
10. Carvalho, V. M., & Gabaix, X. (2013). The great diversification and its undoing. *American Economic Review, 103*(5), 1697–1727.

11. Chen, S. H., & Chie, B. T. (2005). A functional modularity approach to agent-based modeling of the evolution of technology. In A. Namatame et al. (Eds.), *The complex networks of economic interactions: Essays in agent-based economics and econophysics* (Vol. 567, pp. 165–178). Heidelberg: Springer.
12. Chen, S. H., & Chie, B. T. (2007). Modularity, product innovation, and consumer satisfaction: An agent-based approach. In *International Conference on Intelligent Data Engineering and Automated Learning*.
13. Chiarli, T., Lorentz, A., Savona, M., & Valente, M. (2010). The effect of consumption and production structure on growth and distribution: A micro to macro model. *Metroeconomica, 61*(1), 180–218.
14. Chiarli, T., Lorentz, A., Savona, M., & Valente, M. (2016). The effect of demand-driven structural transformations on growth and technical change. *Journal of Evolutionary Economics, 26*(1), 219–246.
15. Chie, B. T., & Chen, S. H. (2014). *Non-price competition in a modular economy: An agent-based computational model*. Manuscript, U. Trento.
16. Chie, B. T., & Chen, S. H. (2014). Competition in a new industrial economy: Toward an agent-based economic model of modularity. *Administrative Sciences, 2014*(4), 192–218.
17. Dawid, H., Gemkow, S., Harting, P., van der Hoog, S., & Neugart, M. (2018). Agent-based macroeconomic modeling and policy analysis: The Eurace@Unibi model. In S. H. Chen, M. Kaboudan, & Ye-Rong Du (Eds.), *Handbook of computational economics and finance*. Oxford: Oxford University Press.
18. Delli Gatti, D., Gaffeo, E., Gallegati, M., & Giulioni, G. (2008). *Emergent macroeconomics: An agent-based approach to business fluctuations*. Berlin:Springer.
19. Dosi, G., Fagiolo, G., & Roventini, A. (2005). An evolutionary model of endogenous business cycles. *Computational Economics, 27*(1), 3–34.
20. Fagiolo, G., & Roventini, A. (2017). Macroeconomic policy in DSGE and agent-based models redux: New developments and challenges ahead. *Journal of Artificial Societies and Social Simulation, 20*(1).
21. Foster, L., Haltiwanger, J. C., & Syverson, C. (2008). Reallocation, firm turnover, and efficiency: Selection on productivity or profitability? *American Economic Review, 98*(1), 394–425.
22. Gabaix, X. (2011). The granular origins of aggregate fluctuations. *Econometrica, 79*, 733–772.
23. Georges, C. (2011). *A hedonic approach to product innovation for agent-based macroeconomic modeling*. Manuscript, Hamilton College.
24. Grossman, G. M., & Helpman, E. (1991). *Innovation and growth in the global economy*. Cambridge, MA: The MIT Press.
25. Hidalgo, C. A., Klinger, B., Barabasi, A. L., & Hausmann, R. (2007). The product space conditions the development of nations. *Science, 317*, 482–487.
26. Hommes, C., & Iori, G. (2015). Introduction to special issue crises and complexity. *Journal of Economic Dynamics and Control, 50*, 1–4.
27. Horvath, M. (2000). Sectoral shocks and aggregate fluctuations. *Journal of Monetary Economics, 45*, 69–106.
28. Hottman, C., Redding, S. J., & Weinstein, D. E. (2014). *What is 'firm heterogeneity' in trade models? The role of quality, scope, markups, and cost*. Manuscript, Columbia U.
29. Jackson, L. F. (1984). Hierarchic demand and the engel curve for variety. *Review of Economics and Statistics, 66*, 8–15.
30. Klette, J., & Kortum, S. (2004). Innovating firms and aggregate innovation. *Journal of Political Economy, 112*(5), 986–1018.
31. Lancaster, K. J. (1966). A new approach to consumer theory. *Journal of Political Economy, 74*(2), 132–157.
32. Lancaster, K. J. (1966). Change and innovation in the technology of consumption. *American Economic Review, 56*(1/2), 14–23.

33. Marengo, L., & Valente, M. (2010). Industrial dynamics in complex product spaces: An evolutionary model. *Structural Change and Economic Dynamics, 21*(1), 5–16.
34. Pintea, M., & Thompson, P. (2007). Technological complexity and economic growth. *Review of Economic Dynamics, 10*, 276–293.
35. Strotz, R. H. (1957). The empirical implications of a utility tree. *Econometrica, 25*(2), 269–280.
36. Sutton, J. (2007) Market share dynamics and the "Persistance of leadership debate. *American Ecmomic Review, 97*(1), 222–241.
37. Valente, M. (2012). Evolutionary demand: A model for boundedly rational consumers. *Journal of Evolutionary Economics, 22*, 1029–1080.
38. Windrum, P., & Birchenhall, C. (1998). Is product life cycle theory a special case? Dominant designs and the emergence of market niches through coevolutionary-learning. *Structural Change and Economic Dynamics, 9*(1), 109–134.

Part II
New Methodologies and Technologies

Measuring Market Integration: US Stock and REIT Markets

Douglas W. Blackburn and N. K. Chidambaran

Abstract Tests of financial market integration traditionally test for equality of risk premia for a common set of assumed risk factors. Failure to reject the equality of market risk premia can arise because the markets do not share a common factor, i.e. the underlying factor model assumptions are incorrect. In this paper we propose a new methodology that solves this joint hypothesis problem. The first step in our approach tests for the presence of common factors using canonical correlation analysis. The second step of our approach subsequently tests for the equality of risk premia using Generalized Method of Moments (GMM). We illustrate our methodology by examining market integration of US Real Estate Investment Trust (REIT) and Stock markets over the period from 1985 to 2013. We find strong evidence that REIT and stock market integration varies through time. In the earlier part of our data period, the markets do not share a common factor, consistent with markets not being integrated. We also show that during this period, the GMM tests fail to reject the equality of risk premia, highlighting the joint hypothesis problem. The markets in the latter half of our sample show evidence of both a common factor and the equality of risk premia suggesting that REIT and Stock markets are integrated in more recent times.

Keywords Market integration · Canonical correlation · Financial markets · REIT · Stock · GMM · Risk premia · Common factor · Factor model · Principal components

---

D. W. Blackburn · N. K. Chidambaran (✉)
Fordham University, New York, NY, USA
e-mail: blackburn@fordham.edu; chidambaran@fordham.edu

1 Introduction

Financial market integration has significant impact on the cost of capital, the pricing of risk, and cross-market risk sharing. Understanding whether markets are or are not integrated is of great interest and importance to academics, practitioners, and policymakers. Though definitions vary, two markets are said to be integrated if two assets from two different markets with identical exposure to the same risk factors yield the same expected returns (see, e.g., [4, 10, 24]). That is, financial market integration requires that the same risk factors explain prices of assets in both markets and that the risk-premia associated with the factors are equal across the two markets. Empirical tests for determining whether markets have equal risk premia for shared risk factors are econometrically challenging. First, the tests suffer from a joint hypothesis problem as they require the right factor model specification with common factors correctly identified. Second, financial market integration is a changing and volatile process introducing noise in the data. In this paper, we present a new approach to testing for market integration that untangles these issues and we illustrate the methodology by examining integration of US Real Estate Investment Trust (REIT)[1] and stock markets.

Our approach for testing financial market integration is based on the Canonical Correlation (CC) technique and is in two steps. In the first step, we determine the factor model describing returns in each market. We begin with determining a set of principal components that explain variation of returns in each dataset and search for the presence of a common factor within the factor space of each market using canonical correlations, as in [2].[2] In the absence of a common factor, tests of market integration are meaningless and this first step therefore represents a necessary condition for integration tests. If a common factor is present, we then proceed to the second step and develop economic proxies for the common factor and run tests for equality of risk premia with respect to the economic factors. For each step of our procedure, we develop tests for statistical significance and carefully examine the implications for market integration.

---

[1] REITs are companies that invest in real estate. REITs are modeled after closed-end mutual funds with REIT shareholders able to trade their shares as do mutual fund shareholders. In addition to being able to sell and divest their shares, REIT shareholders receive dividend payouts representing income generated by the real estate properties owned by REITs.

[2] The Canonical Correlation approach determines two factors, one from each data set, that have the highest correlation compared to any other two factors across the two data sets. In determining the set of factors with the highest correlation, the methodology recognizes that the principal components are not a unique representation of the factor model and that a rotation of the principal components is also a valid representation of the factor model. A rotation of the principal component is a set of orthogonal linear combination of the principal components and are called canonical variates. The search for a set of factors that has the highest correlation also includes all canonical variates, i.e. all possible rotations of the principal components. A canonical variate in the first data set that is highly correlated with a canonical variate in the second data set represents a common factor, if the correlation is statistically significant.

Prior integration studies tend to fall into one of two camps. Either equality of risk premia is tested relative to an assumed set of factors or implications of integration are tested while ignoring the requirement that risk-premia be equal for shared risk factors. In regard to the first, researchers assume a particular factor model of returns and examine whether estimated risk premia of assumed common factors are equal across markets using econometric techniques such as GMM or the [8] methodology. Liu et al. [15] assume the Capital Asset Pricing Model (CAPM) and Ling and Naranjo [13] use a multi-factor model in their study of equity and REIT market integration. This is a direct test of market integration as it explicitly determines the consistency of pricing across two markets relative to an assumed pricing model. Bekeart and Harvey [4] assume a nested world and local CAPM model in their investigation of world market integration. While these tests work very well if the model and the assumptions are valid, results are subject to a joint hypothesis problem. A rejection of integration may be a result of the markets not being integrated or a result of using the incorrect model. It is also possible for tests to not reject integration even when markets do not share a common factor in the first place. By separately testing for the presence of a common factor, our approach resolves the dual hypothesis issue.

The second strand of research studies market integration by examining criteria other than equality of risk premia. Such tests are indirect measures of market integration as they do not explicitly compare the pricing of assets across markets. Two common measures, for example, are R-square (see [20]) and comovement (see [3]). Studies based on these statistics seek to imply trends in integration by measuring changes in the explanatory power of the assumed common factors or changes in the economic implications of integration such as the degree of diversification benefits, respectively. However, trends in R-Square and comovement only suggest an increase in the pervasiveness of a set of common factors but it is not clear what level of R-Square or comovement is required to imply market integration or the existence of a common factor. Without knowing whether markets are or are not integrated, interpreting these measures is difficult. As our tests directly determine whether markets are integrated in each year, our approach allows us to better interpret trends in measures such as comovement and R-Square.

In addition to the joint hypothesis problem, another significant challenge in testing for financial market integration arises from the fact that integration is not static but is instead a changing and volatile process. Markets can move from being segmented to integrated, or from being integrated to segmented, because of changes in risk premia or changes in risk factors. This volatility introduces noise into the analysis especially when using data spanning periods when markets are integrated and periods when they are not. We test for equality of risk premia only for the time-period when a common factor exists, thereby improving the signal-to-noise ratio and improving the power of our statistical tests. The increased power makes it possible for us to test over a shorter time period, which allows for economically meaningful analysis that accounts for changes in the integration process and time varying risk premia. A related important aspect of the changes in integration over time is that the pattern may be non-monotonic. While the literature has mostly assumed, and looked for, monotonic trends, we do not pre-suppose a monotonic transition of markets from being segmented to being integrated.

We illustrate our methodology and contrast it to the standard approach through a careful study of US REIT and stock market integration over the period from 1985 to 2012. We begin with the standard approach by assuming a single factor model with the Center for Research in Security Prices (CRSP) value-weighted portfolio (MKT) as the common factor. GMM is used to estimate the risk premia of both markets every quarter using daily data. Using this standard approach, we are unable to reject equal risk premia over 1985–1992 and 2000–2012. It is only during the middle period 1993–1999 that equality of risk premia is consistently rejected. Further analysis however reveals problems with these results. Specifically, we show that MKT is not even a REIT factor throughout the 1985–1992 time period. For example, in 1991, the average R-square from regressing REIT returns on MKT is only 1.7%. Additionally, the R-square from regressing MKT on the first eight REIT principal components is a dismal 3.9% suggesting MKT is not in the REIT factor space. This illustrates the surprising result that we may find support for integration even when the assumed factor is not present in the factor models of one, or both, markets.

To contrast the previous result, we first test for the presence of a common factor prior to testing for integration. Using canonical correlation analysis on the factors for REITs and Stocks confirms that there are no common factors for nearly all quarters prior to 1998. From 1998 to 2005, the canonical correlations vary above and below 0.87, the critical level shown by simulations to be necessary for the presence of a common factor. It is only after 2005 that the canonical correlations are consistently higher than 0.87. Tests for unequal premia are only consistently meaningful after 2005 when it is known that the two markets share a common factor. Testing for integration prior to 2005 is therefore meaningless suggesting that the inability to reject equal premia is a spurious result. Using GMM, we confirm the equality of risk premia each quarter from 2005 to 2012 and over the entire sub-sample from 2005 to 2012.

Our choice of the US REIT and stock markets is interesting for a number of reasons. First, there is mixed evidence in the literature as to whether the markets are integrated. On the one side, early studies by Schnare and Struyk [21], Miles et al. [18], and Liu et al. [15] conclude that the two markets are not integrated while on the other side, Liu and Mei [16], Gyourko and Keim [11], and Ling and Naranjo [13] find evidence that the two markets are at least partially integrated.[3] Ling and Naranjo [13] are unable to reject integration after 1990 fewer times than before 1990 suggesting an increase in integration. Our results add to this debate. Second, though the real estate market is very large, the number of REITs is small making integration tests less robust. Ling and Naranjo [13], for example, use a sample of five real estate portfolios and eleven equity portfolios and are unable to

---

[3]Liu et al. [15] find that equity REITs but not commercial real estate securities in general are integrated with the stock market.

reject a difference in risk premia even though the difference is economically large—1.48% for real estate and 5.91% for equities. Third, REIT data is noisy. Titman and Warga [25] discuss the lack of power in statistical tests brought about by the high volatility of REITs. By first identifying regimes over which integration is feasible, we decrease the adverse effects of noise brought about by these data issues. Last, US Stock and REIT markets present a nice setting to study a new methodology for testing financial market integration without the confounding effects arising from differences in legal regimes, exchange rates, time zone differences, or accounting standards that are present in the study of international stock and bond markets.

The paper proceeds as follows. Section 2 describes the US REIT and Stock return data used in the study. Section 3 presents the traditional approach used to study integration as a first pass. Section 4 presents our new two-step methodology and Section 5 concludes.

2 Data and Relevant Literature

In this section we describe the data and the structure of our tests, informed by the relevant literature in the area.

Financial market integration requires that returns on assets in the two markets be explained by the same set of risk-factors, i.e. a set of common factors, and that assets exposed to these risk factors have the same expected returns. Testing for market integration therefore involves estimating the risk-premia for a set of common factors and is econometrically challenging. Tests inherently involve a dual hypothesis—that the econometrician has the right factor model and that the estimated risk-premia for the assumed risk factors are equal across the two data sets. A rejection of integration may be a result of the markets not being integrated or a result of using the incorrect model. It is also possible for tests to not reject integration even when markets do not share a common factor in the first place.

Two approaches have been used in prior studies to test for market integration. In the first approach, researchers assume a particular factor model of returns motivated by financial theory applicable to one single market such as the US or to non-segmented markets. They then test whether risk-premia, estimated using GMM or other statistical methods, of the assumed common factors are equal across markets. Liu et al. [15], for example, assume the Capital Asset Pricing Model (CAPM) as the factor model and Ling and Naranjo [13] use a multi-factor model in their study of equity and REIT market integration. Bekeart and Harvey [4] assume a nested world and local CAPM as the factor model in their investigation of world market integration. The problem with this approach is that the tests for equality of risk premia for an assumed common factor has little statistical power when the factor is not a common factor. Failure to reject unequal risk-premia may simply be because the assumed common factor is not common to the two markets. Similarly, tests that fail to reject equality of risk-premia may also simply reflect an incorrect common factor assumption.

In the second approach, researchers examine trends in criteria other than equality of risk premia. The underlying argument in this approach is that integrated markets differ from segmented markets with respect to specific statistical measures. Two measures that have been extensively used are R-Square, a measure representing the explanatory power of the assumed common factors, and comovement of asset returns, which impacts the degree of diversification benefits and is a measure of the economic implications of integration. If the R-Square or comovement is sufficiently high, one can conclude that the markets are integrated. The issue, however, is that it is not clear what level of R-Square or comovement is required to imply market integration or the existence of a common factor. Further, noise in the data arising from autocorrelation or heteroskedasticity can make measurements difficult. Researchers, therefore, examine changes in R-Square and comovement over time and interpret the time trends in integration as implying a movement towards or away from integration. Pukthuanthong and Roll [20] focus on the time trend of the degree to which asset returns are explained by R-square and Bekaert et al. [3] examine the trend in the comovement of returns. These tests can inherently only suggest an increase in the pervasiveness of a set of common factors but cannot really address whether expected returns of assets facing similar risks are equal across the two markets. It is plausible for financial markets to be completely segmented yet each market exhibiting an increasing R-square relative to an assumed common factor. Consider, for example, an economy that is comprised of two perfectly segmented markets with returns of each market exhibiting an upward R-square trend relative to its own market-specific value weighted portfolio. A researcher who assumes the common factor is the value-weighted portfolio of all securities from both markets will find that both markets have an increasing R-square relative to this common factor. This is because the returns of each market are increasingly explained by the part of the common factor from the same market while being orthogonal to the part of the common factor from the other market. Without knowing whether markets are or are not integrated, interpreting these measures is difficult. Theoretical asset pricing literature has also suggested that the risk premia and factor sensitivities vary over time (see [12]), and the changing risk premia may result in markets moving in and out of integration. Previous empirical studies have found that markets can become more and less integrated through time (see, e.g., [4, 6]).[4]

---

[4]Bekeart and Harvey [4] study international market integration over 1977 to 1992 using a conditional regime-switching model that nests the extreme cases of market segmentation and integration and find that most countries are neither completely integrated nor segmented but are instead somewhere in-between fluctuating between periods of being more integrated and periods of being more segmented. India and Zimbabwe are particularly interesting cases. Both countries exhibit extreme and sudden regime-shifts—in 1985, India instantaneously moves from being integrated to segmented, and likewise, Zimbabwe suddenly changes from being integrated to segmented in 1986 and then just as suddenly switches back to being integrated in 1991. Carrieri et al. [6] document similar results of a volatile integration process characterized by periods of increasing and decreasing integration.

Our approach differs from these standard approaches by first testing for the presence of a common factor and then checking for equality of risk-premia only when a common factor is present. By directly determining whether markets are integrated in each year, our approach allows us to better interpret trends in measures such as comovement and R-Square.

We use daily return data in our tests. We measure the integration level of the US REIT and Stock Markets each quarter from 1985 to 2013. This choice of annual time steps stems from a number of considerations. On the one hand, since we want to study the time variation in integration, we need to develop a sequence of integration measures with each measurement estimated over a short time-step. This allows us to observe variation in the factors and factor structures over time. On the other hand, we use principal components to compute the factor structure and GMM to measure risk-premium, techniques which require a longer data series for robust estimation leading to stronger results. Our choice of using daily data over each quarter strikes a reasonable balance yielding a time series of 116 quarters with a time series of 60 days used to compute each measurement.

Daily stock return (share codes 10 and 11) and REIT return (share codes 18 and 48) data for the sample period 1985–2013 are from CRSP dataset available through Wharton Research Data Services (WRDS). To be included in the data set for a particular quarter, stocks and REITs must have daily returns for all days during that quarter. Depending on the year, the number of stocks (REITs) used in our analysis ranges from a low of 3429 (55) to a high of 7060 (210). Additionally, we use the Fama-French-Carhart factors also available through WRDS. These factors are the excess return on the CRSP value-weighted portfolio (MKT) and the three well-known factor mimicking portfolio SMB, HML, and UMD.

Our statistical tests are structured as follows. We begin with a simulation to identify the level of canonical correlation expected when a common factor is present. The simulation uses the actual stock and REIT returns and embeds a common factor into the returns. Performing the simulation in this way ensures that the returns, and more importantly the errors, maintain their interesting statistical properties. We can say that the two markets share a common factor when the canonical correlations rise above the simulated benchmark level. Once a regime characterized by the existence of common factors is identified, factor mimicking portfolios based on canonical correlations of the principal components are compared with economically motivated factors that are believed to explain returns. We can thus identify economically meaningful common factors and run tests for the equality of risk premia to identify market integration.

3 Market Integration: A First Pass

We begin our investigation by examining integration using a pre-specified factor model for assets returns. As in earlier studies, we assume a model of expected returns and test whether the assumed risk factors are priced equally across the markets. Liu et al. [15], one of the first studies of stock and REIT market integration, assume the CAPM and test for integration relative to six different proxies for the market factor while [13] assume a multi-factor model that includes such factors as MKT, a term structure premium, a default premium, and consumption growth. Following these studies, we select the single factor CAPM and use MKT as the proxy for the market factor for our first series of tests. MKT is a value-weighted portfolio that includes both stocks and REITs providing theoretic as well as economic support for its use.

We begin by dividing all stocks and all REITs into 15 stock portfolios and 15 REIT portfolios according to each securities beta coefficients relative to MKT. Portfolio betas and factor risk premia are simultaneously estimated using a GMM framework and standard errors are computed using [19] with three lags. Coefficients estimated using GMM are equal to those found using Fama-MacBeth. By using GMM, however, we avoid the error-in-variable problem and standard errors are appropriately estimated as in [22] with the added adjustment for serially correlated errors. This estimation is performed each year from 1985 to 2013 using daily return data. Estimating risk premia at the annual frequency allows for some time variation in betas and in risk premia. A Wald test is used to test the hypothesis of equal stock and REIT risk premia.

Table 1 shows the results of these tests. The second and third columns in the table list the risk premia estimated from REITs and their corresponding t-statistics. The fourth and fifth columns are the premia and t-statistics estimated for the stock market. The premia are listed as daily percent returns. The sixth column is the Wald statistic. For 1992, for example, the daily risk premia for REITs is 0.028% and for stocks is 0.092% which roughly correspond to 0.62% and 2.02% monthly premia for REITs and stock, respectively. Though the premia are economically significant, only the stock premium is statistically significant, most likely due to noisy data. Despite the large difference in premia, we are unable to reject the hypothesis that the premia are equal suggesting the possibility that the stock and REIT markets are integrated. In contrast, in 1995, the REIT and stock markets have daily risk premia of 0.030% and 0.111%, respectively, which correspond to monthly premia of 0.66% and 2.44%. Again, only the stock premia is significant; however, we do reject the hypothesis of equal risk premia. We conclude, therefore, that for 1995 either markets are not integrated or the assumed model is incorrect.

Overall, we reject integration in only six of the 29 years in the sample leading to the belief that the markets are, by in large, integrated. This conclusion is questionable upon further investigation. We perform two tests to check whether the initial modeling assumptions are reasonable. The first test is based on the R-square measure proposed by Pukthuanthong and Roll [20] who argue that a reasonable measure of market integration is the proportion of returns that can be explained by a common set of factors. If the level is low, then returns are primarily influenced by local sources of risk. We construct this measure by taking the average R-square from regressing each REIT on MKT. The average R-square is listed in column 7 of Table 1 labeled REIT R2. With the exception of 1987, the year of a big market crash, R-squares are consistently below 10% from 1985 to 2002. R-squares then

Table 1 Mean canonical correlations

Test for equal risk premia							
	REIT Premium	REIT t-stat	Stock Premium	Stock t-stat	Wald	REIT R2	PC R2
1985	0.041	0.56	0.090	1.89	1.06	0.009	0.068
1986	0.125	1.49	0.049	0.75	1.34	0.021	0.110
1987	−0.076	−0.61	0.012	0.09	0.41	0.111	0.618
1988	0.015	0.14	0.104	1.57	1.82	0.016	0.074
1989	0.008	0.08	0.070	1.35	1.43	0.014	0.064
1990	−0.105	−0.85	−0.041	−0.55	0.71	0.018	0.031
1991	0.242	2.27**	0.201	2.91***	0.34	0.017	0.039
1992	0.028	0.37	0.092	2.13**	2.17	0.008	0.051
1993	0.178	2.01**	0.084	2.40**	7.12***	0.011	0.046
1994	0.013	0.20	0.021	0.48	0.03	0.025	0.082
1995	0.030	0.47	0.111	3.09***	5.07**	0.011	0.039
1996	0.255	3.72***	0.087	1.48	8.23***	0.028	0.399
1997	0.163	1.56	0.113	1.43	0.39	0.049	0.490
1998	−0.187	−1.18	0.052	0.48	4.95**	0.078	0.427
1999	−0.117	−0.65	0.243	2.86***	17.97***	0.012	0.131
2000	0.133	0.88	−0.038	−0.29	1.72	0.029	0.336
2001	0.163	1.01	0.072	0.67	0.73	0.074	0.476
2002	0.124	0.73	−0.054	−0.49	2.55	0.116	0.456
2003	0.231	2.71***	0.231	3.00***	0.00	0.137	0.581
2004	0.130	1.54	0.056	1.03	1.89	0.107	0.482
2005	0.003	0.06	0.018	0.41	0.11	0.213	0.646
2006	0.088	1.74*	0.043	0.95	1.01	0.187	0.696
2007	−0.087	−1.08	−0.019	−0.31	1.19	0.291	0.689
2008	−0.047	−0.29	−0.127	−0.81	0.26	0.384	0.796
2009	0.130	1.42	0.183	1.73*	0.25	0.425	0.826
2010	0.121	1.48	0.098	1.27	0.09	0.397	0.810
2011	0.011	0.13	−0.008	−0.09	0.05	0.509	0.874
2012	0.123	1.87*	0.063	1.12	1.15	0.244	0.677
2013	0.049	0.76	0.136	3.02***	3.76*	0.235	0.561

For each year, the risk premium relative to the excess return on the CRSP Value-Weighted Portfolio (MKT) is estimated from the set of all REITs and the set of all US common stocks. Premia are listed as percent daily returns. A Wald test is used to test for equal premia. REIT R2 is the average R-Square from regressing each REIT on MKT, and PC R2 is the R-square from regressing the MKT on the first eight principal components estimated from REIT returns

begin to increase after 2002 and reach a maximum of 0.509 in 2011. With such small R-squares early in the sample, relative to our choice of common factors, the REIT market must be primarily influenced by local factors up until 2002.

The second test determines whether MKT lies in the REIT factor space. We follow Bai and Ng [2] who argue that a proposed economic factor should lie in the factor space spanned by the set of principal components.[5] For each year, we select the first eight principal components to represent the REIT factor space. The choice of eight principal components is determined using the information criteria of [1]. MKT is regressed on the eight principal components and the R-square is reported in column 8 of Table 1, PC R2. In their study, Bai and Ng [2] state that when an economic factor lies in the factor space, we should observe a canonical correlation (square root of R-square in this case) near one. We find many R-squares below 10% in the early part of the sample. R-square does eventually reach 70% in year 2007. During the early part of the sample, it is clear that MKT does not lie in the REIT market's factor space.

These tests show that despite failing to reject equal premia, there is absolutely no evidence that MKT is a REIT factor for the vast majority of the sample period, especially prior to 2002. Hence, the GMM tests are misspecified in the early part of the sample and the failure to reject integration is meaningless.

Our analysis above suggests that the critical issue in market integration is in determining the appropriate model for stock and REIT returns and there are several questions to ask with respect to the factor model. Do the two markets even share a common factor or are the two markets completely segmented? If the markets do share a common factor, what are the reasonable proxies for the common factors? Our proposed technique, presented in the next section, addresses these questions prior to estimating and testing risk premia.

4 Two-Step Approach to Testing Integration

The first step in our two-step testing approach is to determine the factor model for returns in each of the markets. As is standard in the investigation of multiple markets, we allow returns of stocks and REITs to follow different linear factor models. Returns may depend on either factors common to both markets—common factors—or factors that are unique to a particular market—local factors.

For the $i = 1, 2, \ldots, M$ stocks and $j = 1, 2, \ldots, N$ REITs, the return generating functions for stocks and REITs are given by

$$\begin{aligned} \text{Stock}: R_{it}^S &= E_{t-1}[R_{it}^S] + \sum_{l=1}^{g} \beta_{il}^S F_{lt}^C + \sum_{k=1}^{h} \gamma_{ik}^S F_{kt}^S + \epsilon_{it}^S \\ \text{REIT}: R_{jt}^R &= E_{t-1}[R_{jt}^R] + \sum_{l=1}^{g} \beta_{jl}^R F_{lt}^C + \sum_{k=1}^{l} \gamma_{jk}^R F_{kt}^R + \epsilon_{jt}^R. \end{aligned} \quad (1)$$

where g, h, and l are the number of common factors, local stock factors, and local REIT factors, respectively. For stocks (REITs), R_{it}^S (R_{jt}^R) are the asset returns, $E[R_{it}^S]$ ($E[R_{jt}^R]$) are the expected returns, β_{il}^S (β_{jl}^R) are the sensitivities to the

---

[5]See also [23].

common factors F_{lt}^C, and γ_{ik}^S (γ_{jk}^R) are the factor sensitivities to the local factors F_{kt}^S (F_{kt}^R). We assume that all local and common factors are mutually orthogonal.

Expected returns are given by

$$\begin{aligned}\text{Stock}: E_{t-1}[R_{it}^S] &= \lambda_{0,t}^S + \sum_{l=1}^{g} \beta_{il}^S \lambda_{lt}^C + \sum_{k=1}^{h} \gamma_{ik}^S \lambda_{kt}^S + \epsilon_{it}^S \\ \text{REIT}: E_{t-1}[R_{jt}^R] &= \lambda_{0,t}^R + \sum_{l=1}^{g} \beta_{jl}^R \lambda_{lt}^C + \sum_{k=1}^{l} \gamma_{jk}^R \lambda_{kt}^R + \epsilon_{jt}^R.\end{aligned} \quad (2)$$

where $\lambda_{0,t}^S$ ($\lambda_{0,t}^R$) is the riskless rate of return at time t for the stock (REIT) market. λ_{lt}^C are the risk premia for the g common sources of risk, and λ_{kt}^S (λ_{kt}^R) are the risk premia for the h (l) local sources of risk. In many studies of market integration it must be tested whether the riskless rates are equal across markets. For this study, since REITs and stocks both trade in the USA, it is reasonable to assume the same riskless rate applies to both markets, $\lambda_{0,t}^S = \lambda_{0,t}^R$.

Perfectly integrated markets require that the same expected return model applies to securities in all markets. Expected returns are determined only by common factors and not local factors, and the risk premia on the common factors must be equal across markets. The presence of local risk factors, therefore, is sufficient to show markets are not perfectly integrated. The literature does discuss partial integration. For example, [7] have an equilibrium model that allows for mild or partial segmentation, where both common and local factors are present.

We next turn to determining the number of common factors g and proxies for the common factors. As noted before, once the correct model and appropriate proxies for the common factors are identified, we can test for equal premia using GMM.

4.1 Existence of Common Factors

The approach we use to identify the appropriate model for each market builds on [1, 2]. It is well known that the factor space for a particular market can be consistently constructed using principal components [23]. Therefore, we first determine the factor structure and factor proxies for each market separately using [1] to find the number of principal components required to explain the returns of assets in each market—$g + h$ for equities and $g + l$ for REITs. Canonical correlation analysis is used to compare the factor spaces spanned by each market's principal components in order to identify the number common factors g. Canonical variates associated with sufficiently large canonical correlations are related to proposed economic factors to identify economically meaningful sources of risk. A test for equal risk premia is conducted over periods when $g > 0$ using the identified factor proxies.[6]

---

[6] Studies that have used a similar approach are [9] in their comparison NYSE and Nasdaq stocks, and Blackburn and Chidambaran [5] who study international return comovement derived from common factors.

We first identify the number of factors (principal components) required to explain the returns in each market individually. Using [1] on daily data over each quarterly we find that the number of factors varies from quarter to quarter.[7] For stocks, the number of factors ranges from one to four while for REITs, the number of factors ranges from five to eight. In every quarter, the number of REIT factors is strictly larger than the number of stock factors implying $l < h$.[8] At this point it is clear that the two markets cannot be perfectly integrated since REITs always have $h > 0$ local factors. Nonetheless, it is still interesting to investigate further to determine the extent to which the two markets are partially integrated.

To simplify the analysis going forward, we hold the number of principal components constant for each market. We use eight principal components for REITs and four for stocks. These choices have support in the literature. Liow and Webb [14] use seven principal components to describe REIT returns—equal to the number of eigenvalues greater than one, and the four-factor Fama-French-Carhart model is commonly used for equities in the asset pricing literature. In addition to using the first four principal components to proxy for stock factors, we also use the four Fama-French-Carhart factors.[9] Since the use of the four-factor equity model has strong support in the literature, it is reasonable to assume that these factors are prime candidates for being common factors.

While it is the case that there are many quarters when the number of factors is less than four for stocks and eight for REITs, the use of too many principal components does not present a problem. What is essential is that the factor space be completely spanned by the set of principal components. Additional principal components simply add noise. For the cases when the true number of factors is less than what we assume, the true factor space will be a subspace of the one empirically estimated.

Canonical correlation analysis is used to determine the number of common factors g from the set of $g + h$ stock latent factors and the $g + l$ REIT latent factors by finding the linear combination of the stock factors that has the maximum correlation with a linear combination of REIT factors [2]. The pair of factors with maximum correlation are called first canonical variates and the correlation between them is the first canonical correlation. The process is then repeated to determine a pair of second canonical variates that are characterized as having the maximum correlation conditional on being orthogonal to the first canonical variates. The correlation between the pair of second canonical variates is called the

---

[7]This analysis was also performed on an annual basis. Results are similar.

[8]It is common to use variables other than the Fama-French factors in studying REIT returns. For example, [17] use the excess returns on the value-weighted market portfolio, the difference between the 1-month T-bill rate and inflation, the 1-month T-bill rate relative to its past 12-month moving average, and the dividend yield (on an equally weighted market portfolio).

[9]While it would be ideal to also include a standard set of economically motivated REIT factors, unlike the literature on explaining the cross-section of stock returns, the REIT literature has not yet agreed on such standard set of factors.

second canonical correlation, and so on. Essentially, canonical correlation analysis determines whether and to what degree the two factor spaces overlap.

Other researchers have also used a principal component approach to identify factor models while studying integration. Instead of estimating principal components from returns in each market separately as we do here, the common approach is to combine returns of assets from all markets into one large data set and then calculate the principal components of the combined set of returns (see, e.g., [3, 20]). This traditional approach presents several problems that can lead to incorrectly identifying a strong local factor in one market as a global factor. First, when markets do not share any common factors the first set of principal components will be comprised of some rotation of local factors. Second, suppose, as is the case here, that the two markets are of unequal size, with market one having a large number of assets while market two having only few securities. If market one has a particularly strong local factor relative to a weak common factor, then principal components will identify those factors that explain the greatest proportion of return variation—the strong local factor. In both cases, the first set of principal components that represent local factors are assumed to be proxies for common factors. Our approach avoids these pitfalls by first identifying the factor structure for each market separately and then comparing the two factor spaces.

4.1.1 Benchmark Canonical Correlation Level

An important issue in canonical correlation analysis not rigorously addressed in the literature is the magnitude of the canonical correlation required to claim existence of one or more common factor. Previous research has tended to arbitrarily select a benchmark value [2, 9]. Since our approach requires making strong statements about the existence of common factors, we need to be more precise. To do so, we perform a simulation to determine the appropriate benchmark canonical correlation level for the stock and REIT data.

From the estimated four and eight principal components for stocks and REITs, respectively, we follow the canonical correlation methodology to calculate the first four canonical variates for both stocks and REITs. We begin with one common factor. The following steps are then repeated for two, three, and four common factors. For one common factor, each stock return is regressed on the first stock canonical variate and each REIT return is regressed on the first REIT canonical variate to determine their respective alphas, betas, and errors:

$$\text{Stock}: R_{it}^S = \alpha_i^S + \beta_i^S V_t^S + \epsilon_{it}^S \\ \text{REIT}: R_{jt}^R = \alpha_j^R + \beta_j^R V_t^R + \epsilon_{jt}^R. \quad (3)$$

where β_i^S is the loading of stock i on the first stock canonical variate V_t^S and β_j^R is the factor loading on the first REIT canonical variate V_t^R. The errors are represented by ϵ_{it}^S for stocks and ϵ_{jt}^R for REITs. We do not require any particular level of canonical correlation between canonical variates in this step.

We now create simulated returns with the feature of having a common factor. We do this by replacing the canonical variates V_t^S and V_t^R in Eq. (3) with a proxy for the common factor F_t^C:

$$\text{Stock} : \hat{R}_{it}^S = \alpha_i^S + \beta_i^S F_t^C + \epsilon_{it}^S \\ \text{REIT} : \hat{R}_{jt}^R = \alpha_j^R + \beta_j^R F_t^C + \epsilon_{jt}^R. \quad (4)$$

The alphas, betas, and errors remain the same as in Eq. (3) only the factor is changed. Constructing the simulated returns in this way preserves many of the statistical features of the true stock returns such as heteroskedasticity, serially correlated errors, and cross-correlated errors. However, the simulated returns now share a common factor. The proxy used for the common factor is derived from the sum of the stock and REIT canonical variates:

$$F_t^C = \frac{V_t^S + V_t^R}{\sqrt{Var(V_t^S) + Var(V_t^R) + 2Cov(V_t^S, V_t^R)}} \quad (5)$$

where the denominator normalizes the factor F_t^C to have unit variance.[10] This proxy for the common factor maintains many of the desirable statistical features observed in the canonical variates.

Using the simulated returns $\hat{R}_{it}^S$ and $\hat{R}_{jt}^R$, principal components are estimated and canonical correlation analysis is used to determine the simulated canonical correlations when common factors exist and the number of common factors is known. Since we have embedded common factors into the simulated returns, we should expect canonical correlations much closer to one; however, estimation error in calculating principal components caused by finite samples, heteroskedasticity, and autocorrelation may cause the first canonical correlation to be large, but less than one. We perform this simulation for each quarter of our sample period using the same set of returns—same time-series length and cross-section—as used in the study. The benchmark levels are unique to this particular study.

Results are plotted in Fig. 1. Figure 1a is the plot of the first four canonical correlations when there is only one common factor. The bold, solid line is the first canonical correlation, and as expected, is close to one with a time-series average, as shown in Table 2-Panel A, of 0.954. The standard error, 0.045, is listed below the mean. All other canonical correlations are much smaller with time-series averages of 0.521 for the second canonical correlation (light, solid line), 0.367 for the third canonical correlation (dashed line), and 0.215 for the fourth canonical correlation

---

[10] We also used the stock canonical variates and REIT canonical variates as common factor proxies and the results did not change.

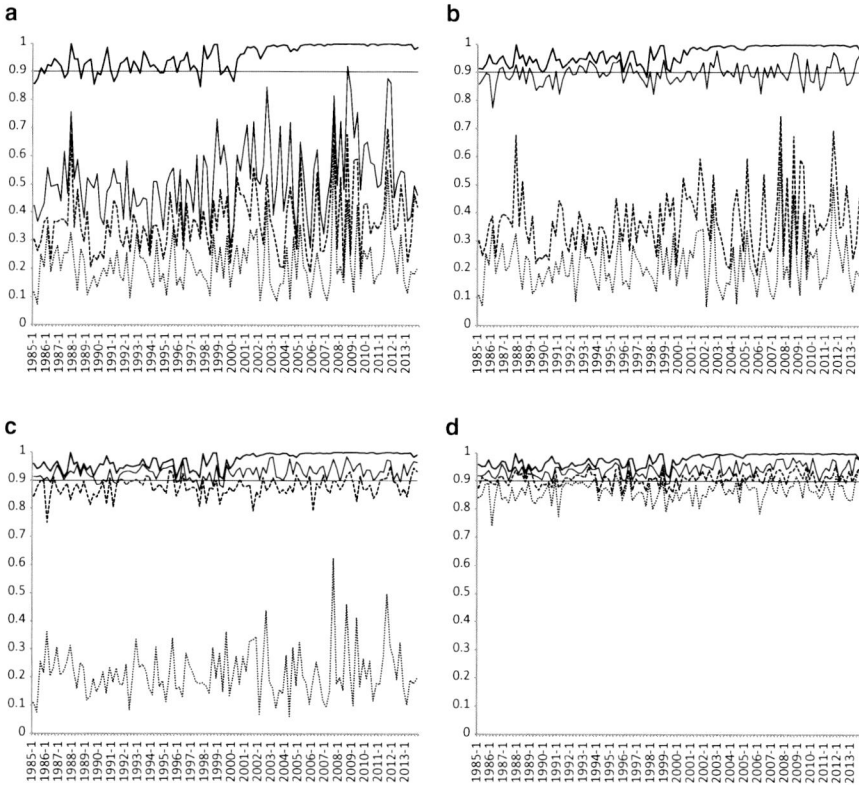

Fig. 1 Simulated canonical correlations. The first four canonical correlations between REIT and stock principal components are estimated each quarter from 1985 to 2013 using simulated returns. The simulated returns are modeled to include one, two, three, and four factors common to both markets. The bold solid line is the first canonical correlation, the thin solid line is the second canonical correlation, the dashed line is the third, and the dotted line is the fourth canonical correlation

(dotted line). This is exactly what is expected. With only one common factor, it is expected that we find one canonical correlation close to one with the remaining much lower.

In Fig. 1b, we plot the time-series of canonical correlation when we simulate returns with two common factors. We notice two different affects. First, the second canonical correlation, with a time-series average of 0.896, is much larger than in the one common factor case when its average was 0.521. Second, the average first canonical correlation is slightly larger and has noticeably less variance. Meanwhile, the time-series average for the third and fourth canonical correlations is nearly unchanged. The presence of two common factors increases the signal-to-noise ratio in the system thus making it easier to identify common factors.

Figure 1c for three common factors and Fig. 1d for four common factors tell similar stories. Common factors are associated with high canonical correlations.

Table 2 Mean canonical correlations

Panel A: average canonical correlations from simulated returns				
	CC 1	CC 2	CC 3	CC 4
One common factor	0.954	0.521	0.367	0.215
	0.045	0.135	0.114	0.094
Two common factors	0.967	0.896	0.368	0.216
	0.032	0.037	0.114	0.091
Three common factors	0.972	0.928	0.875	0.216
	0.027	0.026	0.034	0.091
Four common factors	0.977	0.941	0.909	0.864
	0.022	0.024	0.025	0.035
Panel B: average canonical correlations from actual returns				
Eight REIT PCs vs Four stock PCs	0.752	0.514	0.371	0.216
Eight REIT PCs vs FFC four-factors	0.731	0.452	0.316	0.199

We use canonical correlation analysis [2] to measure the presence of common factors between the US stock and REIT markets. In Panel A, we simulate return data to have one to four common factors to determine a canonical correlation level expected when common factors exist. The standard error is listed below each mean canonical correlation. In Panel B, we calculate the time-series mean canonical correlation from actual stock and REIT returns each quarter from 1985 to 2013. For REITs, we use the first eight principal components to represent REIT factors. Two different proxies for stock factors are used—the first four principal components (PC) and the four Fama-French-Carhart factors (FFC)

From Table 2-Panel A, we see the dramatic jump in average canonical correlation from 0.216 to 0.864 when moving from three to four common factors. These results clearly demonstrate that for our particular system, common factors are associated with high canonical correlations. The benchmarks we use in this paper to claim statistical evidence for the presence of one or more common factors is the 95% lower confidence level below the time series canonical correlation mean. The benchmark threshold levels for the first, second, third, and fourth canonical correlations, used in this paper are 0.87, 0.82, 0.81, and 0.79, respectively.

4.1.2 Evidence for a Common Factor

The canonical correlation method is now used on the real stock and REIT return data to find regimes when a common factor exists. Results are plotted in Fig. 2a–d. Each figure shows two time-series of canonical correlations. The solid line is the canonical correlation estimated from the comparison of eight REIT principal components and four stock principal components. The dotted line is estimated from eight REIT principal components and the four Fama-French-Carhart factors. The two dotted horizontal lines represent the simulated time series mean canonical correlations and its 95% lower confidence. We say that a common factor exists when the canonical correlation rises above the 95% lower confidence interval (the lower of the two horizontal dotted lines).

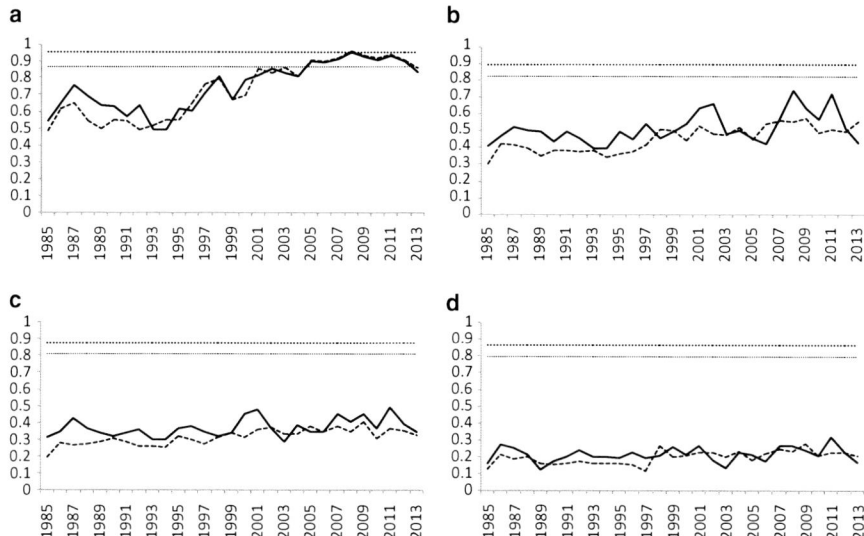

Fig. 2 Canonical correlations. The first four canonical correlations between the principal components of daily REIT returns and proxies for stock factors are plotted for each quarter from 1985 to 2013 [2]. Two proxies for stock factors are considered—principal components from stock returns (given by the solid line), and the three Fama-French plus Carhart factors (given by the dotted line). The first canonical correlations is given in (**a**), second canonical correlations in (**b**), third in (**c**), and the fourth in (**d**). The two dotted horizontal lines represent the simulated canonical correlation time series mean and its 95% lower confidence interval

Figure 2a depicts the quarterly first canonical correlations values. We first notice that the choice of stock factors, whether the first four principal components or the four Fama-French-Carhart factors, does not greatly affect our results. The solid and dotted curves track each other closely. Second, the time series average canonical correlation is lower than the 95% lower confidence benchmark of 0.87. Table 2-Panel B lists a time-series average first canonical correlation of 0.752; however, there is substantial time variation in the canonical correlation suggesting significant variation in integration between the two markets. The first canonical correlation initially increases from 0.6 to 0.97 in the fourth quarter of 1987, concurrent with the 1987 market crash. It then trends downward until 1993 when it begins a long trend upward until 2012 before it drops down again.

The time series of canonical correlations describes the integration process from markets having completely separate return models with only local factors to models comprised of both local and common factors. Using 0.87 as the threshold required for a common factor, we observe the process as it moves toward and away from the threshold. With the exception of the fourth quarter of 1987, the first canonical correlation remains below 0.87 until 1998. After 2000, the canonical correlation remains close to the benchmark level fluctuating above and below the 0.87 line. Finally, in 2005 the canonical correlation breaches the threshold and for the most part remains above 0.87. The canonical correlation drops back down in the last two quarters of 2013. From 2005 to 2012, the canonical correlation is above the threshold 29 out of 32 quarters providing strong support for the existence of a single factor over this time period.

Figure 2b–d shows plots of the second, third and fourth canonical correlations, respectively. The second canonical correlation rises above its threshold level of 0.82 three times but each time it drops immediately back down. Spikes tend to be associated with market crashes when returns of all assets tend to be highly correlated. The third and fourth canonical correlations never rise above their respective benchmarks.

Our evidence shows that only one common factor exists and only during the latter part of the sample. Both choices of stock factors, the latent factors and the Fama-French-Carhart factors, support this result. Conservatively, we can claim the existence of a common factor between 2005 and 2012; but, more liberally, there is some support for a common factor over the longer period from 2000 to 2013. We find no evidence for any significant common factor prior to this thus calling into question the research that found evidence for stock and REIT integration using samples from the 1980s and 1990s.

4.1.3 Economic Identity of Common Factors

Having identified a regime over which a common factor exists, we relate the single statistically determined latent factor with the Fama-French-Carhart factors to identify an economically meaningful proxy for the common factor. While the literature has posited a large number of possible risk factors, it has also consistently agreed on the use of the four-factor model comprised of the MKT, SMB, HML, and UMD. Therefore, it is reasonable to assume that since these factors perform well in explaining stock returns that any common factors are likely to be selected from this list. In the end, it is an empirical matter as to how well the latent factor can be explained by these four factors.

For the statistical proxy for the common factor, we use F^C from Eq. (5) constructed each quarter from 1985 to 2013. We look over the entire time period in order to contrast the regime when integration is believed to be possible with the regime over which integration is believed to be impossible. Figure 3a plots the R-square from regressing F^C on MKT each quarter and Fig. 3b is the R-square from regressing F^C on all four Fama-French-Carhart factors. In both plots, we find a very noisy and inconsistent relationship between the statistically determined common factor and the economic factors. It is not until 2005 that the R-squares are consistently high. This is particularly true for the R-squares in Fig. 3b where R-square is on average 0.93 over 2005–2013. For the results using MKT, R-square is 0.82 on average over the same time period. The market factor, MKT, is clearly the primary driver of the high R-square in Fig. 3b. In unreported results, we show that SMB accounts for much of the remaining difference in R-square while HML and UMD contribute very little. As previously mentioned, the value-weighted MKT

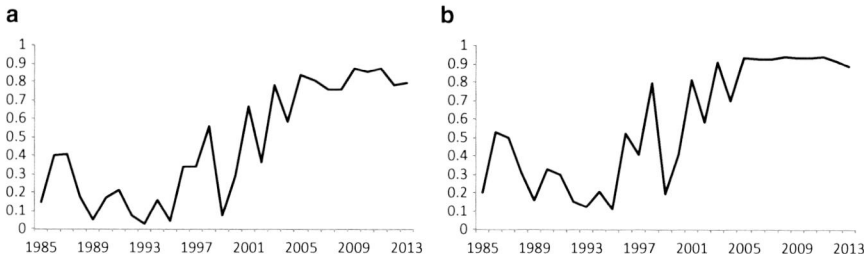

Fig. 3 Economic factors. We relate the statistically determined common factor shared by both the REIT and equity markets to four commonly used factors—the excess CRSP value-weighted portfolio, the Fama-French hedge portfolios SMB and HML, and Carhart's UMD portfolio. The time-series of the R-squares from regressing the statistical common factor on MKT (**a**) and on all four factors (**b**) is plotted. Regressions are performed each quarter using daily data from 1985 to 2013

portfolio comprised of stocks and REITs, among other assets, has strong economic and theoretical support. We feel that we are on both strong statistical and economic footing by selecting MKT as our proxy for the common factor towards the end of our sample period.

4.1.4 Test for Equal Risk Premia

As our final step, we test for equal risk premia over the identified regime using the previously described GMM approach. In Table 3-Panel A, we list the estimated risk premia for both the stock and REIT markets and the result of the Wald test for the more conservative regime 2005–2012 and the more liberal regime 2000–2013. In both cases, we are unable to reject differences in risk premia. Moreover, the daily percent risk premia are indeed very similar. For the 2005–2012 regime, the daily REIT and stock premia are 0.032% and 0.037%, respectively, equivalent to 0.72% and 0.81% per month. For the 2000–2013 regime, the daily premia are 0.057% and 0.058%, equivalent to 1.24% and 1.27% monthly, for REITs and stock, respectively. The highly similar premia provide strong evidence supporting our proposed approach to studying integration.

To contrast these results, we take one last view over the entire sample period 1985–2013. In Panel B of Table 3, risk premia are estimated over 5-year rolling windows. An interesting result emerges. Rejection of equal premia appears cyclical. This is clearly seen in Fig. 4 where we plot the daily risk premia from Table 3-Panel B. The solid line represents the REIT premia and the dotted line represents stock premia. The stock risk premia (dotted line) varies through time but the variation is rather stable—no large jumps or swings. The REIT premia, on the other hand, are highly cyclical and characterized by large swings. The premia is at its minimum in 1989–1993 but is at its maximum over the period 1993–1997. The grey shaded regions in Fig. 4 indicate the time periods when statistical tests reject equal premia.

Table 3 Mean canonical correlations

Panel A: test for equal risk premia						
	REIT Premium	REIT t-stat	Premium	Stock t-stat	Stock Wald	P-value
2005–2012	0.033	1.12	0.037	1.17	0.02	0.89
2000–2013	0.057	2.37**	0.058	2.28**	0.00	0.96
Panel B: test for equal risk premia: 5-year rolling window						
1985–1989	0.026	0.75	0.066	1.58	0.91	0.340
1986–1990	−0.012	−0.32	0.033	0.75	1.03	0.311
1987–1991	0.003	0.07	0.090	1.90*	3.34	0.068*
1988–1992	−0.013	−0.29	0.153	3.92***	18.00	0.001***
1989–1993	−0.041	−0.72	0.126	3.68***	23.80	0.001***
1990–1994	0.094	2.25**	0.148	4.47***	2.65	0.103
1991–1995	0.160	5.53***	0.162	5.74***	0.01	0.928
1992–1996	0.180	6.06***	0.159	6.69***	0.73	0.394
1993–1997	0.214	5.80***	0.137	4.77***	7.10	0.008***
1994–1998	0.079	1.95*	0.109	3.16***	0.74	0.390
1995–1999	0.074	1.66*	0.146	3.61***	3.19	0.074*
1996–2000	0.057	1.12	0.110	2.17**	1.09	0.295
1997–2001	0.029	0.53	0.109	1.97**	2.09	0.149
1998–2002	0.028	0.48	0.071	1.21	0.55	0.457
1999–2003	0.134	2.67***	0.114	2.06**	0.12	0.727
2000–2004	0.153	3.07***	0.067	1.29	2.81	0.094*
2001–2005	0.104	2.48**	0.074	1.78*	0.52	0.472
2002–2006	0.093	2.58***	0.068	1.97**	0.53	0.465
2003–2007	0.038	1.31	0.068	2.53**	1.25	0.263
2004–2008	−0.004	−0.10	−0.002	−0.05	0.00	0.960
2005–2009	0.015	0.38	0.028	0.65	0.09	0.764
2006–2010	0.032	0.78	0.044	0.99	0.08	0.784
2007–2011	0.024	0.56	0.031	0.66	0.02	0.886
2008–2012	0.052	1.22	0.048	1.03	0.01	0.935
2009–2013	0.075	2.36**	0.096	2.71***	0.33	0.563

Risk premia relative to the excess return on the CRSP Value-Weighted Portfolio (MKT) is estimated from the set of all REITs and the set of all US common stock using GMM. Premia are listed as percent daily returns. A Wald test is used to test for equal premia. T-statistics testing the hypothesis of a significant risk premium is provided next to each premium. Panel A measures the risk premia over the regime determined from the canonical correlation methodology. Panel B estimates risk premia using a 5-year rolling window over the entire sample period

Rejection is nearly always associated with the peaks and troughs of the REIT premia. As the REIT premia move through the cycle and cross over the stock premia, statistical tests are unable to reject equal premia. This is not because markets are integrated, but because of the time-varying, cyclical nature of the REIT premia.

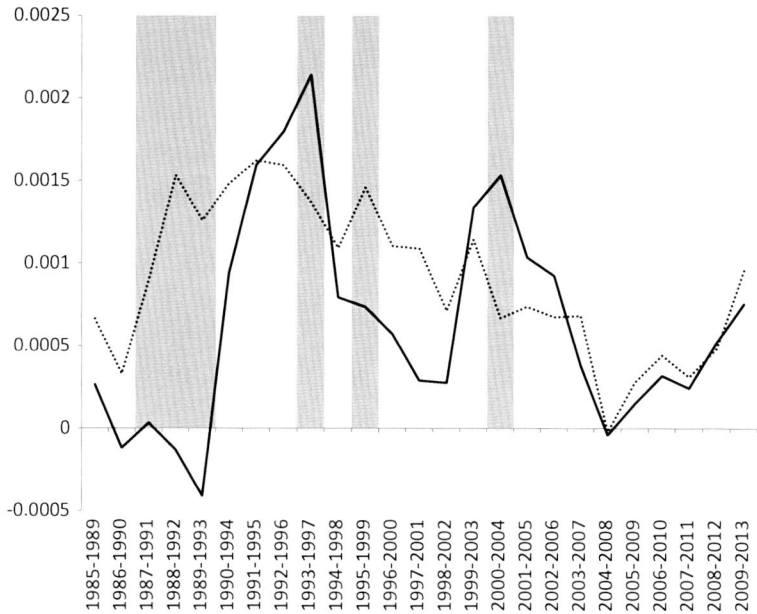

Fig. 4 Stock and REIT risk premia by year. Risk premia relative to the CRSP value-weighted market portfolio for REITs (solid line) and stocks (dotted line) are estimated over 5-year rolling windows using GMM. Premia are estimated using daily data and are plotted as daily premia. The grey shaded regions represent the intervals when equality of risk premia is rejected

As additional support for our approach, Table 3-Panel B and Fig. 4 clearly show the two premia converging in the later part of the sample, beginning around 2003. After this point, the stock and REIT risk premia are nearly equal, tracking each other closely as the rolling window advances through time. This is the precise regime the canonical correlation methodology identifies as having a common factor—the regime over which integration of the two markets is a possibility.

4.2 Measures of Integration

Due to the complications involved in testing for integration, a number of papers instead indirectly test integration by examining its implications, such as changes in diversification benefits. In this section, we discuss these alternative tests and relate them to time-varying canonical correlations underlying our procedure. As in the case of the results of GMM tests, the results of these alternate tests are also difficult to interpret. Specifically, while such measures are very informative for understanding the changing opportunities and investment environment, they are difficult to interpret as measures of integration.

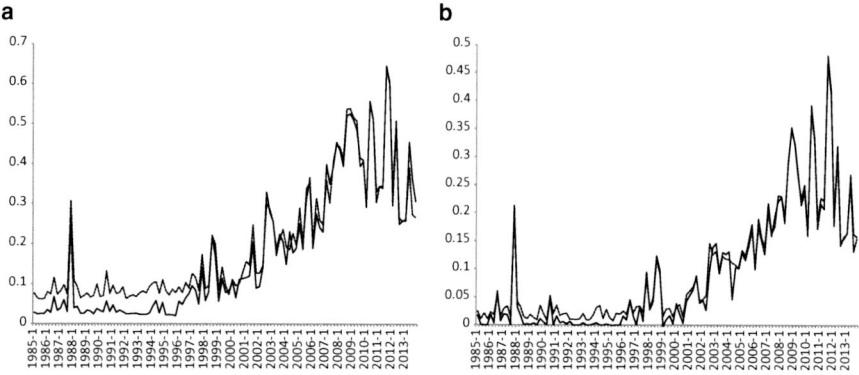

Fig. 5 Measures of integration. We calculate the R-square measure of integration [20] and the comovement measure of Bekaert et al. [3] using two different models—a single factor model comprised of a factor derived from canonical correlation analysis (solid line) and a four-factor model that includes the Fama-French-Carhart factors (dotted line). Measurements are taken using daily REIT and stock return data over each quarter from 1985 to 2013

In a recent study, Pukthuanthong and Roll [20] argue that R-square is a "sensible intuitive quantitative measure of financial market integration", in the sense that if R-square is small, "the country is dominated by local or regional influences. But if a group of countries is highly susceptible to the same global influences, there is a high degree of integration". (p 214–215) We compute the R-square by averaging the individual R-squares obtained from regressing each REIT on proxies for common factors. A plot of the R-square measured each quarter using daily return data is shown in Fig. 5a. The solid line uses F^C from Eq. (5) as the proxy for the common factor while the dotted line assumes the Fama-French-Carhart factors. Both curves tell a similar story. The factors explain a very small proportion of REIT returns throughout the late 1980s and 1990s. At the end of the 1990s and throughout the 2000s, both measurements begin to trend upward, explaining greater proportion of the REIT returns. There is a decrease in R-square at the end of the sample.

Comovement of asset returns across markets is another commonly used measure. Bekaert et al. [3] state, "If the increase in covariance is due to increased exposure to the world markets, as opposed to an increase in factor volatilities, the change in covariance is much more likely to be associated with the process of global market integration". (p2597). We compute the comovement between each REIT and each stock relative to our common factors and the results are plotted in Fig. 5b. Specifically, from Eq. (1) the covariance between the returns on stock i and REIT j as

$$cov\left(R_i^S, R_j^R\right) = \beta_i^{S'} \Sigma_C \beta_j^R + cov\left(\epsilon_i^S, \epsilon_j^R\right)$$

where Σ_C is the covariance matrix of some set of common factors.[11] For M stocks and N REITs, (equal-weighted) comovement is defined as

$$\Gamma_C = \frac{1}{MN} \sum_{i=1}^{M} \sum_{j=1}^{N} \beta_i^{S'} \Sigma_C \beta_j^{R}. \quad (6)$$

From Eq. (6), comovement either increases due to increased factor volatility or by increased factor loadings. Long trends in comovement persisting across multiple periods are likely caused by increased factor loadings and therefore are consistent with increasing integration. The time variation in comovement observed in Fig. 5b is similar to R-square. REITs exhibit low comovement with stocks until early 2000 when it then begins a steady upward trend. The long last trend is most likely due to increased betas as opposed to a persistent upward trend in factor volatility.

Both measures are consistent with our previous findings. There is not a strong relationship between markets in the early part of the sample and the strongest relationship is observed in the 2000s. The plots also demonstrate large time variation in the relationships between the two markets. It is challenging, however, to make any statement about integration from these two measures. Unlike our canonical correlation approach, threshold values do not exist that describe regimes over which integration is possible. Further, markets can be integrated and have moderate or low levels of R-square and comovement.[12] If markets are integrated at R-square or comovement values of say 10%, then what is the interpretation when an increase from 11% to 12% is observed? It cannot be the case that markets are even more integrated than before. Similarly, if markets are not integrated and the measure decreases, then it is not useful to say that markets are now less integrated than they were in the previous period. Hence, in order to interpret these measures of integration, one must know whether markets are already integrated and at what point the markets change from being segmented to integrated—defeating the purpose of the measures.

Two measures that do not suffer from these problems are the regime-switching model of Bekaert and Harvey [4] and the canonical correlation methodology proposed here. Bekaert and Harvey [4] measure of time varying integration allows for easy interpretation by measuring integration on a zero to one scale with values of zero indicating complete segmentation and one indicating complete integration. Changes in this value are readily interpreted as either trending toward one extreme state or the other. Similarly, the time-varying canonical correlations used in this study clearly identify regimes over which common factors exist, a necessary requirement for integration. Time variation in canonical correlations can be used

---

[11] Local factors are assumed to be orthogonal across markets and therefore do not affect covariance.

[12] Pukthuanthong and Roll [20] argue this point as a critique of comovement as a measure of integration. This can also be true for R-square in markets characterized by high idiosyncratic risk.

to track the process from the state when the two markets having two completely different models for returns to the state when there is one market comprised of only common factors.

5 Conclusion

Financial market integration requires that asset returns across distinct markets are determined by a common set of factors *and* the risk premia for these factors be equal across markets. The dual requirement makes econometric testing for market integration challenging, in particular a failure to accept or reject integration could be driven by an incorrect factor model. The approach we have presented in this paper, based on factor analysis and canonical correlation analysis, overcomes the joint hypothesis problem. Our approach is to first determine the correct factor model in each market and determine whether markets share a common factor. We subsequently develop economic proxies for the shared common factor and test for the equality of risk premia conditional on a common factor being present for the data period.

We illustrate the issues involved and the efficacy of our approach using the US stock and REIT markets over the period from 1985 to 2013. We demonstrate that traditional GMM tests are often times not able to reject segmentation thus implying integration. Using canonical correlation analysis on the principal components constructed from each market separately, we are able to identify the factor structure for each market and show that often times this is because markets do not share a common factor. By knowing when common factors exist, we are able to increase the power of our testing methodology by testing only over those regimes when integration is possible. Outside of these periods, time varying risk premia, noise in the data, and changes in integration over time can lead to spurious conclusions.

We find that the REIT and stock markets share a common factor over the 2005–2012 time period. Relative to the early part of the sample, risk premia over the 2005–2012 regime are indeed similar. The inability to reject equal risk premia in the early part of the sample is due to the REIT premia exhibiting a strong cyclical time varying pattern that is much different than the much more stable stock risk premia. Our results strongly support the necessity of first identifying the correct model of returns, both common and local factors, prior to testing for equal premia. We also relate our results to other measures used to imply time variation in integration, such as R-Square from asset pricing regressions and comovement. These measures do provide confirming evidence but are much more difficult to interpret.

We note that U. S. REIT markets are relatively small, especially when compared to the US stock markets. Further, tests have to be run each year as market integration can vary from year to year. These data features reduce the power of standard econometric techniques, such as GMM or Fama-Macbeth regressions that require a large amount of data. Researchers have tried to increase the power of these tests by increasing the data period or by making strong assumptions such as constant betas and risk premia—assumptions that are not valid when market integration is changing over time. Our methodology clearly works in small data sets and across datasets of unequal sizes. We believe, therefore, our approach is generally applicable and can be used in other settings where market capitalizations vary greatly or when a market has a fewer number of assets, e.g. emerging equity and fixed income markets.

Acknowledgements We thank Mark Flannery, Ren-Raw Chen, Shu-Heng Chen, An Yan, Yuewu Xu, Andre de Souza, seminar participants at Fordham University, the Indian School of Business and participants at the 1st Conference on Recent Developments in Financial Econometrics and Applications for their comments and suggestions.

Appendix

We briefly describe the essential elements of Canonical Correlations for identifying common factors in this appendix.

Let the data generating processes for the two dataset be as given below.

$$\mathbf{X}_t^k = \beta^k \mathbf{F}_t + \delta^k \mathbf{H}_t^k + \epsilon_t^k \tag{7}$$

where there are r^c common factors and r^k set-specific factors leading to $\kappa_k = r^c + r^k$ total factors for sets $k = 1, 2$. Our goal is to determine r^c, r^1 and r^2. We determine the number of principal components, κ_1 and κ_2, required to span the factor space for each individual set using [1]. We denote the first κ_1 and κ_2 principal components as $\mathbf{P}_1$ and $\mathbf{P}_2$, respectively.

We next use canonical correlation analysis to relate the two sets of principal components $\mathbf{P}_1$ and $\mathbf{P}_2$ in order to determine the dimension of the subspace at the intersection of the spaces spanned by the two sets. Originally developed by Hotelling (1936), the idea behind canonical correlation analysis is to find matrices that can be used to rotate two sets of factors so that their columns are ordered in terms of decreasing pairwise correlation.[13] Since the common factors are the only factors shared across the two sets, this procedure separates the common factors from the set-specific factors.

Let set $\mathbf{X}^1$ have κ_1 factors and set $\mathbf{X}^2$ have κ_2 factors, and let $r_{\max}^c = \min(\kappa_1, \kappa_2)$ be the maximum possible number of common factors. Canonical correlation analysis finds two $T \times r_{\max}^c$ matrices $\mathbf{U}_1$ and $\mathbf{U}_2$, called canonical variates, such that the ith columns of $\mathbf{U}_1$ and $\mathbf{U}_2$, $\mathbf{U}_1(i)$ and $\mathbf{U}_2(i)$ have maximum correlation.

---

[13] Canonical correlations have been used in prior literature on factor analysis. For example, Bai and Ng [2] used the canonical correlation idea to test the equivalence of the space spanned by latent factors and by some observable time series. Our use of canonical correlations is very much in the flavour of Bai and Ng [2], but for very different purposes. We show that the methodology can be used to determine common and unique factors in panels of data that have intersecting factors and is able to identify the true factor structure in multilevel data.

When there are r^c common factors, then the pairwise correlation between the ith canonical variates from each set will be large for $i = 1, ..., r^c$ and small for $i = r^c + 1, ..., r^c_{max}$.

Define the covariance matrix

$$\Sigma = \begin{bmatrix} \Sigma_{\widehat{P}_1 \widehat{P}_1} & \Sigma_{\widehat{P}_1 \widehat{P}_2} \\ \Sigma_{\widehat{P}_2 \widehat{P}_1} & \Sigma_{\widehat{P}_2 \widehat{P}_2} \end{bmatrix}$$

We search for vectors α_1 and α_2 such that the linear combinations $\widehat{U}_1(1) = \alpha_1^\top \widehat{P}_1$ and $\widehat{U}_2(1) = \alpha_2^\top \widehat{P}_2$ have maximum correlation, and such that $\widehat{U}_1(1)$ and $\widehat{U}_2(1)$ have zero mean and unit variance. We use the *hat* symbol to emphasize that the principal components and rotations of the principal components are estimates of the true factors. The objective function to be maximized is

$$\tilde{\rho}_1 = \text{Max}\left[\alpha_1^\top \Sigma_{\widehat{P}_1 \widehat{P}_2} \alpha_2 - \frac{1}{2}\lambda_1\left(\alpha_1^\top \Sigma_{\widehat{P}_1 \widehat{P}_1} \alpha_1 - 1\right) - \frac{1}{2}\lambda_2\left(\alpha_2^\top \Sigma_{\widehat{P}_2 \widehat{P}_2} \alpha_2 - 1\right)\right],$$

where λ_1 and λ_2 are Lagrange multipliers. The objective function is maximized when vectors α_1 and α_2 satisfy the first order conditions,

$$\begin{bmatrix} -\lambda_1 \Sigma_{\widehat{P}_1 \widehat{P}_1} & \Sigma_{\widehat{P}_1 \widehat{P}_2} \\ \Sigma_{\widehat{P}_2 \widehat{P}_1} & -\lambda_2 \Sigma_{\widehat{P}_2 \widehat{P}_2} \end{bmatrix} \begin{bmatrix} \alpha_1 \\ \alpha_2 \end{bmatrix} = \begin{bmatrix} 0 \\ 0 \end{bmatrix}.$$

The maximum correlation is

$$\tilde{\rho}_1 = \lambda_1 = \lambda_2 = \alpha_1^\top \Sigma_{\widehat{P}_1 \widehat{P}_2} \alpha_2.$$

The correlation $\tilde{\rho}_1$ is said to be the first canonical correlation, and $\widehat{U}_1(1)$ and $\widehat{U}_2(1)$ are referred to as first canonical variates.

The analysis can be repeated to find the second canonical variates $\widehat{U}_1(2)$ and $\widehat{U}_2(2)$ that are orthogonal to $\widehat{U}_1(1)$ and $\widehat{U}_2(1)$, respectively, and have the second canonical correlation $\tilde{\rho}_2 \leq \tilde{\rho}_1$. We apply the process recursively to yield a collection of canonical correlations $\tilde{\rho}_1 \geq \tilde{\rho}_2 \geq \ldots \geq \tilde{\rho}_{r^c_{max}}$ with the canonical variates $\widehat{U}_{1, T \times r^c_{max}}$ and $\widehat{U}_{2, T \times r^c_{max}}$. The squared canonical correlations obtained above are the eigenvalues of

$$\Sigma_{\widehat{F}^x \widehat{F}^x}^{-1} \Sigma_{\widehat{F}^x \widehat{F}^y} \Sigma_{\widehat{F}^y \widehat{F}^y}^{-1} \Sigma_{\widehat{F}^y \widehat{F}^x}. \tag{8}$$

The properties of the squared canonical correlations have been well studied. Anderson (1984) shows that if $\mathbf{P}_1$ and $\mathbf{P}_2$ are observed (not estimated) and are normally distributed, then

$$z_i = \frac{\sqrt{T}\left(\tilde{\rho}_i^2 - \rho_i^2\right)}{2\tilde{\rho}_i(1 - \tilde{\rho}_i^2)} \xrightarrow{d} \mathbf{N}(0, 1) \qquad (9)$$

where $\tilde{\rho}_i$ denotes the sample canonical correlation and ρ_i denotes the benchmark correlation. Muirhead and Waternaux (1980) extend this result to show that if $\mathbf{P}_1$ and $\mathbf{P}_2$ are observed (not estimated) and are elliptically distributed, then

$$z_i = \frac{1}{(1 + \nu/3)} \frac{\sqrt{T}\left(\tilde{\rho}_i^2 - \rho_i^2\right)}{2\tilde{\rho}_i(1 - \tilde{\rho}_i^2)} \xrightarrow{d} \mathbf{N}(0, 1) \qquad (10)$$

where ν is excess kurtosis. Bai and Ng [2] further extend these results to show that if one set is estimated while the second set is observed, then Eqs. (9) and (10) still hold.

References

1. Bai, J., & Ng, S. (2002). Determining the number of factors in approximate factor models. *Econometrics, 70*, 191–221.
2. Bai, J., & Ng, S. (2006). Evaluating latent and observed factors in macroeconomics and finance. *Journal of Econometrics, 131*, 507–537.
3. Bekaert, G., Hodrick, R., & Zhang, X. (2009). International stock return comovement. *Journal of Finance, 64*, 2591–2626.
4. Bekeart, G., & Harvey, C. (1995). Time-varying world market integration. *Journal of Finance, 50*, 403–444.
5. Blackburn, D., & Chidambaran, N. (2014). *Is the world stock return comovement changing?* In Working Paper, Fordham University.
6. Carrieri, F., Errunza, V., & Hogan, K. (2007). Characterizing world market integration through time. *Journal of Financial and Quantitative Analysis, 42*, 915–940.
7. Errunza, V., & Losq, E. (1985). International asset pricing under mild segmentation: Theory and test. *Journal of Finance, 40*, 105–124.
8. Fama, E., & MacBeth, J. (1973). Risk, return, and equilibrium: Empirical tests. *Journal of Political Economy, 81*, 607–636.
9. Goyal, A., Perignon, C. & Villa, C. (2008). How common are common return factors across NYSE and Nasdaq? *Journal of Financial Economics, 90*, 252–271.
10. Gultekin, M., Gultekin, N. B. & Penati, A. (1989). Capital controls and international capital market segmentation: The evidence from the Japanese and American stock markets. *Journal of Finance, 44*, 849–869.
11. Gyourko, J., & Keim, D. (1992). What does the stock market tell us about real estate returns. *Journal of the American Real Estate and Urban Economics Association, 20*, 457–485.
12. Jagannathan, R., & Wang, Z. (1998). An asymptotic theory for estimating beta-pricing models using cross-sectional regression. *Journal of Finance, 53*, 1285–1309.
13. Ling, D., & Naranjo, A. (1999). The integration of commercial real estate markets and stock markets. *Real Estate Economics, 27*, 483–515.
14. Liow, K. H., & Webb, J. (2009). Common factors in international securitized real estate markets. *Review of Financial Economics, 18*, 80–89.
15. Liu, C., Hartzell, D., Greig, W., & Grissom, T. (1990). The integration of the real estate market and the stock market: Some preliminary evidence. *Journal of Real Estate Finance and Economics, 3*, 261–282.

16. Liu, C., & Mei, J. (1992). The predictability of returns on equity REITs and their co-movement with other assets. *Journal of Real Estate Finance and Economics, 5*, 401–418.
17. Mei, J., & Saunders, A. (1997). Have U.S. financial institutions' real estate investments exhibited trend-chasing behavior? *Review of Economics and Statistics, 79*, 248–258.
18. Miles, M., Cole, R., & Guilkey, D. (1990). A different look at commercial real estate returns. *Real Estate Economics, 18*, 403–430.
19. Newey, W., & West, K. (1987). A simple, positive semi-definite, heteroskedasticity and autocorrelation consistent covariance matrix. *Econometrica, 55*, 703–708.
20. Pukthuanthong, K., & Roll, R. (2009). Global market integration: An alternative measure and its application. *Journal of Financial Economics, 94*, 214–232.
21. Schnare, A., & Struyk, R. (1976). Segmentation in urban housing markets. *Journal of Urban Economics, 3*, 146–166.
22. Shanken, J. (1992). On the estimation of beta pricing models. *Review of Financial Studies, 5*, 1–34.
23. Stock, J., & Watson, M. (2002). Forecasting using principal components from a large number of predictors. *Journal of the American Statistical Association, 97*, 1167–1179, 32.
24. Stulz, R. (1981). A model of international asset pricing. *Journal of Financial Economics, 9*, 383–406.
25. Titman, S., & Warga, A. (1986). Risk and the performance of real estate investment trusts: A multiple index approach. *Real Estate Economics, 14*, 414–431.

Supercomputer Technologies in Social Sciences: Existing Experience and Future Perspectives

Valery L. Makarov and Albert R. Bakhtizin

Abstract This work contains a brief excursus on the application of supercomputer technologies in social sciences, primarily, in part of the technical implementation of large-scale agent-based model (ABM). In this chapter, we will consider the experience of scientists and practical experts in the launch of agent-based model on supercomputers. On the example of agent model developed by us of the social system in Russia, we will analyze the stages and methods of effective projection of a computable core of a multi-agent system on the architecture of the acting supercomputer.

Keywords Agent-based models · Parallel calculations · Supercomputer technologies

1 Introduction

Computer modeling is the broadest, most interesting, and intensely developing area of research and is in demand today in many spheres of human activity. The agent-based approach to modeling is universal and convenient for practical researchers and experts because of its visualization, but at the same time sets high requirements for computing resources. It is obvious that significant computing capacities are necessary for direct modeling of sufficiently long-term social processes in a given country or on the planet as a whole.

Due to exponential growth of data volumes, the upcoming trend in agent-based model (ABM) development is the ABM development using supercomputer technologies (including those based on geoinformation systems). This direction is developing rapidly today, and it is already the subject of discussions at world congresses dedicated to ABM.

V. L. Makarov (✉) · A. R. Bakhtizin
CEMI RAS, Moscow, Russia
e-mail: makarov@cemi.rssi.ru

© Springer Nature Switzerland AG 2018
S.-H. Chen et al. (eds.), *Complex Systems Modeling and Simulation in Economics and Finance*, Springer Proceedings in Complexity,
https://doi.org/10.1007/978-3-319-99624-0_13

The relevance of using supercomputer technologies to develop ABM is explained by the fact that an ordinary personal computer no longer has enough memory to fit the number of programming environment items corresponding to, e.g., the population of the Earth, or even some densely populated countries. The launch of original models in specialized environments for ABM design with the number of agents exceeding several million already exceeds the memory amount in a personal computer.

The same situation is observed in terms of performance. Computing the state of a large-scale system with nontrivial behavior logic and interaction between agents requires substantial computing resources comparable to the demands of computational methods used in mathematical physics with the same number of computational cells. However, unlike the movements of particles, the social agents' behavior comprises an element of chance, which makes necessary an additional series of calculations to estimate the probability distribution of the key characteristics in the final stage of the modeled environment.

The factors listed above stipulate the need for large-scale experiments using supercomputer versions of the models in which the agents' population is distributed over a variety of supercomputer nodes and the calculations are performed in parallel. This, in return, requires adaptation of the models built in traditional development environments to supercomputer architecture.

Similarly to the supercomputer programs used to address numerous physical tasks, the potential for parallelizing multi-agent systems is in using the locality of agent interaction. Most of the interactions in the model, just like in real life, occur between subjects located near each other. This allows parallelizing "by space," i.e., ensuring a more uniform distribution of the agents' population over supercomputer nodes given their geographic proximity. Thus, breaking down the territory occupied by the agents into the so-called blocks provides the basic capacity for parallelizing the task. This is the approach most commonly used in practice for cases where distribution of elements in an environment being modeled. Whether we model agents in multi-agent systems or individual computational cells with averaged parameters of the physical environment, they should meet the special localization principle: the connections and exchange of data happen mostly for elements with close coordinates and are processed almost instantly within each computational node.

Supercomputers allow to enormously increase the number of agents and other quantitative characteristics (network nodes, and the size of territory) of models, originally developed for use on ordinary desktop computers. For this reason, supercomputer modeling is a logical and desirable step for those simplified models, which have already passed practical approbation on conventional computers. Unfortunately, the specific architecture of modern computers does not guarantee that the software of a computer model will immediately work on a supercomputer. The paralleling of the computable core is required as a minimum as well as frequently its deep optimization. In the absence of these adjustments, the use of expensive supercomputer calculation will most likely not pay off.

2 Experience of Some Scientists and Practical Experts

In September of 2006, a project on the development of a large-scale ABM of the European economy—**EURACE**, i.e., Europe ACE (Agent-based Computational Economics), was launched, with a very large number of autonomous agents, interacting within the socioeconomic system [4]. Economists and programmers from eight research centers in Italy, France, Germany, Great Britain, and Turkey are involved in the project, including an advisor from Columbia University, USA, the Nobel Prize winner Joseph Stiglitz.

According to the developers, virtually all existing ABMs either cover only a single industry or a relatively restricted geographical area and accordingly, small populations of agents, while the EURACE presents the entire European Union, so the scope and complexity of this model is unique, and its numerical computation requires the use of supercomputers as well as special software.

The information about 268 regions in 27 countries was used to fill the model with necessary data, including some geoinformation maps.

In the model, there are three types of agents: households (up to 10^7), enterprises (up to 10^5), and banks (up to 10^2). They all have a geographical reference, and are also linked to each other through social networks, business relationships, etc.

EURACE was implemented using a flexible scalable environment for simulating agent-based model—**FLAME** (Flexible Large-scale Agent Modeling Environment), developed by Simon Coakley and Mike Holcombe[1] initially to simulate the growth of cells under different conditions. With the help of the developed model, several experiments were conducted in order to study the labor market. Without going into detail about the obtained numerical results, we note that, according to the authors, the main conclusion of the research is that the macroeconomic measures of two regions with similar conditions (resources, economic development, etc.) during a long period (10 years and more) may vary significantly, due to an initial heterogeneity of the agents.[2]

In ABM **EpiSims**, developed by researchers from the Virginia Institute of bioinformatics (Virginia Bioinformatics Institute), the movement of agents is studied as well as their contacts within an environment as close as possible to reality and containing roads, buildings, and other infrastructure objects [10]. To develop this model, a large array of data was necessary, including information about the health of individual people, their age, income, ethnicity, etc.

The original goal of the research was to construct an ABM of high dimension to be launched on the supercomputer, which could be used to study the spreading of diseases in society. However, afterwards, in the course of work, another task was also being resolved, regarding the creation of specialized **ABM++** software, which allows to carry out the development of ABM in the C++ language, and also

---

[1] For more thorough information, see www.flame.ac.uk.

[2] More thorough information can be found on the website of the project: www.eurace.org.

contains functions, facilitating the allocation of the program code in use among the cluster nodes of the supercomputer. Apart from that, ABM++ provides for the possibility of dynamic redistribution of currents of calculations as well as the synchronization of ongoing events.

ABM++, the first version of which appeared in 2009, is the result of the modernization of the instrument, developed in 1990–2005 in the Los Alamos National Laboratory during the process of constructing large-scale ABMs (EpiSims, TRANSIMS, and MobiCom).

Specialists of another research team from the same Bioinformatics Institute of Virginia created an instrument for the study of the particularities of the spreading of infectious diseases within various groups of society—**EpiFast**, among the assets of which is the scalability and high speed of performance. For example, the simulation of social activity of the population of Greater Los Angeles Area (agglomerations with a population of over 17 million people) with 900 million connections between people on a cluster with 96 dual-core processors POWER5 took less than 5 min. Such fairly high productivity is provided by the original mechanism of paralleling presented by the authors [2, 7].

Classic standard models of spread of epidemics were mostly based on the use of differential equations; however, this tool complicates the consideration of connections between separate agents and their numerous individual particularities. ABM allows to overcome such shortcomings. In 1996, Joshua Epstein and Robert Axtell published a description of one of the first ABMs, in which they reviewed the process of the spread of epidemics [6]. Agent models, which differ from each other in their reaction to the disease, which depends on the state of their immune system, are spread out over a particular territory. At that, in this model, agents, the number of which constitutes a mere few thousand, demonstrate fairly primitive behavior.

Later on, under the supervision of Joshua Epstein and Jon Parker at the Center on Social and Economic Dynamics at Brookings, one of the largest ABMs was constructed, which included data about the entire population of the USA, that is around 300 million agents [9]. This model has several advantages. First of all, it allows to predict the consequences of the spread of diseases of various types. Second of all, it focuses on the support of two environments for calculations: one environment consists of clusters with an installed 64-bit version of Linux, and the other of servers with quad-core processors and an installed Windows system (in this regard, Java was chosen as the language of the programming, although the developers did not indicate which particular version of Java they used). Third of all, the model is capable of supporting from a few hundred million to six billion agents.

The model in question (the US National Model) includes 300 million agents, which move around the map of the country in accordance with the mobility plan of 4000×4000 dimensions, specified with the help of a gravity model. A simulation experiment was conducted on the US National Model, imitating the 300-day long process of spreading a disease, which is characterized by a 96-h incubation period and a 48-h infection period. In the course of the study, among other things, it was determined that the spreading of the disease was declining, after 65% of the infected got better and obtained immunity. This model has repeatedly been used by the specialists of the Johns Hopkins University as well as by the US Department of National Security, for research, dedicated to the strategy of rapid response to various types of epidemics [5].

In 2009, a second version of the US National Model was created, which included 6.5 billion agents, whose actions were specified taking into consideration the statistical data available. This version of the model was used to imitate the spreading of the A(H1N1/09) virus all over the planet.

Previously, this kind of model was developed by the Los Alamos National Laboratory (USA), and the results of the work with this model were published on April 10, 2006 [1]. One of the most powerful computers which existed at the time known by the name of "Pink," which consisted of two 1024 processors with a 2.4 GHz frequency and a 2 GB memory each was used for the technical realization of the model. This large-scale model, composed of 281 million agents, was used to study scenarios of the spreading of various viruses, including the H5N1, and took into consideration several possible operational interventions such as vaccinations, closing of schools, and introducing of quarantines in some territories.

Researchers at Argonne National Laboratory have been successfully using a new modeling paradigm agent-based modeling and simulation (ABMS)—to address challenges and gain valuable insights in such key areas as energy, biology, economics, and social sciences. To maximize potential, they are developing a next-generation ABMS system that can be extended to exascale computing environments to achieve breakthrough results in science, engineering, and policy analysis.

Argonne researchers have developed and used large-scale agent-based model to provide important information to policymakers that would not be available using other modeling approaches. One outstanding example—Electricity Markets Complex Adaptive Systems (**EMCAS**)—was used to model the Illinois electric power industry under deregulation conditions in an effort to anticipate the likely effects of deregulation on electricity prices and reliability.[3]

Interdisciplinary project on modeling of technological, social, and economic systems of the world was launched in 2012 and has involved scientists from almost all developed countries. Its implementation period is 10 years and the initial funding is 1 billion Euros.[4] Project managers emphasize the use of advanced information technologies (first of all, the agent-based model on the basis of geographic information systems). According to one of the leaders of the project—Dirk Helbing, despite the relevancy of developing such multilevel systems and the existence of some of their components, an integrated product is still missing due to institutional barriers and lack of resources. In this regard, **FuturICT** promises to become the first of its kind.

---

[3]Electricity Market Complex Adaptive System (EMCAS), Argonne National Laboratory, http://www.dis.anl.gov/projects/emcas.html.

[4]www.futurict.eu.

FuturICT is organized as a network of national centers, each of which is represented by several scientific institutions within the same country. In addition to this network, there is a problem-oriented network, which uses the main national science centers for solving specific issues. Thus, FuturICT integrates organizations on institutional and problem-solving level.

Every scientific community at the country level has a certain degree of autonomy, but also a set of obligations. Developers offer the FuturICT platform, which includes the three following components:

1. the nervous system of the planet (Planetary Nervous System);
2. simulator of a living planet (Living Earth Simulator);
3. global unified platform (Global Participatory Platform).

A set of models, forming a "simulator of a living planet," through "observatories" will allow for the detection of crises and finding solutions for mitigation. These models will be verified and calibrated using data collected in real time using the "nervous system of the planet." Ultimately, decision-makers will interact with the global unified platform, which will combine the results of the first two parts of FuturICT.

Various "observatories" (financial, economic, energy, and transport) are arranged in four groups of main directions of research in the field of public relations, technological and economic development as well as in monitoring of the environmental state.

"The nervous system of the planet" can be represented in the form of a global network of sensors that collect information on socioeconomic, technological, and ecological systems of the world. For its construction, the project coordinators of FuturICT work closely with the team of Sandy Pentland from the Massachusetts Institute of technology (MIT) in order to "connect the sensors to modern gadgets."

In the framework of the "simulator of the living planet," an open software platform will be implemented, which, according to the initiators of the project, will remind the well-known App Store in the sense that scientists, developers, and interested people will be able to upload and download other models related to the various parts of the planet. The basic modeling approach will be based on an agent-based paradigm. In the future, unified model components are expected to be implemented with the use of supercomputer technologies.

"Global unified platform," will also serve as an open platform for discussion of the forecasts of development of the world in regard to citizens, government officials, and the business community obtained using FuturICT.

There are several tools for high-performance computing for ABM.

Microsoft Axum is a domain-specific concurrent programming language, based on the Actor model that was under active development by Microsoft between 2009 and 2011. It is an object-oriented language based on the .NET Common Language Runtime using a C-like syntax which, being a domain-specific language, is intended for the development of portions of a software application that is well suited to concurrency. But, it contains enough general-purpose constructs that one need not switch to a general-purpose programming language (like C#) for the sequential parts of the concurrent components.[5]

The main idiom of programming in Axum is an Agent (or an Actor), which is an isolated entity that executes in parallel with other Agents. In Axum parlance, this is referred to as the agents executing in separate isolation domains; objects instantiated within a domain cannot be directly accessed from another.

Agents are loosely coupled (i.e., the number of dependencies between agents is minimal) and do not share resources like memory (unlike the shared memory model of C# and similar languages); instead, a message passing model is used. To coordinate agents or having an agent request the resources of another, an explicit message must be sent to the agent. Axum provides Channels to facilitate this.

The Axum project reached the state of a prototype with working Microsoft Visual Studio integration. Microsoft had made a CTP of Axum available to the public, but this has since been removed. Although Microsoft decided not to turn Axum into a project, some of the ideas behind Axum are used in TPL Dataflow in .Net 4.5.

Repast for High-Performance Computing (**Repast HPC**) 2.1, released on 8 May 2015, is a next-generation agent-based modeling and simulation (ABMS) toolkit for high-performance distributed computing platforms.

Repast HPC is based on the principles and concepts developed in the Repast Simphony toolkit. Repast HPC is written in C++ using MPI for parallel operations. It also makes extensive use of the boost (http://boost.org) library.

Repast HPC is written in cross-platform C++. It can be used on workstations, clusters, and supercomputers running Apple Mac OS X, Linux, or Unix. Portable models can be written in either standard or Logo-style C++.

Repast HPC is intended for users with:

- Basic C++ expertise
- Access to high-performance computers
- A simulation amenable to a parallel computation. Simulations that consist of many local interactions are typically good candidates.

Models can be written in C++ or with a Logo-style C++.[6]

CyberGIS Toolkit is a suite of loosely coupled open-source geospatial software components that provide computationally scalable spatial analysis and modeling capabilities enabled by advanced cyberinfrastructure. CyberGIS Toolkit represents a deep approach to CyberGIS software integration research and development and is one of the three key pillars of the CyberGIS software environment, along with CyberGIS Gateway and GISolve Middleware.[7]

---

[5] https://www.microsoft.com/en-us/download/details.aspx?id=21024.

[6] More thorough information on the S/W can be found in the User Manual [3]; a new version of the 1.0.1. package (dated March 5, 2012) can be downloaded from the following website: http://repast.sourceforge.net/repast$_$hpc.html.

[7] http://cybergis.cigi.uiuc.edu/cyberGISwiki/doku.php/ct.

The integration approach to building CyberGIS Toolkit is focused on developing and leveraging innovative computational strategies needed to solve computing- and data-intensive geospatial problems by exploiting high-end cyberinfrastructure resources such as supercomputing resources provided by the Extreme Science and Engineering Discovery Environment and high-throughput computing resources on the Open Science Grid.

A rigorous process of software engineering and computational intensity analysis is applied to integrate an identified software component into the toolkit, including software building, testing, packaging, scalability and performance analysis, and deployment. This process includes three major steps:

1. Local build and test by software researchers and developers using continuous integration software or specified services;
2. Continuous integration testing, portability testing, small-scale scalability testing on the National Middleware Initiative build and test facility; and
3. XSEDE-based evaluation and testing of software performance, scalability, and portability. By leveraging the high-performance computing expertise in the integration team of the NSF CyberGIS Project, large-scale problem-solving tests are conducted on various supercomputing environments on XSEDE to identify potential computational bottlenecks and achieve maximum problem-solving capabilities of each software installation.

Pandora is a novel open-source framework designed to accomplish a simulation environment corresponding to the above properties. It provides a C++ environment that automatically splits the run of an ABM in different computer nodes. It allows the use of several CPUs in order to speed up the decision-making process of agents using costly AI algorithms. The package has also support for distributed execution and serialization through HDF5, and several analysis techniques (spatial analysis, statistics and geostatistics, etc.).[8]

In addition, pyPandora is a Python interface to the framework, designed to allow people with minimal programming background a tool to develop prototypes. The ability to develop ABMs using Python is an important feature for social scientists, because this programming language is also used in other common modeling tools, like Geographical Information Systems (i.e., ESRI ArcGIS).

PyPandora allows researchers to create ABMs using a language that they already know and enables the design of a common framework where one can combine simulation, spatial analysis, and geostatistics. Finally, the package is complemented by Cassandra, a visualization tool created to detect spatiotemporal patterns generated by the simulation. This application allows any user to load the data generated by a Pandora parallel execution into a single computer and analyze it.

SWAGES, a distributed agent-based Alife simulation and experimentation environment that uses automatic dynamic parallelization and distribution of simulations in heterogeneous computing environments to minimize simulation times.

---

[8]https://www.bsc.es/computer-applications/pandora-hpc-agent-based-modelling-framework.

SWAGES allows for multi-language agent definitions, uses a general plug-in architecture for external physical and graphical engines to augment the integrated SimWorld simulation environment, and includes extensive data collection and analysis mechanisms, including filters and scripts for external statistics and visualization tools.[9] Moreover, it provides a very flexible experiment scheduler with a simple, web-based interface and automatic fault detection and error recovery mechanisms for running large-scale simulation experiments.

A Hierarchical Parallel simulation framework for spatially explicit Agent-Based Models (**HPABM**) is developed to enable computationally intensive agent-based model for the investigation of large-scale geospatial problems. HPABM allows for the utilization of high-performance and parallel computing resources to address computational challenges in agent-based model.[10]

Within HPABM, an agent-based model is decomposed into a set of sub-models that function as computational units for parallel computing. Each sub-model is comprised of a subset of agents and their spatially explicit environments. Sub-models are aggregated into a group of super-models that represent computing tasks. HPABM based on the design of super- and sub-models leads to the loose coupling of agent-based model and underlying parallel computing architectures. The utility of HPABM in enabling the development of parallel agent-based model was examined in a case study.

Results of computational experiments indicate that HPABM is scalable for developing large-scale agent-based model and, thus, demonstrate efficient support for enhancing the capability of agent-based modeling for large-scale geospatial simulation.

The growing interest in ABM among the leading players in the IT industry (Microsoft, Wolfram, ESRI, etc.) definitely shows the relevance of this instrument and its big future, while exponential growth of overall data volumes related to human functioning and the need for analytical systems to obtain new-generation data needed to forecast social processes call for the use of supercomputer technologies.

There are currently several international associations uniting groups of researchers from the largest institutes and universities working in this direction. The most famous ones are: (1) North American Association for Computational Social and Organizational Sciences (NAACSOS); (2) European Social Simulation Association (ESSA); and (3) Pacific Asian Association for Agent-Based Approach in Social Systems Science (PAAA). Each association holds regular conferences on social modeling in America, Europe, and Asia, respectively. A World Congress on the topic is also conducted every 2 years.

In Russia, ABMs are a relatively new scientific development, pioneered by the Central Economics and Mathematics Institute of the Russian Academy of Sciences.

---

[9]http://www.hrilab.org/.

[10]https://clas-pages.uncc.edu/wenwu-tang.

The content and results of the studies conducted by this institution are described in the second part of this chapter.

3 Adaptation of Agent-Based Models for the Supercomputer: Our Approach

3.1 Experiment #1

In March of 2013, the model was launched on the supercomputer Lomonosov which simulated the socioeconomic system of Russia for the next 50 years. This ABM is based on the interaction of 100 million agents, which hypothetically represent the socioeconomic environment of Russia. The behavior of each agent is set by a number of algorithms, which describe its actions and interaction with other agents in the real world.

Five people participated in the project: two specialists of the Central Economic and Mathematical Institute of the Russian Academy of Sciences (V.L. Makarov and A.R. Bakhtizin) and three researchers of the Moscow State University (V.A. Vasenin, V.A. Roganov, and I.A. Trifonov) [8]. The data for the modeling was provided by the Federal Service of State Statistics and by the Russian Monitoring of the Economic Conditions and Health of the Population. A model for a standard computer was developed in 2009, and in 2013 it was converted into a supercomputer version. Below, we will examine the main stages and methods of effective projection of a computable kernel of a multi-agent system on the architecture of a modern supercomputer, which we have developed during the process of resolving the issue in question.

3.1.1 The Problem of Scaling

It is important to understand that the scaling of programs for supercomputers is a fundamental problem. Although the regular and supercomputer programs carry out the same functionalities, the target functions of their development are usually different.

During the initial development of the complex application software, the first and foremost goal is to try to minimize the costs of programming, personnel training, enhance the compatibility between platforms, etc., and to leave optimization for later. This is quite reasonable, since at the early stages, the priority of development is exclusively the functionality.

However, when the developed software is already being implemented, it is often discovered that there is not enough productivity for real massive data. And, since modern supercomputers are not simple computers, but work thousands of times faster than personal computers, in order to launch the program on a supercomputer it is necessary to introduce significant alterations first. Doing this effectively without special knowledge and skills is by far not always successfully achieved.

Doing work properly and correctly usually leads to significant increase in efficiency on the following three levels:

1. multisequencing of calculation;
2. specialization of calculative libraries by task;
3. low-level optimization.

3.1.2 Specialization and Low-Level Optimization

Before seriously talking about the use of supercomputers, the program must be optimized to the maximum and adapted to the target hardware platform. If this is not done, the parallel version will merely be a good test for the supercomputer, but the calculation itself will be highly inefficient.

To use a supercomputer without optimization and adaptation of the program to the target hardware platform is the same as sending a junior regiment on a combat mission: first of all, it is necessary to teach the recruits how to properly carry out their tasks (specialization, and optimization of software), as well as how to efficiently handle weapons (low-level optimization of software), and only then considerations of effective use of resources can really enter into account.

In the universal systems of modeling of the AnyLogic type, the procedures presented are universal. And, a universal code can often be optimized for a particular family of tasks.

3.1.3 Selection of a Modeling Support System

Certainly, ABM can be programmed without a special environment, in any object-based language. In addition, the main shortcoming of the existing products for creating ABM except for RepastHPC is its inability to develop projects that would run on a computing cluster (i.e., there is no mechanism for paralleling the process of executing the program code).

However, a more reasonable approach would be to use one of the proven systems for ABM because of the unified implementation of standard ways of interacting agents. In this chapter, we will limit our scope to the ADEVS system.[11]

ADEVS is a set of low-level libraries for discrete modeling done in C++ language. Some of the advantages worth mentioning are the following:

- ease of implementation;
- high efficiency of models;

---

[11] The ADEVS system and its description can be downloaded from here: http://www.ornl.gov/~1qn/adevs/.

- support of basic numerical methods used for modeling;
- built in paralleling of simulation with the help of OpenMP;
- the possibility of using standard means of paralleling;
- fairly rapid development of libraries in the current time;
- Cross-platform software;
- low-level software (current functionality does not impose any restrictions on the model);
- independence of the compiled code on unusual libraries;
- open-source code.

However, significant shortcomings of this product is a complete lack of means of presentation and quite complicated modeling, when compared to, for example, AnyLogic. Therefore, this product cannot be used to build models on the consumer level, however, is an effective platform for implementing parallel simulations.

The main elements of the program when using the ADEVS library to build an ABM are the following:

- adevs simulator:: Simulator$< X >$;
- primitive of adevs agents:: Atomic$< X >$;
- model of adevs (container of agents)::::: Digraph$< VALUE, PORT >$.

It was decided to develop the supercomputer version of the program described below, based on the ADEVS system, in view of its advantages listed above. Within the framework of this project, an MPI version of the ADEVS simulator was created, as well as a visualization system of the calculation process based on the Qt library—a cross-platform set of tools for creating software written in the C++ programming language.

Next, we turn to a brief description of the developed model and the procedures for its subsequent run on a supercomputer.

3.1.4 The Initial Agent-Based Model

The first stage of development of the ABM described below is constructing a tool that effectively solves the problem of research on conventional computers as well as adjusting the parameters of the model. The model is then tested on a conventional computer with a small number of agents (depending on the complexity of agents, conventional computers are able to perform calculations at a satisfactory rate and with good productivity for approximately 20 thousand agents). During the first stage, the AnyLogic product was used, the technical capabilities of which allowed to debug the model at a satisfactory speed and to configure its settings. After successful testing, we proceeded to the second stage of development: to convert the model so that it could be used on a supercomputer. The model for an ordinary computer was built in 2009, and it was converted into a supercomputer version in 2013.

Figure 1 shows the operating window of the developed ABM (dots—agents). During the operation of the system, current information can be obtained on the

Supercomputer Technologies in Social Sciences

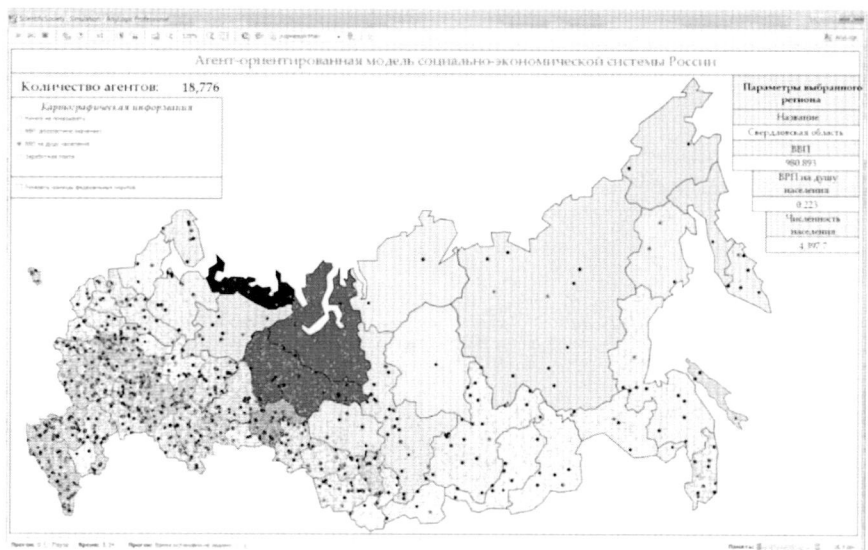

Fig. 1 Operating window of the developed ABM

socioeconomic situation in all regions of Russia, including the use of cartographic data, changing in real time depending on the values of the endogenous parameters.

The specification of the agents of the model was carried out taking into consideration the following parameters: age, life expectancy, specialization of parents, place of work, region of residence, income, and others.

The specification of regions was carried out, taking into consideration the following parameters: geographic borders, population, number of workers (by type), GRP, GRP per capita, volume of investments, volume of investments per capita, average salary, average life expectancy, index of population growth, etc.

Statistics manuals of Rosstat as well as sociological databases of RLMS were used to fill the model with the necessary data.

Agents in the model are divided into two groups (types) with different reproductive strategies. Agents of the first type follow the traditional strategy, known for high birth rate, while the second group follows the modern strategy with a much lower birth rate. The model starts working by initializing its starting environment conditions and creating a population of agents with personal characteristics (age, gender, reproductive strategy type, and the number of children sought) assigned so as to reproduce the age, gender, and social structure of the population in the region being modeled.

Next, the natural population movement processes—mortality and birth—are imitated using the age movement methods and probability mechanisms. Agents die following the mortality rates that are differentiated by age and gender but corresponding to the actual figures for the entire population. Meanwhile, the creation of new agents in the model (childbirth) is a result of other agents' actions.

First, individual agents interact to form couples and agree on how many children they want to have together. Next, the couples agree on when each child will be born, depending on them belonging to one or another type.

The model was used to conduct experiments forecasting changes in the population of agents inhabiting a certain region, the age structure of this population, and the correlation of agent numbers in various types for main age groups and the population in general. Experiments were conducted with the following parameter values: total population—20,000; share of agents with traditional reproductive strategy type—10%. The outcomes show that the model adequately imitates processes observed in real life as a reduction in the overall population numbers, as well as its aging—reduction of the younger population groups and increase in older groups.

3.1.5 Conversion of the Model into a Supercomputer Program

Earlier, we had already discussed the problems of using ABM development tools for the realization of projects, carried out on the computing clusters of the supercomputer. Due to the difficulties in separating the computing part from the presentational part as well as to the realization of the code using a high level of the JAVA language, the productivity of the implementation of the code is significantly lower for AnyLogic than for ADEVS. Apart from that, it is extremely difficult to reprocess the generated code into a concurrently executed program.

Below is the algorithm of the conversion of the AnyLogic model into a supercomputer program.

Translation of the Model

The models in the AnyLogic project are kept in the format of an XML file, containing the tree diagram of the parameters necessary for the generation of the code: classes of agents, parameters, elements of the presentation, and descriptions of the UML diagrams of the behavior of agents.

During the work of the converter, this tree diagram is translated into code C++ of the program, calculating this model. The entry of the tree is executed depth wise. At that, the following key stages are marked, and their combination with the translation process is carried out.

1. *Generating the main parameters.* The search for the root of the tree and the reading of the parameters of daughter nodes, such as the name of the model, address of assembly, type of model, and type of presentation.
2. *Generating classes.* Generating classes (more detailed):

    (a) configuration of the list of classes;
    (b) reading of the main class parameters;
    (c) reading of the variables;
    (d) reading of the parameters;

(e) reading of the functions;
(f) generating a list of functions;
(g) reading the functions code;
(h) conversion of the functions code Java $->C++$;
(i) reading of the figures and elements of control that are being used;
(j) generating the code of initialization of figures and elements of control;
(k) generating constructor, destructor, and visualizer codes;
(l) generating the class structure;
(m) generating header code and source files.

3. *Generating the simulator.* Search for the peak, containing the information about the process of simulation (controlling elements, significance of important constants, elements of presentation, etc.).
4. *Generating shared files of the project* (main.cpp, mainwindow.h, mainwindow.cpp, etc.).

Import of Incoming Data Data from the geoinformation component of the initial model (map of Russia), containing all of the necessary information, is imported into the model as input data.

Generating Classes and the Transformation of the Functions Code When generating the functions from the tree, the following information is read: name of function, return type, parameters, and its body.

Based on the list of classes constructed earlier, changes are introduced into the arguments of the functions, replacing the heavy classes, i.e., all generated classes, classes of figures, and other classes, which do not form a part of the standard set with corresponding indicators. The purpose of this is to save memory space and avoid mistakes when working with it. After that, the titles of the functions are generated, which are later inserted into the title and source files. In the course of such reading, the body of the function, by the means of the relevant function (listing 5), is transformed from a Java-based code into an analogical C++ code (this is possible due to the fairly narrow class of used functions; as for more complicated functions, manual modification of the translated code is required), after which it is added to the list of bodies for this class.

In the course of the translation, the need for the transformation of the initial functions code from the Java language to the C++ often arises. It can be presented in the form of sequential replacements of constructions, for example:

- *Transformation of cycles*: Java format.
- *Transformation of indicators.* Java, unlike C++, does not contain such an obvious distinction between the object and the object indicator, hence the structure of the work with them does not differ. That is why, a list of classes is introduced, in which it is important to use operations with object indicators, and not with the object itself, and all of the variables of such classes are monitored with the subsequent replacement of addresses to the objects with corresponding addresses to the object indicators within the framework of the given function.

- *The opening of black boxes.* In Java, and in the AnyLogic library in particular, there is a certain number of functions and classes, which do not have analogues in the C++ itself, nor in the ADEVS library. Due to this fact, additional libraries, such as, shapes.h and mdb-work.h, had been created, which compensate for the missing functions.
- *During the generating stage of the main parameters of the lists of classes* the name of the main class and the names of the modulated agent classes are obtained. The procedure of adding an agent into the visibility range of the simulator is introduced into the code of the main class.

Generating Outside Objects In the process of generating outside objects, a separate function $Main :: initShapes()$ is created, which contains all of the graphic information, i.e., the initialization of all figures, the classes of which had also been implemented in the shapes.h., and is carried out within the framework of the function. The relevant example is presented in the following code fragment.

Generating Classes and the Code of the Title and Source Files Based on all the data that has been read and generated, the title and source files of the corresponding class are created.

Generating Simulation For the generation of simulation, it turned out to be enough to have the main.cpp, mainwindow.cpp, mainwindow.h files, written beforehand, in which the templates define the type of the main class and the added title files. When compiling the initial code, the templates are replaced with the names of the classes received earlier (at the generating stage). This is enough for the double-flow simulation, which can later be replaced with a corresponding module for a multiprocessor simulation.

Additional Attributes At the stage of analyzing the tree (see above), a tree, similar in structure, is formed for the generation of a C++ code, with the help of which the necessary attributes of compilation can be set (visualization of certain parts, visual validation of code recognition, additional flags of assembly etc.), during the stage of preparation for translation. After that, at receiving the command for transformation, the final compilation takes place, taking into consideration all of these attributes.

Assembly of the Ready Project For the assembly of the translated project, the QtCreator is used—cross-platform shareware integrated environment for work with the Qt framework.

Agent Code With the help of the translator described above, an initial code (except for the behavior pattern of agent) has been generated from the data of the files of the AnyLogic project (model.alp and others).

The behavior pattern of the agent must be generated from the diagram of conditions; however, currently the automation of this process has not yet been implemented. Therefore, a certain volume of the code had to be added to the generated code.

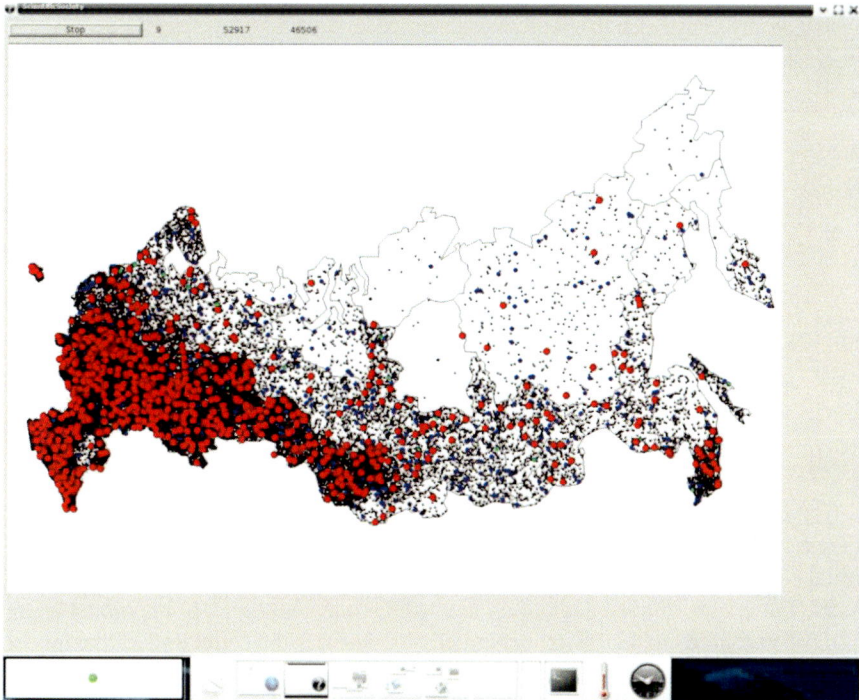

Fig. 2 Result of the work of the supercomputer program in graphic format

After the introduction of the necessary changes, a cross-platform application, repeating the main functionality of the given model, was achieved as a result of the compilation.

Statistics and Visualization of Time Layers Given the noninteractive mode of the launching of the program on big supercomputers, the collection of data and visualization were separated (this has to do with the load imbalance on clusters at various times of the day; as for exclusive access, it is simply impossible). After the recalculation of the model, the information obtained can once again be visualized, for example, in the following manner (Fig. 2).

Supercomputers Available for Calculations At the moment of making the calculations, three supercomputers were available to us (Table 1), which were in the top five of the supercomputer rating of the Top-50 supercomputers in the CIS countries.

For our calculations, we used two supercomputers—the "Lomonosov" and the MVS-100K.

Table 1 Supercomputers available for research group

Position in Top-50	Supercomputers	Nodes	CPU	Kernels	RAM/node	TFlops
1	Lomonosov (Moscow State University)	5130	10,260	44,000	12 GB	414
4	MVS-100K (Interagency Supercomputing Center of the RAS)	1256	2332	10,344	8 GB	123
5	Chebyshev (Moscow State University)	633	1250	5000	8 GB	60

3.1.6 Results

By using supercomputing technologies and optimizing the program code, we were able to achieve very high productivity.

The optimization of the code and the use of C++ instead of Java allowed for the increase in speed of execution of the program. The model was tested in the following initial conditions: (1) number of agents, 20 thousand; and (2) forecasting period, 50 years. The results of the calculations showed that the count time of the model using ADEVS amounted to 48 s using one processor, whereas the count time of the model using AnyLogic and a single processor amounted to 2 min and 32 s. It means that the development framework was chosen correctly.

As has already been noted above, an ordinary personal computer with high productivity is able to carry out calculations with satisfactory speed with a total number of 20 thousand agents, provided that the behavior of each agent is defined by approximately 20 functions. The average count time of one unit of model time (1 year) amounted to around 1 min. When dealing with a larger number of agents, for example, 100 thousand, the computer simply freezes.

Using the 1000 processors of the supercomputer and executing the optimized code allowed to increase the number of agents to 100 million and the number of model years to 50. All this enormous volume of research was carried out in a period of time which approximately equaled 1 min and 30 s (this indicator may slightly vary depending on the type of processors used).

The results of the modeling showed that if the current tendencies persist, the population of Russian Siberian and Far-Eastern Federal Districts will significantly decrease, while the Southern Federal District, on the contrary, will see a significant increase in population. In addition to that, the model suggests a gradual decrease of Russian GDP as well as of several other macroeconomic indicators.

The results of the experiments carried out using the developed ABM revealed that the scaling of the model in itself has certain meaning. For example, when launching the same version of the model for 50 model years using the same parameters, except for the number of agents (in the first case, there were 100 million agents, and in the second, 100 thousand agents), the results received diverged by approximately 4.5%.

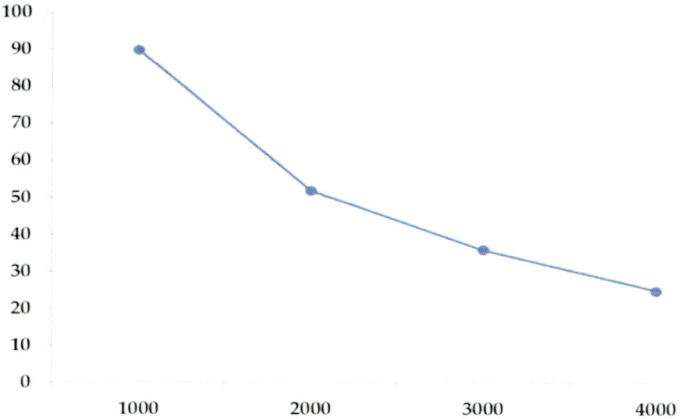

Fig. 3 Model performance (X axes number of processors, Y axes time in seconds)

It can be assumed that in complex dynamic systems the same parameters (birth rate, life expectancy, etc.) may produce different results depending on the size of the community.

Then, we continued to increase the number of processors under the same model parameters in order to establish a dependency of the time for computation from the resources involved. Here are the results (Fig. 3).

3.2 Experiment #2

That model version did not have interagent communication that, on the one hand, did not allow to use the agent approach to the full extent, and on the other, significantly simplified paralleling of the source code. In the current version, agents communicate with each other that resulted in changing of the model paralleling technology as well as other software libraries. Six people participated in the project: three specialists of the Central Economic and Mathematical Institute of the Russian Academy of Sciences (V.L. Makarov, A.R. Bakhtizin, and E.D. Sushko) and three researchers of the Moscow State University (V.A. Vasenin, V.A. Roganov, and V.A. Borisov).

Besides a common paralleling scheme that determines the model source code translation method, it is necessary to meet the requirements baseline of up-to-date supercomputers for which the de facto standard is focused on MPI (Message Passing Interface).

The use of Java standard environment is undesirable, because it is seldom available in supercomputer. That is why, all specificity peculiar for the source environment of AnyLogic version and standard Java environment, it shall be adapted to supercomputer, and communication between computational nodes shall be based on MPI standard.

Already tested during paralleling of demographic model of Russia, the library for multi-agent modeling ADEVS declared itself quite well. Last *ADEVS* versions have some support of Java as well being an advantage too. However, ADEVS designers have not implemented the parallel operation on supercomputer (except for OpenMP technology for multiprocessors that required significant finalization in part of MPI support in our previous work) up to now.

Also, during paralleling of a previous, quite simple model, it was rewritten using C++ entirely, being superfluous: pre- and post-processing of data, and creation of the initial status of multi-agent environment are not critical operations in terms of time.

For a supercomputer, paralleling of an algorithm to compute kernel is usually enough, i.e., in this case—population status recalculation phase.

Analysis of advanced software technologies showed that integrated tooling for Java programs execution has been developed for the last time actively. It uses the so-called *AOT (Ahead-Of-Time)* compilation.

At the same time, the result of AOT-compiler operation is a usual autonomous executable module that contains a machine code for a target platform. Such approach is used in new versions of *Android* operating system that, in our opinion, is not accidental—both for embedded systems, and for a supercomputer the code execution effectiveness is a key factor. Experiments with one such product—*AOT-compiler Avian*—made it possible to draw the following conclusions:

- Avian allows obtaining an autonomous executable module in the form of MPI application for a supercomputer, at the same time an undirected additional code is implemented using C++ easily, including initialization and reference to MPI communication library.
- Operating speed of the obtained software module (e.g., in classic game by Conway "Game of Life") approximately corresponds to the operating speed of ADEVS.
- It allowed transferring a considerable part of work on AOT-compiler and to use C++ for implementation of the most necessary one leaving for ADEVS a niche of accelerated stages support with complicated interagent communication.

3.2.1 Technology of Parallel Version Obtaining and Launching

A source, developed in AnyLogic environment model description, represents an XML file with extension *.ALP (XML file in standard AnyLogic)* that contains description of agents' essences of models, their parameters, and rules of agents' status recalculation during evolution. Except for ALP description, the model is attached with a data file in Excel format, where numeric parameters are specified used both at the population formation stage and during recalculation of population status.

The whole process of creation of a paralleled supercomputer version of the program is performed in several stages listed below.

1. *Input ALP file* with model description is being read by *converter* that redesigns an object representation for all described essences and rules and, having performed their necessary processing, forms a package of software modules (Java class of the total volume of several thousand lines) containing all significant information. At the same time, *agents' variables* are transformed into *classes' fields*, and rules of agents' response to events—*into the corresponding methods*.
2. *Input Excel file* with source parameters of the model is converted into the source Java text (also having the volume of several thousand lines) to provide for the immediate access to model parameters at every node of a supercomputer. In other words, external set of model parameters becomes a constituent part of the executable module for a supercomputer. *Formed software modules* together with a developed code of emulation of used AnyLogic environment functions are compiled and configured into machine code for a target supercomputer.
3. *Executable module*, at the same time, is completely autonomous, and during launching it takes several key parameters only in the command line. For example, launch of the model *with a million of agents for 20 years on 12 computational nodes* is executed by the following command:

$ mpirun -np 12 -x Population=1000000 rt.Main

4. *Output data*, at the same time, is collected from computational nodes in the process of calculation, and population key characteristics and total calculation time are preserved in the log and printed upon launch end:

[0]: totalNumberPeople = 990221

[0]: ***Total stages time: 21.173s***

3.2.2 Comparative Results of Models' Operating Speed

For the first experiments with a parallel version of a simplified demographic model, they used a fragment of the laboratory computational cluster with 12 computational kernels and total volume of random access memory 96 GB. With such configuration, RAM accommodates 60 million of agents easily that allows simulating population growth dynamics within a small country.

A parallel version was tested on multicore processor as well. Results of measurements of operating time of the model original and parallel version are shown in Tables 2, 3, and 4.

As you can see in the above tables, operating speed and compatibility in the number of agents of a supercomputer version is considerably higher than readings of an original model. One can see as well that increasing the localing of interagent communication (reducing quarters area), the calculation effectiveness increases.

Table 2 Original version, 1 computational kernel (8 GB RAM)

Agents	Seconds
25,000	2.5
50,000	4.7
100,000	17.6
200,000	52.7
400,000	194
1,000,000	766

Table 3 Parallel version, 4 computational kernels

Kernels	Quarters	Agents	Seconds
4	12	1,000,000	50
4	40	1,000,000	30
4	40	2,000,000	75
4	20	4,000,000	344

Table 4 Fragment of computational cluster (12 computational kernels Core i7, 96 GB RAM)

Kernels	Quarters	Agents	Seconds
12	12	1,000,000	21
12	24	1,000,000	12
12	24	10,000,000	516
12	60	10,000,000	303
12	300	10,000,000	132
12	300	30,000,000	585
12	300	50,000,000	1371
12	300	60,000,000	1833

3.2.3 Conclusions and Development Prospects of the Technology Developed

The main positive moment of the developed approach to paralleling of models designed in AnyLogic environment is *automation of their supercomputer versions creation*. It simplifies the development considerably, because in most cases, after insignificant modification of the source model, it does not require finalization of rules of transformation into executable module for a supercomputer.

This approach is expendable in part of used source language and software–hardware platform. Besides already successfully tested execution platforms *Avian and ADEVS*, we can develop other, lower-level means for agents' status recalculation state acceleration, and in prospect consider an issue of using such hardware accelerators as *Xeon Phi and NVidia CUDA*.

The used technology of internodal communication by means of active communication technology makes it possible, whenever required, to implement easily both interactive simulation and *interactive visualization of simulation process within estimated time*. However, it is possible only if a supercomputer is available in a burst mode, for example, if a compact personal supercomputer is used.

The main research question remaining is a question of maximum achievable effectiveness of paralleling in case of mass communication of agents being in different computational nodes of a supercomputer.

It is quite evident that *if every agent communicates actively with all other agents, the productivity will be low in view of extensional internodal traffic.*

Nevertheless, even in such unfavorable case, a supercomputer makes it possible to accelerate simulation considerably, when a model is launched either repeatedly (for statistics) or with different parameters. In other extreme case, when almost all communications are localized in terms of geographic location of agents, effectiveness of paralleling will be good.

Therefore, *effectiveness* of the parallel version depends directly on that *part of interagent communication* that requires transfer of large data volume between computational nodes.

Acknowledgements This work was supported by the Russian Science Foundation (grant ♯ 14-18-01968).

References

1. Ambrosiano, N. (2006). Avian flu modeled on supercomputer. *Los Alamos National Laboratory NewsLetter, 7*(8), 32.
2. Bisset, K., Chen, J., Feng, X., Kumar, V. S. A., & Marathe, M. (2009). EpiFast: A fast algorithm for large scale realistic epidemic simulations on distributed memory systems. In *Proceedings of 23rd ACM International Conference on Supercomputing (ICS'09), Yorktown Heights, New York* (pp. 430–439)
3. Collier, N. (2012). Repast HPC Manual. [Electronic resource] February 23. Access mode: http://repast.sourceforge.net, free. Screen title. Language. English. (date of request: May 2013)
4. Deissenberg, C., Hoog, S., & van der Herbert, D. (2008, June 24). EURACE: A massively parallel agent-based model of the European economy. Document de Travail No. 2008 (Vol. 39).
5. Epstein, J. M. (2009, August 6). Modeling to contain pandemics. *Nature, 460,* 687.
6. Epstein, J. M., & Axtell, R. L. (1996) *Growing artificial societies: Social science from the bottom up. Ch. V.* Cambridge, MA: MIT Press.
7. Keith, R. B., Jiangzhuo, C., Xizhou, F., Anil Kumar, V. S., & Madhav, V. M. (2009, June 8–12). EpiFast: A fast algorithm for large scale realistic epidemic simulations on distributed memory systems. In *ICS09 Proceedings of the 23rd international conference on supercomputing, Yorktown Heights, New York* (pp. 430–439)
8. Makarov, V. L., Bakhtizin, A. R., Vasenin, V. A., Roganov, V. A., & Trifonov, I. A. (2011). Tools of supercomputer systems used to work with agent-based models. *Software Engineering, 2*(3), 2–14.
9. Parker, J. (2007). A flexible, large-scale, distributed agent based epidemic model. Center on social and economic dynamics. Working Paper No. 52
10. Roberts, D. J., Simoni, D. A., & Eubank, S. (2007). *A national scale microsimulation of disease outbreaks.* RTI International. Research Triangle Park, NC: Virginia Bioinformatics Institute.

Is Risk Quantifiable?

Sami Al-Suwailem, Francisco A. Doria, and Mahmoud Kamel

Abstract The work of Gödel and Turing, among others, shows that there are fundamental limits to the possibility of formal quantification of natural and social phenomena. Both our knowledge and our ignorance are, to a large extent, not amenable to quantification. Disregard of these limits in the economic sphere might lead to underestimation of risk and, consequently, to excessive risk-taking. If so, this would expose markets to undue instability and turbulence. One major lesson of the Global Financial Crisis, therefore, is to reform economic methodology to expand beyond formal reasoning.

Keywords Financial instability · Gödel's incompleteness theorem · Irreducible uncertainty · Lucas critique · Mispricing risk · Quantifiability of risk · Reflexivity · Rice's theorem · Self-reference

1 Introduction

> *It is dangerous to think of risk as a number*
> *— William Sharpe [10]*

Can we estimate uncertainty? Can we quantify risk? The late finance expert Peter Bernstein wrote in the introduction to his book, *Against the Gods: The Remarkable Story of Risk* [9, pp. 6–7]:

S. Al-Suwailem (✉)
Islamic Development Bank Group, Jeddah, Saudi Arabia
e-mail: sami@isdb.org

F. A. Doria
Advanced Studies Research Group, PEP/COPPE, Federal University at Rio de Janeiro, Rio de Janeiro, Brazil

M. Kamel
College of Computer Science, King Abdul-Aziz University, Jeddah, Saudi Arabia

© Springer Nature Switzerland AG 2018
S.-H. Chen et al. (eds.), *Complex Systems Modeling and Simulation in Economics and Finance*, Springer Proceedings in Complexity,
https://doi.org/10.1007/978-3-319-99624-0_14

> The story that I have to tell is marked all the way through by a persistent tension between those who assert that the best decisions are based on quantification and numbers, determined by the patterns of the past, and those who base their decisions on more subjective degrees of belief about the uncertain future. This is a controversy that has never been resolved ...
>
> The mathematically driven apparatus of modern risk management contains the seeds of a dehumanizing and self-destructive technology. ...Our lives teem with numbers, but we sometimes forget that numbers are only tools. They have no soul; they may indeed become fetishes.

We argue in this chapter that developments in science and mathematics in the past century bring valuable insights into this controversy.

Philosopher and mathematician William Byers, in his book *The Blind Spot: Science and The Crisis of Uncertainty* [25], makes an interesting case for why we need to embrace uncertainty. He builds on discoveries in mathematics and science in the last 100 years to argue that:

- Uncertainty is an inevitable fact of life. It is an irreducible feature of the universe.
- Uncertainty and incompleteness are the price we pay for creativity and freedom.
- No amount of scientific progress will succeed in removing uncertainty from the world.
- Pretending that scientific progress will eliminate uncertainty is a pernicious delusion that will have the paradoxical effect of hastening the advent of further crises.

These aspects probably did not receive enough attention in mainstream literature, despite the pioneering efforts of many economists (see, e.g., Velupillai et al. [111]).

We deal here with the question at the roots of uncertainty in the sciences which use mathematics as their main tool. To be specific, by "risk" we mean indeterminacy related to future economic loss or failure. By "quantifiability" we mean the ability to model it using formal mathematical systems rich in arithmetic, in order to be able to derive the quantities sought after. Of course, there are many ways to measure and quantify *historical* risk.[1] The question more precisely, therefore, is: Can we systematically quantify uncertainty regarding future economic losses?

We argue that this is not possible.

This is not to say that we can never predict the future, or can never make quantitative estimates of future uncertainty. What we argue is that, when predicting the future, we will never be able to escape uncertainty in our predictions. More important, we will never be able to quantify such uncertainty. We will see that not even the strictest kind of mathematical rigor can evade uncertainty.

1.1 A Comment: Mainstream Economics and Risk Evaluation

There have been many studies criticizing the mathematization of economics and the formal approach to economic analysis (e.g. Clower [30], and Blaug [14] and [15]).

---

[1] But see Sect. 3 below.

Many of the critiques of the use of mathematical modeling in economics rely on the idea that models are approximations of reality. Models cannot capture all what we know about the world. Economists and investors, therefore, should be less reliant on models and be little more humble when making forecasts and pricing risks.

While this argument is generally valid, for some it might lead to the opposite conclusion. The counterargument holds that approximation can always be improved. There will always be new methods and innovative techniques that help us get more accurate results, even if we cannot get the exact ones. If we work harder and be little smarter, we will come closer and closer to the "ultimate truth", even if we will never be able to reach it.

So the "approximation argument" would likely lead to the opposite of what was intended. Rather than discouraging over-reliance on formal models, it encourages more sophisticated techniques, and larger reliance on computing power.

In this chapter, however, we take a different approach. We examine ideal conditions for employing formal models and supercomputers, and see to what extent they can help us predict the future. As we shall see, ideal models and supercomputers are fundamentally limited in manners that will not be even possibly approximated. This fundamental uncertainty calls for a radically different approach for studying and analysing markets and economic phenomena.

The point is: we envisage "predictability" as "computability", "Turing computability".[2]

Ignoring these limitations leads to systematic mispricing of risk, which in turn encourages unwarranted risk taking. Markets therefore become less stable, and the economy becomes vulnerable to booms and crashes. Blind faith in risk-quantification leads to more, not less, risks.

1.2 A Summary of Our Work

We may summarize the gist of this chapter as follows: we want to look at risk, and risk-evaluation, from the viewpoint of Gödel's incompleteness phenomenon. Therefore the next sections of the chapter deal with Gödel incompleteness, and its rather unexpected consequences for economics. We then apply the concepts we have introduced to formal models in economics, and conclude that Gödel incompleteness appears everywhere (and in crucial situations) in mathematical economics. Here, we arrive at our main question, the quantitative evaluation of risk.

Our discussion is, admittedly, intuitive and nontechnical, as we deliberately sacrifice rigor for understanding.

---

[2] We will later elaborate on that characterization.

2 Formal Models

A more detailed presentation of these ideas can be found in Chaitin et al. [28]. Let us start with the use of formal, axiomatic, mathematical models. Suppose we have an ideal economic world W, that we would like to investigate.[3] The economy W is characterized by the following properties:

- *It is deterministic.* No randomness whatsoever exists in this economy.
- *It is fully observable.* Observation captures the true state of the economy at the current period t. There is no gap between observation and reality at t.

Now suppose we are able to build a formal mathematical model M of the economy W, that has the following ideal characteristics:

- *M is internally consistent.* No two contradictory M-statements can be proved within M.
- *M accurately captures all the knowledge that we have accumulated so far about W.* Any valid statement within M corresponds to a true state of the economy W. The mathematical model M is therefore sound, i.e. it proves only true statements about the economy W.
- *All statements of M are verifiable.* There is no ambiguous or indeterminate M statement. Any M statement about the state of the economy W at a certain time t' can be verified by directly observing W at t'.
- *M lies in a language with enough arithmetic.* That is both a technical condition and a pragmatic one. A pragmatic one: we must calculate if we are to quantify aspects of the economy, and in order to quantify we require arithmetic. A technical condition: "enough" arithmetic is required for incompleteness, as will be discussed shortly.

Given the ideal model of the ideal economy, we ask the following question: can M systematically predict the future states of W?

We will embed our M in a theory S that has the following characteristics:

- The language of S is the first order classical predicate calculus.
- S has a set of theorems that can be enumerated by a computer program.[4]
- PA, Peano Arithmetic, is a subtheory of S and describes its arithmetic portion; M contains the arithmetic portion of S.[5]

---

[3] We use here "model" in the sense of "mathematical model", that is, the mathematical depiction of some phenomenon; we do not use it in the model-theoretic sense.

[4] That is, the set of theorems of S is recursively enumerable.

[5] This is what we mean by "enough arithmetic". Actually we can obtain the same results we require if we add a still weaker arithmetic condition.

2.1 Risk

We informally state a result,[6] later to be formulated in a more rigorous way:

M cannot systematically predict (calculate) all future states of an (ideal) economy W.

The proof of this theorem follows from Kurt Gödel's famous (first) Incompleteness Theorem, published in 1931. In simple words, Gödel's result states (waving hands) that any formal system, rich enough to include arithmetics, cannot be both consistent and complete. To be consistent means there will not be a statement derived from the system that contradicts any of its statements. To be complete means that all true interpretations of formal statements in the system are either axioms or can be derived from the axioms.

Since we have already assumed that model M is consistent, it follows that M cannot be complete, provided that our theory includes enough arithmetics in its framework. Since we are investigating "quantification" of the economy, this condition is readily met. Consequently, there are true M statements about W that cannot be proved within M. These statements are described as "undecidable propositions" [37, p. 119]. We call such statements "Gödel sentences".

One important example of such Gödel sentences is the one that states: "*M* is consistent", i.e., is free from contradictions. As Gödel [51] has shown, this sentence cannot be proved from within M [42, p. 154].

From the economic point of view, there are many important questions that are undecidable, including:

- *Will the economy be stable in the future? Or will it be chaotic?*
- *Will markets clear in the future? Would the economy attain equilibrium?*
- *Will resources be allocated efficiently?*

Moreover, if we define:

A *risky set* is a subset $R \subset W$ that satisfies some property P_R that represents our conceptions of risk,

we can ask

- *Will the risky set R be decidable?*

(A set $A \subset B$ is *decidable* if and only if there is a computer program M_A so that $M_A(x) = 1$ whenever $x \in A$, and 0 otherwise. If there is no such a program, then A is *undecidable*.)

The four questions stated above have been shown to be undecidable (see da Costa and Doria [36] and the references therein). That is, there is no systematic way to provide a mathematical answer, either in the affirmative or in the negative, to any

---

[6]The theorem that implies such a result is a Rice-like theorem proved by da Costa and Doria in theorem 1990; see Chaitin et al. [28].

of these questions. We just don't know for sure. In fact, any interesting feature of dynamical systems is undecidable [100].

If we cannot derive or prove Gödel sentences of M, this means we cannot predict the corresponding states of the economy W. Consequently, the model cannot fully predict all future states of the economy. This implies that the economy might turn out to be at a particular state, call it Gödel-event, that is impossible to predict in the model M.

We can say that this is an indication of the complexity of our theoretical constructs, where events can occur that cannot be derived mathematically within M [76, p. 207].

We can—sort of—say that Gödel's Theorem establishes that mathematics is inexhaustible [46]. The theorem further shows how a well-structured formal system can give rise to unexpected and unintended results. As Dyson [39, p. 14] writes, "Gödel proved that in mathematics, the whole is always greater than the sum of the parts".

According to Gödel (1995, p. 309), the theorem:

> ...makes it impossible that someone should set up a certain well-defined system of axioms and rules and consistently make the following assertion about it: all of these axioms and rules I perceive (with mathematical certitude) to be correct, and moreover I believe that they contain all of mathematics. If someone makes such a statement he contradicts himself.

To prove that a particular proposition is mathematically undecidable is not very easy. It takes a lot of hard work and ingenuity. So, can we assume that a given mathematical statement is provable or decidable in principle, unless it is proved to be otherwise? Put differently, can we follow a rule of thumb: all propositions are decidable unless proved otherwise? Unfortunately, we cannot do so, for two reasons. One is that undecidability seems to be the rule rather than the exception, as we shall see later. Second, there is no finite effective procedure that can decide whether a given statement is provable, in M or in any formal system [12, pp. 107–109]. That is, we cannot make this assumption because this assumption itself is undecidable. Undecidability, therefore, is undecidable!

3 Technicalities Galore

Undecidability, in fact, is everywhere. And undecidability and incompleteness are related: the algorithmical unsolvability of the halting problem for Turing machines[7] means that there is no computer program that decides whether the universal Turing machine stops over a given input. Then, if we formulate Turing machine theory within our axiomatic theory S, there is a sentence in that theory—"There is a x_0 so that the universal Turing machine never stops over it as an input"—which can neither be proved nor negated within S. For if we proved it, then we would be able

---

[7] A simple proof can be found in Chaitin et al. [28].

to concoct a computer program that computed all nonstopping inputs to the universal machine.

We have a more general result proved by da Costa and one of us (Doria) in 1990 [34]. Still sort of waving hands, let us be given M, and let P be a predicate (roughly, the formal version of a property) on the objects of M. Let $x_1 \ne x_2$ be terms in M, that is, the description of objects in M. $P(x)$ can be read as "x has property P". $\neg P(x)$, "x doesn't have P". Then:

> Given any P so that, for $x_1 \ne x_2$, $P(x_1)$ and $\neg P(x_2)$ hold, then there is a term x_3 so that $P(x_3)$ is undecidable.

That is, no algorithm can decide whether either $P(x_3)$ holds or $\neg P(x_3)$ holds. (For a complete discussion, which includes the original reference to this result, see da Costa and Doria [36]).

This also shows that the whole is greater than the sum of the parts. We start with two decidable properties, and end up with a third that is undecidable.

Actually, any scientific discipline that formulates its theories using mathematics will face the problem of undecidability, if it is meant to compute values or produce quantities, that is, if it contains arithmetics.

With more detail: suppose that we know some system S to be at state σ_0 at time t_0. A main task of a scientific discipline is to find some rule ρ to predict the future state of system $\rho(\sigma_0, t)$ for $t > t_0$.

Then, according to the result just quoted, even if we know the exact evolution function of the system, namely rule ρ, and even if we have absolutely precise data about the current state of the system, σ_0, this will not be enough for us to systematically predict an arbitrary future state $\rho(\sigma_0, t)$ of the system. This is in fact a quite unexpected result. (A rigorous statement of the result will soon be given.)

No matter how powerful our mathematical model is, no matter how precise our measurement instruments are, and no matter how comprehensive our data are, we will not be able to systematically predict the future of any natural or social system. Although we may be able to predict the future in many cases, we will never be able to do so for all cases, not even on average.

Wolpert [115] extends the above results to include, in addition to prediction, both observation and retrieval of past measurements. Wolpert argues that physical devices that perform observation, prediction, or recollection share an underlying mathematical structure, and thus call them "inference devices". Inference devices therefore can perform computations to measure data, predict or retrieve past measurements. Based on the work of Gödel and Turing, Wolpert then shows that there are fundamental limits on inference devices within the universe, such that these three functions cannot be performed systematically, even in a classical, non-chaotic world. The results extend to probabilistic inference as well. These limits are purely logical and have nothing to do with technology or resources. The results are totally independent of both the details of the laws of physics and the computational characteristics of the machines. According to Wolpert, these impossibility results can be viewed as a non-quantum mechanical "uncertainty principle".

Hence, not only there are fundamental limits on quantifying the future, there are also fundamental limits on quantifying the observed reality, and recalling the past quantities.

3.1 Island of Knowledge

Mathematical models help us understand and quantify aspects of the universe around us. Scientists clearly have been able to quantify many properties of natural phenomena. However, Gödel's theorem shows that there is a price that we have to pay to get these benefits. While a formal model helps us answer specific questions about the universe, the same model raises more questions and opens the door for more puzzles and uncertainties. Moreover, with different models for different phenomena, we face even more puzzles and uncertainties on how these models link together. Quantum mechanics and general relativity are obvious examples.

So while formal models help us gain knowledge, they make us aware of how ignorant we are. As John Wheeler points out: "As the island of our knowledge grows, so do the shores of our ignorance". William Dampier writes: "There seems no limit to research, for as been truly said, the more the sphere of knowledge grows, the larger becomes the surface of contact with the unknown" (cited in Gleiser [50, p. 288]). Scientist and mathematicians are well aware of this price, and they are more than happy to pay. There will be no end to scientific endeavour and exploration.

Popper [82, p. 162] argues that all explanatory science is incompletable; for to be complete it would have to give an explanatory account of itself, which is impossible. Accordingly:

> We live in a world of emergent evolution; of problems whose solutions, if they are solved, beget new and deeper problems. Thus we live in a universe of emergent novelty; a novelty which, as a rule, is not completely reducible to any of the preceding stages.

Hawking [57] builds on Gödel's Theorem to dismiss a "Theory of Everything". He writes:

> Some people will be very disappointed if there is not an ultimate theory that can be formulated as a finite number of principles. I used to belong to that camp, but I have changed my mind. I'm now glad that our search for understanding will never come to an end, and that we will always have the challenge of new discovery. Without it, we would stagnate. Gödel's theorem ensured there would always be a job for mathematicians. I think M theory [in quantum mechanics] will do the same for physicists. I'm sure Dirac would have approved.

Mainstream economics, in contrast, pays little attention, if any, to these limitations. For mainstream theoreticians, the Arrow–Debreu model [3] depicts essentially a complete picture of the economy. We expect no place in this picture for incompleteness or undecidabilities whatsoever. It is the "Theory of Everything" in the mainstream economics.[8] Limitations arise only in the real world due to transaction costs and other frictions. To this day, this model is the benchmark of Neoclassical economic theory. There is therefore not much to explore or to discover in such a world. In fact, entrepreneurship, creativity and innovation, the engine of economic growth, have no place in Neoclassical theory. Not surprisingly, mainstream economic models have a poor record of predicting major events, not the least of which is the Global Financial Crisis (see, e.g., Colander et al. [31]).

3.2 A Brief Side Comment: Black Swans?

A Gödel sentence describes a Gödel event: a state of the economy that cannot be predicted by M. Nonetheless, after being realized, the state is fully consistent with our stock of knowledge and therefore with M.

This might sound similar to Nassim Taleb's idea of "Black Swan". A Black Swan is an event that:

- Lies outside regular expectations;
- Has an extreme impact, and
- Is retrospectively, but not prospectively, explainable and predictable.

(See Taleb [104, pp. xvii–xviii]). If "retrospective but not prospective" predictability is broadly understood to mean consistency with M but not provability in it, then a Black Swan event might be viewed as an instance of a Gödel statement about the world. However, if a Black Swan event is a "fat tail event",[9] then it is still predictable. In contrast, an event corresponding to a Gödel sentence is fundamentally unpredictable.

This suggests the following characterization:

> An event is predictable *if there is a finite set of inputs and a computer program that computes it.*

We thus equate "prediction" to "effective procedure".

3.3 More Side Comments: Ambiguity

We have assumed that each statement of our M model corresponds unambiguously to a certain state of the economy. But this is not true. When it comes to interpretation, mathematical concepts are "relative" [7]. They can be interpreted to mean different things, and these different interpretations are not always isomorphic, so to say.

---

[8] Yet the Arrow–Debreu theory is undecidable; that follows from a theorem by Tsuji et al. [107].
[9] For a summary discussion, see http://http://www.fattails.ca.

A formal system may have unintended interpretations despite the system being intended to have a particular interpretation [38, pp. 180–187]. This may be seen as another example of an emergent phenomenon where the whole goes beyond the parts.

According to DeLong [38, p. 185], Gödel's incompleteness theorem and the work of the Thoralf Skolem (see Gray [55]) show that there exists an infinite number of interpretations (or models, now in the model–theoretic sense) of any formal system. In general, no formal system is necessarily categorical, i.e., has only isomorphic models (in the model–theoretic sense). Hence, mathematical reality cannot be unambiguously incorporated in axiomatic systems [62, pp. 271–272].

This means that even if an M-statement is predictable (provable), we cannot tell exactly what would be the corresponding interpretation, seen as a state of the economy. Unintended interpretations would arise beyond the intended setup of the model. This ambiguity adds another layer of uncertainty that defeats the best efforts to quantify our knowledge.

4 Technicalities, II

But there is still more to come: how many Gödel sentences can there be? According to Gödel [52, p. 272], there are "infinitely many undecidable propositions in any such axiomatic system". This is obvious: they must be infinite; if they were finite in number, or could be generated by a finite procedure, we would just add them to our axiom system.

Calude et al. [26] prove this result formally, and show that the set of Gödel sentences is very large, and that provable (i.e. predictable) statements are actually rare (see also Yanofsky [117, p. 330]). Almost all true sentences are undecidable [103, p. 17]. Almost all languages are undecidable (Lewis [69, p. 1]; Erickson [40, pp. 5, 7]). Most numbers are "unknowable", i.e. there is no way to exhibit them ([24], pp. 170–171).

We give here a brief discussion, based on an argument of A. Bovykin.[10] Besides Calude's paper there are several folklore-like arguments that show that undecidability is the rule, in formal systems. One of them is based on Chaitin's Ω number, and the fact that it is a normal number.[11] Briefly, one concludes that while the set of well-formed formulae (wff) of a theory with associated halting probability Ω grows as x, the set of *provable wff* grows as $\log x$. (There is an intriguing and still unexplored connection with the Prime Number Theorem here.)

---

[10]Personal communication to FAD.

[11]In binary form, zeros and ones are evenly distributed.

Yet another quite simple argument that leads to the same conclusion is due to A. Bovykin. It applies to the theories we are dealing with here:

- We deal with theories that have a recursively enumerable set of theorems. The sentences in such theories are supposed to be a recursive subset of the set of all words in the theories' alphabet, and therefore they can be coded by a 1–1 map onto ω, the set of all natural numbers.
- The set of theorems of one such theory, say T, is therefore coded by a recursively enumerable subset of ω.
  As a consequence there is a Turing machine M_T that lists all theorems of T. Or, a machine M_T which proves all theorems of T (each calculation is a proof).
- The calculation/proof is a sequence of machine configurations.
- Now consider all feasible, even if "illegal," machine configurations at each step in a calculation. Their number is, say, n, with $n \geq 2$. For a k-step computation there will be n^k possible such legal and illegal sequences, of which only one will be a computation of M_T.
- That is to say, just one in n^k possible sequences of machine configurations, or proof sequences.

This clarifies why among the set of all such strings the set of all proofs is very small. The argument is sketchy, but its rigorous version can be easily recovered out of the sketchy presentation above.

In general, the describable and decidable is countably infinite, but the indescribable and undecidable may even be uncountably infinite. And the uncountably infinite is much, much larger than the countably infinite [117, pp. 80, 345].

4.1 Is There a Limit to Uncertainty?

Quantifying a phenomenon requires a mathematical model of it based on available information. Since the ideal model M by assumption captures all the knowledge—available to us—about the world, it follows that the uncertainty that arises from M cannot be formally modeled, and therefore cannot be quantified.

Suppose we create a "meta model" M_2 that somehow quantifies Gödel sentences arising from the model M. Then the new model will have its own Gödel sentences which cannot be quantified neither in M nor M_2. If we create yet another meta–meta model, say M_3, then another set of Gödel sentences will arise, *ad infinitum*.

For every formal model, there will be (many, many) undecidable propositions and unprovable statements. If we try to model these statements using a more powerful formal model, many more undecidable statements will arise. With each attempt to quantify uncertainty, higher uncertainty inevitably emerges. We will never be able to tame such uncertainty. Uncertainty is truly fundamental, and it breaches the outer limits of reason. It follows that uncertainty is beyond systematic quantification.

A frequently cited distinction between "risk" and "uncertainty" is based, mainly, on Knight [63]. The distinction holds that "risk" is quantifiable indeterminacy, while "uncertainty" is not quantifiable. From the previous discussion, this distinction appears to be less concrete. Since "quantification" must be based on formal mathematical modeling, then uncertainty, as reflected in infinite undecidable Gödel statements that creep up from the model, inevitably arises. This "irreducible uncertainty" evades quantification by all means.

4.2 Rationality

So far we have been discussing the economy at a macro level. Let us now zoom in to economic agents and see if they also face uncertainty.

One of the most controversial assumptions about economic agents in mainstream neoclassical economics is the assumption of rationality. There has been, and still ongoing, a long debate on this assumption, but we shall not address it here; we will simply state a few conditions and ask our agents to satisfy it. We then proceed to see what, if any, kind of uncertainty agents face. In particular, we assume that ideal agents R in the economy W enjoy the following ideal characteristics:

- Agents are well organized and very systematic in planning and executing their plans. They are as systematic as digital computers. In other words, an agent is a "Turing machine", i.e. an ideal computer that is capable of performing any task that can be structured in a clear sequence of steps, i.e. every economic task is computable.
- Agents have (a potentially) unlimited memory and ultra-fast computing power. They face no resource constraints on their computing capabilities.
(That identification can be taken as a definition of rationality, if needed. Putnam [85] and Velupillai [110] show the formal equivalence of economic rationality to Turing machines.)
- Agents act deterministically, and are not subject to random errors.
- Agents have full information about the current and past states of the economy W.
- Agents are honest and truthful (this is to avoid artificial uncertainty due to deliberate misrepresentation).

To be practical, an agent would decide his economic choice based on an ideal plan P: a set of instructions of how to make an economic choice. For example, a consumption plan would have a clear set of instructions on how to screen goods and services, how to evaluate each, how to choose, over how many periods, etc. The same applies to production and investment. Agents formulate plans identical to software programs that specify each step needed to perform consumption, production and investment.

R agents need not be utility or profit maximizers. They need not be optimizers of any sort. But they are well organized and systematic that they formulate their plans in a clear sequence of instructions, just as a computer program.

Now we ask the following question: would R agents face any uncertainty in their course of economic activity?

We informally state another result, whose rigorous formulation will be given below:

Ideal agents R cannot systematically predict the outcome of their ideal plans P.

The proof of this theorem follows quite trivially from Alan Turing's (**(year?)**) famous theorem on the "halting problem". In words, it says that there is no program (or a Turing machine) that can always decide whether a certain program Pr will ever halt or not. A program halts when it completes executing its instructions, given a certain input. It will fail to halt if it is stuck in an infinite loop where it fails to complete the execution of its instructions.

A generalization of Turing's theorem is provided by Rice [88]. Rice's Theorem states that there is no program that can decide systematically whether a given program Pr has a certain non-trivial property y. A non-trivial property is a property that may be satisfied by some programs but not by others.

A good example is found in computer viruses, which are softwares that damage the performance of the operating system. Is there a program that can always decide whether a certain program is (or has) a virus or not? In other words, could there be an "ideal" anti-virus program?

According to Turing and Rice, the answer is no [28]. There is no ideal anti-virus program. Nor there is an ideal debugger program. The reason is intuitively simple. Suppose there is a program that is believed to be an ideal anti-virus. How to know that it is in fact the ultimate anti-virus? To verify this property, we need to design another program to test our "ideal anti-virus". Then the same problem arises for the testing program, *ad infinitum*. We will never be able to be fully sure if our software can catch all viruses, or spot all bugs. Risk is inevitable. The claim of inventing the "ultimate anti-virus" is no less fraudulent than that of a "perpetual motion machine".

For our economic agents, this means that agents cannot always know in advance what will be the full output of ideal plans P.

For example, for a given production or investment plan, we ask the following question: Is there a general procedure that can tell us whether the plan has no "bugs" that could lead to loss or adverse outcomes? Rice's Theorem says no. There is no general procedure or program that can perform this function. Accordingly, an investor contemplating to invest in a given business opportunity is unable to know for sure what would be the output of the business plan, even in a fully deterministic world, free from cheating and lying, and where agents have full information and enjoy super-computing capabilities. Agents can never be sure that they will not lose money or face inadvertent outcomes. Risk is unavoidable.

Again, this risk is not quantifiable. The argument is the same as before. If we were able to design a program to compute the risks associated with a given investment or production plan, how can we verify that the program has no "bug"? We will never be able systematically to do so.

(More on Rice's Theorem at the end of this section.)

4.3 Is Knowledge Quantifiable?

We arrive here at an important question: can we quantify our knowledge? Gödel's theorem makes it clear that there are statements that are true but not provable in prescribed formal systems, and that we cannot simply add those statements to our axioms, as new undecidable statements immediately creep up. If the formal system is consistent, then we know that the statement is true, but it is not provable. As Franzen [46, p. 240] points out, Gödel's theorem "is not an obstacle to knowledge, but an obstacle to the formalization of knowledge". Does that mean that truth extends beyond deductive logic? If so, it is not quantifiable.

Curiously enough, cognitive scientists and neuroscientists agree that "we know more than we can say" [54, p. 99]. Human beings are equipped with sensors that are independent of verbal or even visual communications. The flow of such signals is a transfer of information via non-expressible means. Detecting and analysing these signals is what amounts to "social intelligence". The human brain has regions for handling social intelligence that are different from those for handling deductive and logical reasoning [70].

The knowledge accumulated via non-deductive, informal, means is what is called "tacit knowledge", as Polanyi [81] describes it (from Latin *tacere*, to shut up). Knowing how to ride a bicycle or to drive a car, for example, is tacit knowledge that cannot be fully expressed or articulated. Much of our valuable knowledge that guide our daily and social activities are of this sort of inarticulate or irreducible knowledge (see Lavoie [67]; and Collins [33]). According to Mintzberg [77, pp. 258–260], "soft information", i.e. information not communicated through formal means, account for 70–80% of information used for decision making.

As already pointed out, what we can describe and formalize is countably infinite, but the indescribable may even be uncountably infinite, which is much, much larger than the countably infinite.

This implies that the problem of informational asymmetry is particularly relevant only for formal or expressible knowledge. However, if tacit knowledge is taken into account, the problem could be reduced to a greater extent, although it will never disappear. Main stream economics pays little attention to non-formal or tacit knowledge. This "epistemic deficit", as Velupillai [112, p. xix] describes it, only reinforces the problem of informational asymmetry. Formal requirements of disclosure, for example, will help only up to a point. Beyond that point, we need to enhance social relations to improve social intelligence and communication of tacit knowledge.

4.4 Back to Our Goal: Technicalities Galore; Rice's Theorem and the Rice–da Costa–Doria Result

It is now the moment to go technical again, and add some more mathematical rigor to our presentation.

We sketch here a very brief primer on Rice's Theorem.

Suppose that the theory S we deal with has also the ι symbol (briefly, $\iota_x Q(x)$ reads as "the object x so that $Q(x)$ is true"). Let P be a predicate symbol so that there are two terms $\xi \neq \zeta$ and $S \vdash P(\xi)$ and $S \vdash \neg P(\zeta)$ (that is, $P(\xi)$ and not-$P(\zeta)$ are provable in S; we call such P, *nontrivial predicates*). Then, for the term:

$$\eta = \iota_x[(x = \xi \wedge \alpha) \vee (x = \zeta \wedge \neg \alpha)],$$

where α is an undecidable sentence in S:

Proposition 1 $S \nvdash P(\eta)$ and $S \nvdash \neg P(\eta)$.

This shows that incompleteness is found everywhere within theories like S, that is, which include PA, have as its underlying language the first order classical predicate calculus, and have a recursively enumerable set of theorems.

Notice that this implies Rice's Theorem [73, 89] in computer science. Briefly: suppose that there is an algorithm that settles $P(n)$, for each $n \in \omega$. Then by the representation theorem we may internalize that algorithm into S, and obtain a proof of $P(n)$ for arbitrary $n \in \omega$, a contradiction given Proposition 1.

In 1990 (published in 1991), da Costa and Doria extended Rice's Theorem to encompass any theory that includes "enough arithmetic" (let's leave it vague; we must only be able to make the usual arithmetical computations) and which is consistent, has a recursively enumerable set of theorems and a model (in the sense of interpretation) with standard arithmetic. Let us add predicate symbols to S, if needed (we will then deal with a conservative extension of S). Then,

Proposition 2 *If P is nontrivial, then P is undecidable, and there is a term ζ so that $S \nvdash P(\zeta)$ and $S \nvdash \neg P(\zeta)$.*

So each non-trivial P leads to an undecidable sentence $P(\zeta)$.

Such results explain why undecidability and incompleteness are everywhere to be found in theories like S, which are the usual kind we deal with. Moreover we can show that $P(\zeta)$ can be as "hard" as we wish within the arithmetical hierarchy [89].

It is as if mathematical questions were very very difficult, but for a small set of sentences... Of course this nasty situation spills out to applied sciences that use mathematics as a main tool [35].

4.5 Know Thyself!

Popper [82, pp. 68–77] was among the early writers who considered the fact that a Turing machine faces an inherent prediction problem of its own calculations. Popper argues that, even if we have a perfect machine of the famous Laplace's demon type [66], as long as this machine takes time to make predictions, it will be impossible for the machine to predict its own prediction. That is, the outcome of its calculation will arrive either at the same time as its primary prediction, or most likely, afterwards. Thus, a predictor cannot predict the *future* growth of its own knowledge (p. 71).

Put differently, a Turing machine, i.e. an ideal computer, cannot systematically predict the result of its own computation [72]. This can be seen as an innocent extension of Rice's Theorem above, but it sheds some additional light here. An ideal agent cannot systematically predict what will be the basket of goods and services that he or she will choose given its own consumption plan. Similarly, an agent cannot predict the inputs that will be used for production, or the asset portfolio that he or she will invested in. Furthermore, suppose agents choose their consumption or investment plans from a large set of predetermined plans. Could an agent systematically predict what plan it will choose? The answer, again, is no, for the same reasons.

Can an ideal agent predict the choice of other agents?

> *An ideal agent cannot systematically predict his or her choice, nor the choice of any other ideal agent.*

Since an ideal agent is an ideal computer, it will be able to simulate the decision process of any other ideal agent, assuming the decision process or plan is explicitly provided. So, if an agent cannot systematically predict his, or her, own choice, it follows that an agent cannot systematically predict the choice of any other ideal agent. Since all agents are ideal computers, then no one will be able to systematically predict the choice of any other. As Berto [12, p. 188] points out, "If we are indeed just Turing machines, then we cannot know exactly which Turing machines we are". Paul Benacerraf, moreover, writes: "If am a Turing machine, then I am barred by my very nature from obeying Socrates' profound philosophical injunction: *Know thyself!*" (cited in Berto [12]).[12]

Surprisingly, this uncertainty arises exactly because agents are assumed to possess maximal rationality. As Lloyd [71, p. 36] rightly points out:

> Ironically, it is customary to assign our own unpredictable behavior and that of others to irrationality: were we to behave rationally, we reason, the world would be more predictable. In fact, it is just when we behave rationally, moving logically like a computer from step to step, that our behavior becomes provably unpredictable. Rationality combines with the capacity of self reference to make our actions intrinsically paradoxical and uncertain.

5 Applications, I: Self-reference

Gödel's original example of an undecidable statement arose out of self-reference. Both Gödel and Turing rely on self-reference to prove their theorems [19, 20]. (Even if self-reference isn't essential to prove incompleteness [28], it remains an important tool in the construction of Gödel-like examples.)

---

[12] Actually an injunction engraved at the Apollonic oracle at Delphos.

However, self-reference as such will not always lead to nontrivial results. The statement "This statement is true" does not lead to a paradox, while "This statement is false" does.[13] More important, many concepts and definitions essential to mathematics involve self-reference (like the "maximum" and the "least upper bound"; DeLong [38, p. 84]).

Self-reference is valuable because it allows language to extend beyond the limits of its constituents. It permits the whole to exceed the parts. In physical systems, positive or reinforcing feed-back, which is a form of self-reference, is necessary for emergence and self-organization. In social systems, self-reference allows (social) institutions to be created, which in turn help reduce uncertainty and coordinate expectations (see below). It is also required for "shared understanding", including mutual belief, common knowledge, and public information, all of which rest on self-reference [5, 6].

Self-reference arises from meta-statements: statements about themselves. What Gödel ingeniously discovered was that formal models can be reflected inside themselves [19]. A formal model of consumer's choice, therefore, will have a meta-statement about its own statements. Accordingly, the model describing consumer's preferences will have meta-preferences, specifying the desirability of such preferences [114]. For example, one might have a preference for smoking, but he or she might not like having this particular preference.

Meta-preferences play the role of values and morality that guide consumer's choice [41, p. 41]. Neoclassical economics claims to be "value-neutral", and therefore ignores morality and ethics. But, in the light of Gödel's theorem, this claim is questionable. Meta-statements arise naturally in any (sufficiently rich) formal model, and these statements have the properties of values and morality for models of choice. Simply to ignore meta-statements does not deny their existence and their impact.

5.1 ...and Vicious Self-reference

For self-reference to be "harmful" or vicious, it needs to be "complete" [102, p. 111]. This arises when we try to encapsulate the undescribable within the describable, which is much larger as pointed out earlier. This will necessarily involve contradictions, i.e. the description will be inconsistent. Then to avoid contradiction, the description must be incomplete.

Another way to look at it is that vicious self-reference, as in the Liar paradox for example, has no stable truth value: from one perspective it is true, from the other it is false. We can therefore say that paradoxes are sort of "logically unstable" assertions (see Kremer [64]; Bolander [20]). Vicious self-reference thus creates an

---

[13] Yet one must be careful here, as the so-called "reflection principles" may be interpreted as sentences of the form "X is true".

infinite loop which, if implemented on a physical machine, will exhaust its resources and results in a crash (see Stuart [101, p. 263]). Normal self-reference, in contrast, is a commonplace part of any well-designed program.

Vicious self-reference may occur in various areas of market activity. Financial markets are susceptible to harmful self-reference when they are dominated by speculation. As Keynes [59, p. 156] notes, speculators devote their intelligence to "anticipating what average opinion expects the average opinion to be". It is a "battle of wits", and the objective is to "outwit the crowd". Speculators "spent their time chasing one another's tails", as Krugman [65] remarks. Simon [94, 95] considers the possibility of "outguessing" among economic agents to be "the permanent and ineradicable scandal of economic theory". He points out that "the whole concept of rationality became irremediably ill-defined when the possibility of outguessing was introduced", and that a different framework and methodology must be adopted to understand economic behaviour in such conditions (cited in Rubinstein [91, p. 188]).

Soros [96, 97] develops a theory of "reflexivity", which echoes to some extent the concerns of Simon above. Reflexivity implies that agents' views influence the course of events, and the course of events influences the agents' views. The influence is continuous and circular; that is what turns it into a feedback loop. But positive feedback loops cannot go on forever. As divergence between perceptions and reality widens, this leads to a climax which sets in motion a positive feedback process in the opposite direction. "Such initially self-reinforcing but eventually self-defeating boom-bust processes or bubbles are characteristic of financial markets" [97]. Reflexivity therefore undermines agents' perception and understanding of market dynamics, and at the same time ensures that agents' actions lead to unintended consequences [96, p. 2]. "Financial markets are inherently unstable" [96, pp. 14, 333].

Economists are aware of how agents' beliefs and behaviour actually changes the course of events. The "Lucas Critique" implies that a model may not be stable if it is used to recommend actions or policies that are not accounted for in the model itself (see Savin and Whitman [92]). This is also referred to as "Goodhart's Law" [29]. Lucas' critique therefore was a strong critique to models for policy-making purposes. But there is no reason why the same critique would not apply to models of financial markets. In fact, the Critique is even more relevant to financial markets than policy making given the intensity of the "guessing game" played therein. This became apparent in the years leading to the Global Financial Crisis. Rajan et al. [86] point out that models used to predict risk and probability of defaults are subject to Lucas Critique: Given the estimated probability of default, lenders will take on additional risks, rendering the initial estimates invalid. The authors rightly title their paper: "Failure of models that predict failures".

The problem of vicious self-reference therefore makes models unstable and lead to false predictions. The failure of such models to recognize the possibility of vicious self-reference contributed to additional risk-taking and consequently, market instability. Physicist and fund manager Bouchaud [22, p. 1181] points out: "Ironically, it was the very use of a crash-free model that helped to trigger a crash".

But, how does a vicious circle arise?

There have been many attempts to characterize the nature of paradoxes and vicious self-reference [20, 83, 84, 116].

Bertrand Russell provided a general characterization for paradoxes: "In each contradiction something is said about all cases of some kind, and from what is said a new case seems to be generated, which both is and is not of the same kind..." (cited in Berto [12, p. 37]). Priest [84, ch. 9] capitalizes on the work of Russell, and provides a general scheme of many kinds of paradox. Let Ω be a set with property ϕ, such that:

1. Define: $\Omega = \{y : \phi(y)\}$.
2. If x is a subset of Ω, then:

    a. $\delta(x) \notin x$.
    b. $\delta(x) \in \Omega$.

The function δ is the "diagonalizer". It plays the role of the diagonalization technique to produce non-x members.

As long as x is a proper subset of Ω, no contradiction arises. However, when $x = \Omega$, while other conditions still hold, we end up with a paradox: by condition (a) above, $\delta(\Omega) \notin \Omega$, while condition (b) requires that $\delta(\Omega) \in \Omega$. This is essentially Russell's Paradox.

Intuitively, trying to treat the part as the whole is the heart of the paradox. There must be limits on the subset x to avoid contradiction and paradox.

From an economic perspective, loosely speaking, the financial market is supposed to be a proper subset, x, of the economy, ω, that facilitates trade and production. However, when financial claims generate profits purely from other financial claims, the financial market becomes decoupled from the economy, and so x becomes a subset of itself, i.e. x becomes Ω. As finance becomes detached from the real economy, we end up with vicious circles of self-reference.

Money and language share many features: both are abstract social conventions, as Leonard [68] points out. Each is form, not substance. The two are good examples of complex systems [1]. As we can extrapolate from the work of Gödel and many others, language cannot define itself; it is dependent on an external reference [102, p. 111]. The same is true for money. Money and pure financial claims, therefore, cannot stand on their own; the financial sector must be part of the real economy.

5.2 Quantifying Vicious Self-reference or Reflexivity

According to Sornette and Cauwels [99], most of the financial crashes have fundamentally an endogenous, or internal, origin and that exogenous, or external, shocks only serve as triggering factors. Detecting these internal factors however is not always very obvious. Filimonov and Sornette [45] use techniques originally introduced to model the clustered occurrences of earthquakes, to measure endogeneity of price changes in financial markets. In particular, they aim to quantify how much of price changes are due to endogenous feedback processes, as opposed to exogenous news. They apply the techniques to the E-mini S&P 500 futures contracts traded in the Chicago Mercantile Exchange from 1998 to 2010. They find that the level of endogeneity has increased significantly from 1998 to 2010: 30% of price changes were endogenous in 1998, while more than 70% were endogenous since 2007. At the peak, more than 95% of the trading was due to endogenous triggering effects rather than genuine news. The measure of reflexivity provides a measure of endogeneity that is independent of the rate of activity, order size, volume or volatility.

Filimonov et al. [44] quantify the relative importance of short-term endogeneity for financial markets (financial indices, future commodity markets) from mid-2000s to October 2012. They find an overall increase of the reflexivity index since the mid-2000s to October 2012, which implies that at least 60–70% of financial price changes are now due to self-generated activities rather than novel information, compared to 20–30% earlier.

These results suggest that price dynamics in financial markets are mostly endogenous and driven by positive feedback mechanisms involving investors' anticipations that lead to self-fulfilling prophecies, as described qualitatively by Soros' concept of reflexivity.

It has been observed that while the volatility of the real economy has been declining (in industrial countries) since World War II, volatility of financial markets has been on the rise [2]. Bookstaber [17] comments on this phenomenon saying:

> The fact that total risk of the financial markets has grown in spite of a marked decline in exogenous economic risk to the country is a key symptom of the design flaws within the system. Risk should be diminishing, but it isn't.

6 Applications, II: Mispricing Risk

Systematically ignoring irreducible uncertainty might lead to underestimation of risk, and therefore, to a systematic downward bias in pricing risk.

This can be seen if we realize that ignored uncertainty is not necessarily symmetric across possibilities of loss and gain. The error in mispricing may not necessarily cancel out through averaging. Assuming that ignored uncertainty is symmetric across all possible arrangements of economic factors, it follows that chances of loss will be underestimated more than chances of gain. The reason is that there are many more arrangements that end up in loss than in gain. Gain, like order in physical systems, requires very special arrangements, while loss, like entropy, has many more possibilities. Consequently, systematically ignoring the limits of formal models might lead to underestimation of possibilities of loss more than those of gain. Hence, risk might be systematically underpriced.

With underpricing of risk, agents would take on more risk than they should. Accordingly, markets become more exposed to instabilities and turbulences. As early as 1996, this was visible to people like Bernstein [9]. He points out that despite the "ingenious" tools created to manage risks, "volatilities seem to be proliferating rather than diminishing" (p. 329).

6.1 Shortsightedness

There is another reason for mispricing risk that can lead to greater risks. Theory of finance, as Bernstein [10] also points out, revolves around the notion that risk is equivalent to volatility, measured usually by beta, standard deviation, and related quantities. These quantities by nature measure risk in the short run, not the long run. This causes investors to focus more on the shorter run. Major and extreme events, however, usually build up over a longer horizon. This means that signs of danger might not be detectable by shortsighted measures of risk. By the time the signs are detected, it is too late to prevent or escape the crash.

With shortsightedness, rationality of individual agents may lead to "collective irrationality". In the short run, it is reasonable to assume that certain variables are exogenous and are thus not affected by the decisions of market players. But in the medium to long-run, these variables are endogenous. By ignoring the endogeneity of these variables, agents ignore the feedback of their collective actions. This feedback however might invalidate the short-term estimates of risk that agents are using. When everyone is excessively focusing on the short-run, they collectively invalidate their individual estimates, which is one way to see the rationale of Lucas Critique discussed earlier. Hence, agents will appear to be individually rational, when collectively they might not be so.

Former president of the European Central Bank, Trichet [106], argues that excessive focus on short-term performance resulted in "excessive risk-taking and, particularly, an underestimation of low probability risks stemming from excessive leverage and concentration". He therefore calls for "a paradigm change" to overcome the shortsightedness dominating the financial sector.

6.2 Emotions and Herd Behaviour

Emotions play a major role in financial markets, leading to mispricing of risk. Psychoanalyst David Tuckett conducted in 2007 detailed research interviews with 52 experienced asset managers, working in financial centres around the globe [108]. The interviewed managers collectively controlled assets of more than $500 billion in value. Tucker points out that it was immediately clear to him that financial assets are fundamentally different from ordinary goods and services (p. xvi). Financial assets, he argues, are abstract objects whose values are largely dependent on expectations of traders of their future values (pp. 20–25). This created an environment for decision-making that is completely different from that in other markets. It was an environment in which there is both inherent uncertainty and inherent emotional conflict (p. 19). Traders' behaviour therefore is predominantly influenced by emotions and imaginations. These emotions are continuously in flux, which adds to the volatility of markets. Traders seek protection by developing *groupfeel* and shared positions (p. 173). This explains the well-documented herd behaviour in financial markets that plays a major role in contagion and instability [49, 93, 98]. In short, logical indeterminacy arising from self-reference makes emotions and herding dominate financial markets, causing higher volatility and instability that cannot be accounted for by formal models.

Coleman [32, p. 202] warns that "The real risk to an organization is in the unanticipated or unexpected–exactly what quantitative measures capture least well". Quantitative tools alone, Coleman explains, are no substitute for judgment, wisdom, and knowledge (p. 206); with any risk measure, one must use caution, applying judgement and common sense (p. 137). David Einhorn, founder of a prominent hedge fund, wrote that VaR was "relatively useless as a risk-management tool and potentially catastrophic when its use creates a false sense of security among senior managers and watchdogs" (cited in Nocera [79]).

Overall, ignoring the limits of quantitative measures of risk is dangerous, as emphasized by Sharpe and Bernstein [10, 11], among others [74]. This leads to the principle of Peter Bernstein: a shift away from risk measurement to risk management [16]. Risk management requires principles, rules, policies, and procedures that guide and organize financial activities. This is elaborated further in the following section.

7 Applications, III: Dealing with Uncertainty

If uncertainty is irreducible and fundamentally unquantifiable, even statistically, what to do about it? In an uncertain environment, the relevant strategy is one that does not rely on predicting the future. "Non-predictive strategies" refer to strategies that are not dependent on the exact course the future takes. Such strategies call for co-creating the future rather than trying to predict it and then act accordingly [105, 113]. Non-predictive strategies might imply following a predefined set of rules to guide behaviour without being dependent on predicting the future.

Heiner [58] argues that genuine uncertainty arises due to the gap between agents' competency and environment's complexity. In a complex environment, agents overcome this gap by following simple rules (or "coarse behavior rule"; see Bookstaber and Langsam [18]) to reduce the complexity they face. As the gap between agents' competency and environment's complexity widens, agents tend to adhere to more rigid and less flexible rules, so that their behaviour becomes stable and more predictable. This can be verified by comparing humans with different kinds of animals: as the competency of an animal becomes more limited, its behaviour becomes more predictable. Thus, the larger the gap between competency and complexity, the less flexible will be animal's behaviour, and thus the more predictable it becomes. The theorems of Gödel and Turing show that the gap between agents' competency and environment's complexity can never be closed; in fact, the gap in some respects is infinite. The surprising result of this analysis, as Heiner [58, p. 571] points out, is that:

> ...genuine uncertainty, far from being un–analyzable or irrelevant to understanding behavior, is the very source of the empirical regularities that we have sought to explain by excluding such uncertainty. This means that the conceptual basis for most of our existing models is seriously flawed.

North [80] capitalizes on Heiner's work, and argues that institutions, as constraints on behaviour, develop to reduce uncertainty and improve coordination among agents in complex environments. Institutions provide stable structure to every day life that guide human interactions. Irreducible uncertainty therefore induces predictable behaviour as people adopt rules, norms and conventions to minimize irreducible uncertainty (see Rosser [90]).

Former Federal Reserve chairman, Ben Bernanke, makes the same point with respect to policy (cited in Kirman [61]):

> ...it is not realistic to think that human beings can fully anticipate all possible interactions and complex developments. The best approach for dealing with this uncertainty is to make sure that the system is fundamentally resilient and that we have as many failsafes and back-up arrangements as possible.

The same point is emphasized by former governor of Bank of England, King [60, p. 120]: "rules of thumb – technically known as heuristics – are better seen as rational ways to cope with an unknowable future".

Constraints in principle could help in transforming an undecidable problem into a decidable one. Bergstra and Middleburg [8, p. 179] argue that it is common practice in computer science to impose restrictions on design space to gain advantages and flexibility. They point out that:

> In computer architecture, the limitation of instruction sets has been a significant help for developing faster machines using RISC (Reduced Instruction Set Computing) architectures. Fast programming, as opposed to fast execution of programs, is often done by means of scripting languages which lack the expressive power of full-blown program notations. Replacing predicate logic by propositional calculus has made many formalizations decidable and for that reason implementable and the resulting computational complexity has been proved to be manageable in practice on many occasions. New banking regulations in conventional finance resulting from the financial crisis 2008/2009 have similar characteristics. By making the financial system less expressive, it may become more stable and on the long run more effective. Indeed, it seems to be intrinsic to conventional finance that seemingly artificial restrictions are a necessity for its proper functioning.

Another approach to avoid undecidability is to rely more on *qualitative* rather than *quantitative* measures. This can be viewed as another kind of restrictions that aim to limit uncertainty. According to Barrow [4, pp. 222, 227], if we have a logical theory that deals with numbers using only "greater than" or "less than", without referring to absolute numbers (e.g. Presburger Arithmetic), the theory would be complete. So, if we restrict our models to qualitative properties, we might face less uncertainty. Feyerabend [43, pp. xx, 34] points out that scientific approach does not necessarily require quantification. Quantification works in some cases, fails in others; for example, it ran into difficulties in one of the apparently most quantitative of all sciences, celestial mechanics, and was replaced by qualitative considerations. Frydman and Goldberg [47, 48] develop a qualitative approach to economics, "Imperfect Knowledge Economics", which emphasizes non-routine change and inherently imperfect knowledge as the foundations of economic modelling. The approach rejects quantitative predictions and aims only for qualitative predictions of market outcomes. Historically, several leading economists, including J.M. Keynes, were sceptical of quantitative predictions of economic phenomena (see Blaug [13, pp. 71–79]).

Certain restrictions therefore are needed to limit uncertainty. The fathers of free market economy, Adam Smith, John Stuart Mill and Alfred Marshall, all realized that banking and finance needs to be regulated in contrast to markets of the real economy [27, pp. 35–36, 163].

In fact, the financial sector usually is among the most heavily regulated sectors in the economy, as Mishkin [78, pp. 42–46] points out. Historically, financial crises are associated with deregulation and liberalization of financial markets. Tight regulation of the banking sector following the Word War II suppressed banking crises almost completely during 1950s and 1960s [21]. According to Reinhart and Rogoff [87], there were only 31 banking crises worldwide during the period 1930–1969, but about 167 during the period 1970–2007. The authors argue that financial liberalization has been clearly associated with financial crises.

Deregulation has been visibly clear in the years leading to the Global Financial Crisis. The famous Glass-Steagal Act has been effectively repealed in 1999 by the Gramm-Leach-Bliley Act. Derivatives were exempted from gaming (gambling) regulations in 2000 by the Commodity Futures Modernization Act (see Marks [75]). Within a few years, the world witnessed "the largest credit bubble in history", as Krugman [65] describes it.

Sornette and Cauwels [99] argue that deregulation that started (in the US) approximately 30 years ago marks a change of regime from one where growth is based on productivity gains to one where growth is based on debt explosion and financial gains. The authors call such regime "perpetual money machine" system, that was consistently accompanied by bubbles and crashes. "We need to go back to a financial system and debt levels that are in balance with the real economy" (p. 23).

In summary, the above discussion shows the need and rationale in principle for institutional constraints of the financial sector. The overall objective is to tightly link financial activities with real, productive activities. Regulations, as such, might not be very helpful. Rather than putting too much emphasis on regulations per se, more attention should be directed towards good governance and values that build trust and integrity needed to safeguard the system (see Mainelli and Giffords [74]).

8 Conclusion

In his lecture at Trinity University in 2001 celebrating his Nobel prize, Robert Lucas [23] recalls:

> I loved the [Samuelson's] *Foundations*. Like so many others in my cohort, I internalized its view that if I couldn't formulate a problem in economic theory mathematically, I didn't know what I was doing. I came to the position that mathematical analysis is not one of many ways of doing economic theory: It is the only way. Economic theory *is* mathematical analysis. Everything else is just pictures and talk.

The basic message of this chapter is that mathematics cannot be the *only* way. Our arguments are based on mathematical theorems established over the last century. Mathematics is certainly of a great value to the progress of science and accumulation of knowledge. But, for natural and social sciences, mathematics is a tool; it is "a good servant but a bad master", as Harcourt [56, p. 70] points out. The master should be the wisdom that integrates logic, intuition, emotions and values, to guide decision and behaviour to achieve common good. The Global Financial Crisis shows how overconfidence in mathematical models, combined with greed and lack of principles, can have devastating consequences to the economy and the society as a whole.

Uncertainty is unavoidable, not even statistically. Risk, therefore, is not in principle quantifiable. This would have a substantial impact not only on how to formulate economic theories, but also on how to conduct business and to finance enterprises. This chapter has been an attempt to highlight the limits of reason, and how such limits affect our abilities to predict the future and to quantify risk. It is hoped that this contributes to a better reform of economics and, subsequently, to better welfare of the society.

Acknowledgements We are grateful to the editors and an anonymous referee for constructive comments and suggestions that greatly improved the readability of this text. FAD: wishes to acknowledge research grant no. 4339819902073398 from CNPq/Brazil, and the support of the Production Engineering Program, COPPE/UFRJ, Brazil. SA: wishes to acknowledge the valuable discussions with the co-authors, particularly FAD. The views expressed in this chapter do not necessarily represent the views of the Islamic Development Bank Group.

References

1. Al-Suwailem, S. (2010). Behavioural complexity. *Journal of Economic Surveys, 25*, 481–506.
2. Al-Suwailem, S. (2014). Complexity and endogenous instability. *Research in International Business and Finance, 30*, 393–410.
3. Arrow, K., & Debreu, G. (1954). The existence of an equilibrium for a competitive economy. *Econometrica, 22*, 265–89.
4. Barrow, J. (1998). *Impossibility: The limits of science and the science of limits*. Oxford: Oxford University Press.

5. Barwise, J. (1989). *The situation in logic*. Stanford, CA: Center for the Study of Language and Information.
6. Barwise, J., & Moss, L. (1996). *Vicious circles*. Stanford, CA: Center for the Study of Language and Information.
7. Bays, T. (2014). Skolem's paradox. In E. Zalta (Ed.), *Stanford Encyclopedia of Philosophy*. Stanford, CA: Stanford University, https://plato.stanford.edu/.
8. Bergstra, J. A., & Middelburg, C. A. (2011). Preliminaries to an investigation of reduced product set finance. *Journal of King Abdulaziz University: Islamic Economics, 24*, 175–210.
9. Bernstein, P. (1996). *Against the Gods: The remarkable story of risk*. New York: Wiley.
10. Bernstein, P. (2007). *Capital ideas evolving*. New York: Wiley.
11. Bernstein, P. (2007). Capital ideas: Past, present, and future. In *Lecture delivered at CFA Institute Annual Conference*.
12. Berto, F. (2009). *There's something about Gödel: The complete guide to the incompleteness theorem*. Hoboken: Wiley-Blackwell.
13. Blaug, M. (1992). *The methodology of economics*. Cambridge: Cambridge University Press.
14. Blaug, M. (1999). The disease of formalism, or what happened to neoclassical economics after war. In R. Backhouse & J. Creedy (Eds.), *From classical economics to the theory of the firm* (pp. 257–80). Northampton, MA: Edward Elgar.
15. Blaug, M. (2002). Is there really progress in economics? In S. Boehm, C. Gehrke, H. Kurz, & R. Sturm (Eds.), *Is there progress in economics?* European Society for the History of Economic Thought (pp. 21–41). Northampton, MA: Edward Elgar.
16. Blythe, S. (2007). Risk is not a number. http://www.advisor.ca/investments/alternative-investments/risk-is-not-a-number-26881.
17. Bookstaber, R. (2007). *A demon of our own design*. New York: Wiley.
18. Bookstaber, R., & Langsam, J. (1985). On the optimality of coarse behavior rules. *Journal of Theoretical Biology, 116*, 161–193.
19. Bolander, T. (2002). Self-reference and logic. Φ *News, 1*, 9–44. http://phinews.ruc.dk/.
20. Bolander, T. (2014). Self-reference. In E. Zalta (Ed.), *Stanford Encyclopedia of Philosophy*. Stanford, CA: Stanford University, https://plato.stanford.edu/.
21. Bordo, M., Eichengreen, B., Klingebiel, D., & Martinez-Peria, M. S. (2001). Is the crisis problem growing more severe? *Economic Policy, 32*, 53–82.
22. Bouchaud, J. P. (2008). Economics needs a scientific revolution. *Nature, 455*, 1181.
23. Breit, W., & Hirsch, B. (Eds.). (2009). *Lives of the laureates* (5th ed.). Cambridge, MA: MIT Press.
24. Byers, W. (2007) *How mathematicians think: Using ambiguity, contradiction, and paradox to create mathematics*. Princeton: Princeton University Press.
25. Byers, W. (2011). *The blind spot: Science and the crisis of uncertainty*. Princeton: Princeton University Press.
26. Calude, C., Jurgensen, H., & Zimand, M. (1994). Is independence an exception? *Applied Mathematical Computing, 66*, 63–76.
27. Cassidy, J. (2009). *How markets fail: The logic of economic calamities*. New York: Farrar, Straus, and Giroux.
28. Chaitin, G., Doria, F. A., & da Costa, N. C. A. (2011). *Gödel's way: Exploits into an undecidable world*. Boca Raton, FL: CRC Press.
29. Chrystal, K. A., & Mizen, P. D. (2003). Goodhart's law: Its origins, meaning and implications for monetary policy. In P. Mizen (Ed.), *Central banking, monetary theory and practice: Essays in honour of Charles Goodhart* (Vol. 1). Northampton, MA: Edward Elgar.
30. Clower, R. (1995). Axiomatics in economics. *Southern Economic Journal, 62*, 307–319.
31. Colander, D., Goldberg, M., Haas, A., Juselius, K., Kirman, A., Lux, T., & Sloth, B. (2009). The financial crisis and the systemic failure of the economics profession. *Critical Review, 21*, 249–67.
32. Coleman, T. (2011). *A practical guide to risk management*. Charottesville, VA: CFA Institute.
33. Collins, H. (2010). *Tacit and explicit knowledge*. Chicago: University of Chicago Press.

34. da Costa, N. C. A., & Doria, F. A. (1991). Undecidability and incompleteness in classical mechanics. *International Journal of Theoretical Physics, 30*, 1041–1073.
35. da Costa, N. C. A., & Doria, F. A. (1992). On the incompleteness of axiomatized models for the empirical sciences. *Philosophica, 50*, 73–100.
36. da Costa, N. C. A., & Doria, F. A. (2005). Computing the future. In K. Velupillai (Ed.), *Computability, Complexity and Constructivity in Economic Analysis*. Malden, MA: Blackwell Publishers.
37. Davis, M. (2000). *Engines of logic*. New York: W.W. Norton.
38. DeLong, H. (1970). *A profile of mathematical logic*. Reading, MA: Addison-Wesley.
39. Dyson, F. (2006). *The scientist as rebel*. New York: New York Review Books.
40. Erickson, J. (2015). Models of computation. University of Illinois, Urbana-Champaign. http://jeffe.cs.illinois.edu
41. Etzioni, A. (1988). *The moral dimension: Toward a new economics*. New York: The Free Press.
42. Ferreirós, J. (2008). The crisis in the foundations in mathematics. In T. Gowers (Ed.), *The Princeton companion to mathematics* (pp. 142–156). Princeton: Princeton University Press.
43. Feyerabend, P. (2010). *Against the method* (4th ed.). London: Verso.
44. Filimonov, V., Bicchetti, D., Maystre, N., & Sornette, D. (2013). Quantification of the high level of endogeneity and of structural regime shifts in commodity markets. *The Journal of International Money and Finance, 42*, 174–192.
45. Filimonov, V., & Sornette, D. (2012). Quantifying reflexivity in financial markets: Towards a prediction of flash crashes. *Physical Review E, 85*(5), 056108.
46. Franzen, T. (2004). *Inexhaustibility*. Boca Raton: CRC Press.
47. Frydman, R., & Goldberg, M. (2007). *Imperfect knowledge economics: Exchange rates and risk*. Princeton: Princeton University Press.
48. Frydman, R., & Goldberg, M. (2011). *Beyond mechanical markets: Asset price swings, risk, and the role of state*. Princeton: Princeton University Press.
49. FSA. (2009). *The Turner review*. London: Financial Services Authority.
50. Gleiser, M. (2014). *The island of knowledge: The limits of science and the search for meaning*. New York: Basic Books.
51. Gödel, K. (1931). Über formal unentscheidbare Sätze der Principia Mathematica und verwandter Systeme I (On formally undecidable propositions of Principia Mathematica and related systems I). *Monatshefte für Mathematik und Physik, 38*, 173–198.
52. Gödel, K. (1947). What is Cantor's continuum problem? *American Mathematical Monthly, 54*, 515–525.
53. Gödel, K. (1995). Some basic theorems on the foundations of mathematics and their implications. In S. Feferman (Ed.), *Kurt Gödel Collected Works, Volume III: Unpublished essays and lectures* (pp. 304–323). Oxford: Oxford University Press.
54. Goleman, D. (2006). *Social intelligence*. New York: Bantam Books.
55. Gray, J. (2008). Thoralf Skolem. In T. Gowers (Ed.), *The Princeton companion to mathematics* (pp. 806–807). Princeton: Princeton University Press.
56. Harcourt, G. (2003). A good servant but a bad master. In E. Fullbrook (Ed.), *The crisis in economics*. London: Routledge.
57. Hawking, S. (2002). Gödel and the end of the universe. www.hawking.org.uk.
58. Heiner, R. (1983). The origin of predictable behaviour. *American Economic Review, 73*, 560–595.
59. Keynes, J. M. (1936/1953). *The general theory of employment, interest and money*. San Diego: Harcourt Brace Jovanovich.
60. King, M. (2016). *The end of alchemy*. New York: W.W. Norton.
61. Kirman, A. (2012). The economy and the economic theory in crisis. In D. Coyle (Ed.), *What's the use of economics? Teaching the dismal science after the crisis*. London: London Publishing Partnership.
62. Kline, M. (1980). *Mathematics: The loss of certainty*. Oxford: Oxford University Press.
63. Knight, F. (1921). *Risk, uncertainty, and profit*. New York: Houghton-Mifflin.

64. Kremer, P. (2014). The revision theory of truth. In E. Zalta (Ed.), *Stanford Encyclopedia of Philosophy*. Stanford, CA: Stanford University, https://plato.stanford.edu/.
65. Krugman, P. (2009, September 2). How did economists get it so wrong? *New York Times*.
66. Laplace, P. S. (1902/1951). *A philosophical essay on probabilities*. Translated into English from the original French 6th ed. by F.W. Truscott & F.L. Emory. New York: Dover Publications.
67. Lavoie, D. (1986). The market as a procedure for discovery and conveyance of inarticulate knowledge. *Comparative Economic Studies, 28*, 1–19.
68. Leonard, R. (1994). Money and language. In J. L. Di Gaetani (Ed.), *Money: Lure, lore, and literature* (pp. 3–13). London: Greenwood Press.
69. Lewis, H. (2013). *Lecture notes*. Harvard University, http://lewis.seas.harvard.edu/files/harrylewis/files/section8sols.pdf.
70. Lieberman, M. (2013). *Social: Why our brains are wired to connect*. New York: Crown Publishers.
71. Lloyd, S. (2006). *Programming the universe*. New York: Vintage Books.
72. Lloyd, S. (2013). *A Turing test for free will*. Cambridge, MA: MIT Press. http://arxiv.org/abs/1310.3225v1.
73. Machtey, M., & Young, P. (1979). *An introduction to the general theory of algorithms*. New York: North–Holland.
74. Mainelli, M., & Giffords, B. (2009). The road to long finance: A systems view of the credit scrunch. London: Centre for the Study of Financial Innovation. Heron, Dawson & Sawyer. www.csfi.org.uk.
75. Marks, R. (2013). Learning lessons? The global financial crisis five years on. *Journal and Proceeding of the Royal Society of New South Wales, 146*, 3–16, A1–A43.
76. McCauley, J. (2012). Equilibrium versus market efficiency: Randomness versus complexity in finance markets. In S. Zambelli & D. George (Eds.), *Nonlinearity, complexity and randomness in economics* (pp. 203–210). Malden, MA: Wiley-Blackwell.
77. Mintzberg, H. (1994). *The rise and fall of strategic planning*. New York: The Free Press.
78. Mishkin, F. (2007). *The economics of money, banking, and financial markets* (Alternate edition). New York: Pearson.
79. Nocera, J. (2009, January 2). Risk mismanagement. *The New York Times*.
80. North, D. (1990). *Institutions, institutional change and economic performance*. Cambridge, MA: Cambridge University Press.
81. Polanyi, M. (1966). *The tacit dimension*. New York: Doubleday.
82. Popper, K. (1956/1988). *The open universe: An argument for indeterminism*. London: Routledge.
83. Priest, G. (1994). The structure of the paradoxes of self-reference. *Mind, 103*, 25–34.
84. Priest, G. (2000). *Beyond the limits of thought* (2nd ed.). Oxford: Oxford University Press.
85. Putnam, H. (1967). Psychophysical predicates. In W. Capitan & D. Merrill (Eds.), *Art, mind, and religion*. Pittsburgh: University of Pittsburgh Press.
86. Rajan, U., Seru, A., & Vig, V. (2015). The failure of models that predict failures: Distance, incentives, and defaults. *Journal of Financial Economics, 115*, 237–260.
87. Reinhart, C., & Rogoff, K. (2009). *This time is different: Eight centuries of financial folly*. Princeton: Princeton University Press.
88. Rice, H. G. (1953). Classes of recursively enumerable sets and their decision problems. *Transactions of the American Mathematical Society, 74*, 358–366.
89. Rogers, H., Jr. (1967). *Theory of recursive functions and effective computability*. New York: McGraw-Hill.
90. Rosser, J. B. (2001) Alternative Keynesian and post Keynesian perspectives on uncertainty and expectations. *Journal of Post Keynesian Economics, 23*, 545–66.
91. Rubinstein, A. (1998). *Modeling bounded rationality*. Cambridge, MA: MIT Press.
92. Savin, N., & Whitman, C. (1992). Lucas critique. In P. Newman, M. Milgate, & J. Eatwell (Eds.), *New Palgrave dictionary of money and finance*. New York: Palgrave.
93. Shiller, R. (2009). *Irrational exuberance*. Princeton: Princeton University Press.

94. Simon, H. (1978). Rationality as process and as product of thought. *American Economic Review, 68*, 1–16.
95. Simon, H. (1978/1992). Rational decision-making in business organizations. In A. Lindbeck (Ed.), *Nobel Lectures, Economics 1969–1980* (pp. 343–371). Singapore: World Scientific Publishing Co. https://www.nobelprize.org.
96. Soros, G. (2003). *Alchemy of finance*. Hoboken: Wiley.
97. Soros, G. (2009, October 26). The general theory of reflexivity. *Financial Times*.
98. Sornette, D. (2003). *Why stock markets crash: Critical events in complex financial systems*. Princeton: Princeton University Press.
99. Sornette, D., & Cauwels, P. (2012). *The illusion of the perpetual money machine*. Notenstein Academy White Paper Series. Zurich: Swiss Federal Institute of Technology. http://arxiv.org/pdf/1212.2833.pdf.
100. Stewart, I. (1991). Deciding the undecidable. *Nature, 352*, 664–665.
101. Stuart, T. (2013). *Understanding computation*. Beijing: O'Reilly.
102. Svozil, K. (1993). *Randomness and undecidability in physics*. Singapore: World Scientific.
103. Svozil, K. (1996). Undecidability everywhere? In A. Karlqvist & J.L. Casti (Eds.), *Boundaries and barriers: On the limits to scientific knowledge*. Reading, MA: Addison-Wesley, pp. 215–237. http://arxiv.org/abs/chao-dyn/9509023v1.
104. Taleb, N. (2007). *The black swan*. New York: Random House.
105. Taleb, N. (2012). *Antifragile*. New York: Random House.
106. Trichet, J.-C. (2009). A paradigm change for the global financial system. BIS Review, 4/2009.
107. Tsuji, M., Da Costa, N. C. A., & Doria, F. A. (1998). The incompleteness of theories of games. *Journal of Philosophical Logic, 27*, 553–568.
108. Tuckett, D. (2011). *Minding the market: An emotional finance view of financial instability*. Basingstoke: Palgrave MacMillan.
109. Turing, A. (1936). On computable numbers, with an application to the Entscheidungsproblem. *Proceedings of the London Mathematical Society, Series, 2*(42), 230–265.
110. Velupillai, K. V. (2000). *Computable economics*. Oxford: Oxford University Press.
111. Velupillai, K. V. (2012). Introduction. In K. V. Velupillai et al. (Eds.), *Computable economics*. International Library of Critical Writings in Economics. Northampton, MA: Edward Elgar.
112. Velupillai, K. V., Zambelli, S., & Kinsella, S. (Eds.). (2012). *Computable economics*. International Library of Critical Writings in Economics. Northampton, MA: Edward Elgar.
113. Wiltbank, R., Dew, N., Read, S., & Sarasvathy, S. D. (2006). What to do next? A case for non-predictive strategy. *Strategic Management Journal, 27*, 981–998.
114. Winrich, J. S. (1984). Self-reference and the incomplete structure of neoclassical economics. *Journal of Economic Issues, 18*, 987–1005.
115. Wolpert, D. (2008). Physical limits of inference. *Physica, 237*, 1257–1281.
116. Yanofsky, N. (2003). A universal approach to self-referential paradoxes, incompleteness and fixed points. *The Bulletin of Symbolic Logic, 9*, 362–386.
117. Yanofsky, N. (2013). *The outer limits of reason: What science, mathematics, and logic cannot tell us*. Cambridge, MA: MIT Press.

Index

A

Adaptive expectations, 154
Agent-based artificial stock market, 71
Agent-based macroeconomic model, 205, 206, 215
Agent-based model, 18, 19, 21, 33, 74, 91, 207, 251, 253, 255, 259
Agent-based Scarf economy, 115, 127, 138
Agriculture, 92
Allocative efficiency, 44
Anderson-Darling (AD) test, 105
Artificial stock market, 76, 89, 181
ATS, *see* automated trading systems
Automated trading systems, 39
Average Imitation Frequency, 135–137

B

Barter, 144
Behavioral economics, 193
Bio-fuels, 91
Black swan, 283
Boom-and-bust, 196
Brownian motion, 194
Business cycle, 205, 206, 212, 215, 216

C

Canonical correlation, 223, 224, 229, 235, 242
Canonical correlation analysis, 247
Capital asset pricing model, CAPM, 230
Common agricultural policy, 92
Common factor, 233
Comovement, 225

Consumer search, 210, 211, 214, 215
Consumption set, 118
Continuous double auction, 42, 79
Coordination device, 120
Coordination failure, 118
CRSP value weighted portfolio, 231

D

Dark pools, 17–19
    dark pool usage, 18–20, 22, 25, 29, 30, 33
    midpoint dark pool, 21, 32, 33
    midprice, 22, 25, 32
Decentralized market, 146
Delta profit, 44
Disequilibrium, 114, 118, 136, 138
Double coincidence of wants, 117
DSGE, 141
Dual learning, 164, 171, 173

E

Economic factors, 240, 241
Economic growth, 206
Emergence, 282, 284
    (the whole is greater than the parts), 280, 281, 291
Emotions, 295
Endogenous dynamics, 194, 198
Entropy, 294
Equal risk premia, 224
Equilibrium, 194
Experimental economics, 41
Exploitation, 134, 138
Exploration, 122, 127, 131, 134, 136–138

F

Factor model, 224
Fama-French-Carhart, 234
Flash crash, 36
Forward premium puzzle, 161–166, 169–171, 173, 174, 177
Fractures, 36, 40, 41, 61, 63, 65

G

Gödel–event, 280, 283
Gödel sentences, 279, 283–285
Genetic programming, 71, 76
Geometric Brownian motion, 197
Gibbs-Boltzmann distribution, 121
Global financial crisis, 275, 283, 292, 297–299
GMM, 226
Guru, 191

H

Halting problem, 280, 287
Hedonic, 205, 207–209, 211–213, 215
Herd behavior, 181
Herding, 196
Heterogeneous agent models, 195
Heterogeneous agents, 114, 115, 127–129
Heteroskedasticity, 195
Heuristics, 297
HFT, *see* high-frequency trading
High-frequency trading, 39, 71, 72
Hyperbolic tangent function, 119

I

Imitation, 120, 125, 131, 133, 134, 136
Incompleteness, 276, 277, 279, 282, 289
Informational asymmetry, 288
Information criteria, 232
Innovation, 125, 131, 133, 134, 136
    product, 205–207, 209–212, 214–216
Instability, 292, 296
Institutions, 297, 298
Integration measures, 243, 244
Intensity of choice, 115, 122, 124–137
Interaction effects, 125, 127, 136
Irreducible uncertainty, 276, 286, 294, 296, 297

K

K-means clustering, 134
Kolmogorov-Smirnov (KS) test, 105

L

Law of Effect, 121
Learning, 136
    individual learning, 114, 118, 119, 121, 125, 138
    meta-learning, 114, 115, 121, 122, 124, 125, 127, 131, 133
    social learning, 114, 118, 120, 121, 125, 137, 138
Leontief payoff function, 114, 116
Leptokurtosis, 195
Life cycle assessment (LCA), 92
Limited information, 161, 163, 164, 166–171, 174, 177
Liquidity traders, 21, 22
Lloyd's algorithm, 134
Local interaction, 136
Lucas critique, 292
Luxembourg, 92

M

Market efficiency, 162–165, 167, 168, 174
Market factor, 230
Market fraction, 125, 137
Market integration, 224
Market orders, 19, 25, 30, 32, 33
Market quality, 19, 20
    market volatility, 20, 25, 28–33
    spreads, 18–20, 32, 33
Market share, 205, 206, 211, 215, 216
Market volatility, 18, 19
Mean Absolute Percentage Error, 126, 127, 136
Meta-preferences, 291
Minsky, 202
Mispricing risk, 277, 294
Multiple equilibria, 143, 151

N

Neoclassical economics, 38
Network structures, 181
Non-convergence, 118
Non-Markovian processes, 194
Non-tâtonnement, 114
Normalized Imitation Frequency, 131, 133, 134
Numerairé, 131

O

Order-driven market, 21
    continuous double-auction market, 18

Index 307

limit order book, 19
market impact, 18, 23
price improvement, 18, 20, 25

P
Paradoxes, 291, 293
Persistent learning, 164, 167, 170, 174
Perverse incentives, 196
Positive feedback, 194
Positive mathematical programming (PMP), 92
Power law of practice, 134
Predictability, 277, 279–281, 283
Presburger arithmetic, 297
Price discovery, 98
Price stickiness, 146
Price volatility, 181
Principal components, 224
Private information, 146
Private prices, 116
Production set, 118
Product quality, 205, 206, 211, 213, 214, 217
Product space, 205, 215
Product variety, 206
Profit dispersion, 44

Q
Qualitative predictions, 298
Quantifiability
  of knowledge, 288
  of natural and social phenomena, 275
  of risk, 277, 287, 296
  of the future, 282
  of uncertainty, *see* irreducible uncertainty

R
Rational expectations equilibrium, 142
Rational expectations hypothesis, 142
Rationality, 286, 290, 292, 295
Real business cycles, 143
Recency parameter, 122
Reference point, 122, 123
Reflexivity, 292
Regulations, 297, 298
Reinforcement learning, 114, 121, 122, 131
REIT, 224
Representative agent, 141, 194
Research and development, 207, 209–218
Rice's theorem, 287, 288, 290

Risk premia, 230
Risk premia test, 241
Robot phase transition, 39, 40, 50, 53
  evidence for, 60, 64
RPT, *see* robot phase transition
R-Square, 225

S
Saturation condition, 117
Scarf economy, 115, 118, 123, 127
Self-reference, 290, 291, 293
Separation of timescales, 197
Shortsightedness, 295
Smith's alpha, 43
Social network, 181
Soros, 202
Stochastic differential equation, 194
Stock markets, 224
Subjective prices, 116, 117, 119
Supercomputer technologies, 251, 252, 256, 259
Symmetric rational expectations equilibrium, 150

T
Tacit knowledge, 288
Tâtonnement, 114, 115
Time-decay function, 119
Turing machine, 280, 285–287, 289, 290

U
Uncertainty principle, 281
Undecidability, 280, 285, 297

V
Volume weighted average price, 28

W
Walrasian auctioneer, 114
Walrasian equilibrium, 142
Walrasian general equilibrium, 115

Z
Zero-intelligence, 18, 19, 21, 33

CPI Antony Rowe
Chippenham, UK
2019-01-09 19:50